Specifying Systems

First Printing

Version of 18 June 2002
pdf file recreated 19 March 2020

Specifying Systems

The TLA+ Language and Tools for Hardware and Software Engineers

Leslie Lamport

Microsoft Research

Boston ● San Francisco ● New York ● Toronto ● Montreal
London ● Munich ● Paris ● Madrid
Capetown ● Sydney ● Tokyo ● Singapore ● Mexico City

The publisher offers discounts on this book when ordered in quantity for special sales. For more information, please contact:
  U.S. Corporate and Government Sales
  (800) 382-3419
  corpsales@pearsontechgroup.com

For sales outside of the U.S., please contact:
  International Sales
  (317) 581-3793
  international@pearsontechgroup.com

Visit Addison-Wesley on the Web: *www.awprofessional.com*

Library of Congress Cataloging-in-Publication Data

Lamport, Leslie
    Specifying systems : the TLA+ language and tools for hardware and software engineers / Leslie Lamport.
        p. cm.
    Includes bibliographical references and index.
    ISBN 0-321-14306-X (alk. paper)
        1. System design. 2. Computer systems–Specifications. 3. Logic, symbolic and mathematical. I. Title.

    QA76.9.S88 L35 2003   2002074369
    004.2′1--dc21

ISBN 0-321-14306-X
Text printed on recycled paper
1 2 3 4 5 6 7 8 9 10-MA-0605040302
First printing, July 2002

To Ellen

This whole book is but a draught—nay, but the draught of a draught.

Herman Melville

Contents

    8.5.1 The Liveness Requirement 100
    8.5.2 Another Way to Write It 101
    8.5.3 A Generalization 105
  8.6 Strong Fairness 106
  8.7 The Write-Through Cache 107
  8.8 Quantification 109
  8.9 Temporal Logic Examined 111
    8.9.1 A Review 111
    8.9.2 Machine Closure 111
    8.9.3 Machine Closure and Possibility 113
    8.9.4 Refinement Mappings and Fairness 114
    8.9.5 The Unimportance of Liveness 116
    8.9.6 Temporal Logic Considered Confusing 116

9 Real Time **117**
  9.1 The Hour Clock Revisited 117
  9.2 Real-Time Specifications in General 122
  9.3 A Real-Time Caching Memory 124
  9.4 Zeno Specifications 128
  9.5 Hybrid System Specifications 132
  9.6 Remarks on Real Time 134

10 Composing Specifications **135**
  10.1 Composing Two Specifications 136
  10.2 Composing Many Specifications 138
  10.3 The FIFO 140
  10.4 Composition with Shared State 142
    10.4.1 Explicit State Changes 144
    10.4.2 Composition with Joint Actions 147
  10.5 A Brief Review 150
    10.5.1 A Taxonomy of Composition 151
    10.5.2 Interleaving Reconsidered 151
    10.5.3 Joint Actions Reconsidered 152
  10.6 Liveness and Hiding 152
    10.6.1 Liveness and Machine Closure 152
    10.6.2 Hiding 154
  10.7 Open-System Specifications 156
  10.8 Interface Refinement 158
    10.8.1 A Binary Hour Clock 158
    10.8.2 Refining a Channel 159
    10.8.3 Interface Refinement in General 163
    10.8.4 Open-System Specifications 165
  10.9 Should You Compose? 167

List of Figures and Tables

Figures

Tables

Acknowledgments

I have spent more than two and a half decades learning how to specify and reason about concurrent computer systems. Before that, I had already spent many years learning how to use mathematics rigorously. I cannot begin to thank everyone who helped me during all that time. But I would like to express my gratitude to two men who, more than anyone else, influenced this book. Richard Palais taught me how even the most complicated mathematics could be made both rigorous and elegant. Martín Abadi influenced the development of TLA and was my collaborator in developing the ideas behind Chapters 9 and 10.

Much of what I know about applying the mathematics of TLA to the engineering problems of complex systems came from working with Mark Tuttle and Yuan Yu. Yuan Yu also helped turn TLA$^+$ into a useful tool for engineers by writing the TLC model checker, ignoring my warnings that it would never be practical. While writing the first version of the Syntactic Analyzer, Jean-Charles Grégoire helped me fine tune the TLA$^+$ language.

The following people made helpful comments on earlier versions of this book: Dominique Couturier, Douglas Frank, Vinod Grover, David Jefferson, Sara Kalvala, and Wolfgang Schreiner all pointed out mistakes. Kazuhiro Ogata read the manuscript with unusual care and found a number of mistakes. Kapila Pahalawatta found an error in the *ProtoReals* module. Paddy Krishnan also found an error in the *ProtoReals* module and suggested a way to improve the presentation. And I wish to extend my special thanks to Martin Rudalics, who read the manuscript with amazing thoroughness and caught many errors.

Leslie Lamport
Palo Alto, California
4 March 2002

Introduction

This book will teach you how to write specifications of computer systems, using the language TLA[+]. It's rather long, but most people will read only Part I, which comprises the first 83 pages. That part contains all that most engineers need to know about writing specifications; it assumes only the basic background in computing and knowledge of mathematics expected of an undergraduate studying engineering or computer science. Part II contains more advanced material for more sophisticated readers. The remainder of the book is a reference manual—Part III for the TLA[+] tools and Part IV for the language itself.

The TLA World Wide Web page contains material to accompany the book, including the TLA[+] tools, exercises, references to the literature, and a list of corrections. There is a link to the TLA Web page on

    http://lamport.org

You can also find the page by searching the Web for the 21-letter string

    uidlamporttlahomepage

Do not put this string in any document that might appear on the Web.

What Is a Specification?

> *Writing is nature's way of letting you*
> *know how sloppy your thinking is.*
> —Guindon

A specification is a written description of what a system is supposed to do. Specifying a system helps us understand it. It's a good idea to understand a system before building it, so it's a good idea to write a specification of a system before implementing it.

This book is about specifying the behavioral properties of a system—also called its functional or logical properties. These are the properties that specify what the system is supposed to do. There are other important kinds of properties that we don't consider, including performance properties. Worst-case performance can often be expressed as a behavioral property—for example, Chapter 9 explains how to specify that a system must react within a certain length of time. However, specifying average performance is beyond the scope of the methods described here.

Our basic tool for writing specifications is mathematics. Mathematics is nature's way of letting you know how sloppy your writing is. It's hard to be precise in an imprecise language like English or Chinese. In engineering, imprecision can lead to errors. To avoid errors, science and engineering have adopted mathematics as their language.

The mathematics we use is more formal than the math you've grown up with. Formal mathematics is nature's way of letting you know how sloppy your mathematics is. The mathematics written by most mathematicians and scientists is not really precise. It's precise in the small, but imprecise in the large. Each equation is a precise assertion, but you have to read the accompanying words to understand how the equations relate to one another and exactly what the theorems mean. Logicians have developed ways of eliminating those words and making the mathematics completely formal and, hence, completely precise.

Most mathematicians and scientists think that formal mathematics, without words, is long and tiresome. They're wrong. Ordinary mathematics can be expressed compactly in a precise, completely formal language. It takes only about two dozen lines to define the solution to an arbitrary differential equation in the *DifferentialEquations* module of Chapter 11. But few specifications need such sophisticated mathematics. Most require only simple application of a few standard mathematical concepts.

Why TLA⁺?

We specify a system by describing its allowed behaviors—what it may do in the course of an execution. In 1977, Amir Pnueli introduced the use of temporal logic for describing system behaviors. In principle, a system could be described by a single temporal logic formula. In practice, it couldn't. Pnueli's temporal logic was ideal for describing some properties of systems, but awkward for others. So, it was usually combined with a more traditional way of describing systems.

In the late 1980's, I invented TLA, the Temporal Logic of Actions—a simple variant of Pnueli's original logic. TLA makes it practical to describe a system by a single formula. Most of a TLA specification consists of ordinary, nontemporal mathematics. Temporal logic plays a significant role only in describing those properties that it's good at describing. TLA also provides a nice way to formalize the style of reasoning about systems that has proved to be most effective in practice—a style known as *assertional* reasoning. However, this book is about specification; it says almost nothing about proofs.

Temporal logic assumes an underlying logic for expressing ordinary mathematics. There are many ways to formalize ordinary math. Most computer scientists prefer one that resembles their favorite programming language. I chose instead the one that most mathematicians prefer—the one logicians call first-order logic and set theory.

TLA provides a mathematical foundation for describing systems. To write specifications, we need a complete language built atop that foundation. I initially thought that this language should be some sort of abstract programming language whose semantics would be based on TLA. I didn't know what kind of programming language constructs would be best, so I decided to start writing specifications directly in TLA. I intended to introduce programming constructs as I needed them. To my surprise, I discovered that I didn't need them. What I needed was a robust language for writing mathematics.

Although mathematicians have developed the science of writing formulas, they haven't turned that science into an engineering discipline. They have developed notations for mathematics in the small, but not for mathematics in the large. The specification of a real system can be dozens or even hundreds of pages long. Mathematicians know how to write 20-line formulas, not 20-page formulas. So, I had to introduce notations for writing long formulas. What I took from programming languages were ideas for modularizing large specifications.

The language I came up with is called TLA$^+$. I refined TLA$^+$ in the course of writing specifications of disparate systems. But it has changed little in the last few years. I have found TLA$^+$ to be quite good for specifying a wide class of systems—from program interfaces (APIs) to distributed systems. It can be used to write a precise, formal description of almost any sort of discrete system. It's especially well suited to describing asynchronous systems—that is, systems with components that do not operate in strict lock-step.

About this Book

Part I, consisting of Chapters 1 through 7, is the core of the book and is meant to be read from beginning to end. It explains how to specify the class of properties known as *safety* properties. These properties, which can be specified with almost no temporal logic, are all that most engineers need to know about.

After reading Part I, you can read as much of Part II as you like. Each of its chapters is independent of the others. Temporal logic comes to the fore in Chapter 8, where it is used to specify the additional class of properties known as *liveness* properties. Chapter 9 describes how to specify real-time properties, and Chapter 10 describes how to write specifications as compositions. Chapter 11 contains more advanced examples.

Part III serves as the reference manual for three TLA$^+$ tools: the Syntactic Analyzer, the TLATEX typesetting program, and the TLC model checker. If you want to use TLA$^+$, then you probably want to use these tools. They are available from the TLA Web page. TLC is the most sophisticated of them. The examples on the Web can get you started using it, but you'll have to read Chapter 14 to learn to use TLC effectively.

Part IV is a reference manual for the TLA$^+$ language. Part I provides a good enough working knowledge of the language for most purposes. You need look at Part IV only if you have questions about the fine points of the syntax and semantics. Chapter 15 gives the syntax of TLA$^+$. Chapter 16 describes the precise meanings and the general forms of all the built-in operators of TLA$^+$; Chapter 17 describes the precise meaning of all the higher-level TLA$^+$ constructs such as definitions. Together, these two chapters specify the semantics of the language. Chapter 18 describes the standard modules—except for module *RealTime*, described in Chapter 9, and module *TLC*, described in Chapter 14. You might want to look at this chapter if you're curious about how standard elementary mathematics can be formalized in TLA$^+$.

Part IV does have something you may want to refer to often: a mini-manual that compactly presents lots of useful information. Pages 268–273 list all TLA$^+$ operators, all user-definable symbols, the precedence of all operators, all operators defined in the standard modules, and the ASCII representation of symbols like $\otimes$.

Part I

Getting Started

A system specification consists of a lot of ordinary mathematics glued together with a tiny bit of temporal logic. That's why most TLA$^+$ constructs are for expressing ordinary mathematics. To write specifications, you have to be familiar with this ordinary math. Unfortunately, the computer science departments in many universities apparently believe that fluency in C++ is more important than a sound education in elementary mathematics. So, some readers may be unfamiliar with the math needed to write specifications. Fortunately, this math is quite simple. If exposure to C++ hasn't destroyed your ability to think logically, you should have no trouble filling any gaps in your mathematics education. You probably learned arithmetic before learning C++, so I will assume you know about numbers and arithmetic operations on them.[1] I will try to explain all other mathematical concepts that you need, starting in Chapter 1 with a review of some elementary math. I hope most readers will find this review completely unnecessary.

After the brief review of simple mathematics in the first chapter, Chapters 2 through 5 describe TLA$^+$ with a sequence of examples. Chapter 6 explains some more about the math used in writing specifications, and Chapter 7 reviews everything and provides some advice. By the time you finish Chapter 7, you should be able to handle most of the specification problems that you are likely to encounter in ordinary engineering practice.

---

[1]Some readers may need reminding that numbers are not strings of bits, and $2^{33} * 2^{33}$ equals 2^{66}, not *overflow error*.

Chapter 1

A Little Simple Math

1.1 Propositional Logic

Elementary algebra is the mathematics of real numbers and the operators $+$, $-$, $*$ (multiplication), and $/$ (division). Propositional logic is the mathematics of the two Boolean values TRUE and FALSE and the five operators whose names (and common pronunciations) are

$\wedge$ conjunction (and) $\Rightarrow$ implication (implies)

$\vee$ disjunction (or) $\equiv$ equivalence (is equivalent to)

$\neg$ negation (not)

To learn how to compute with numbers, you had to memorize addition and multiplication tables and algorithms for calculating with multidigit numbers. Propositional logic is much simpler, since there are only two values, TRUE and FALSE. To learn how to compute with these values, all you need to know are the following definitions of the five Boolean operators:

$\wedge$ $F \wedge G$ equals TRUE iff both F and G equal TRUE.

$\vee$ $F \vee G$ equals TRUE iff F or G equals TRUE (or both do).

$\neg$ $\neg F$ equals TRUE iff F equals FALSE.

$\Rightarrow$ $F \Rightarrow G$ equals TRUE iff F equals FALSE or G equals TRUE (or both).

$\equiv$ $F \equiv G$ equals TRUE iff F and G both equal TRUE or both equal FALSE.

iff stands for *if and only if*. Like most mathematicians, I use *or* to mean *and/or*.

We can also describe these operators by *truth tables*. This truth table gives the value of $F \Rightarrow G$ for all four combinations of truth values of F and G:

F	G	$F \Rightarrow G$
TRUE	TRUE	TRUE
TRUE	FALSE	FALSE
FALSE	TRUE	TRUE
FALSE	FALSE	TRUE

The formula $F \Rightarrow G$ asserts that F implies G—that is, $F \Rightarrow G$ equals TRUE iff the statement "F implies G" is true. People often find the definition of $\Rightarrow$ confusing. They don't understand why FALSE $\Rightarrow$ TRUE and FALSE $\Rightarrow$ FALSE should equal TRUE. The explanation is simple. We expect that if n is greater than 3, then it should be greater than 1, so $n > 3$ should imply $n > 1$. Therefore, the formula $(n > 3) \Rightarrow (n > 1)$ should equal TRUE. Substituting 4, 2, and 0 for n in this formula explains why $F \Rightarrow G$ means F *implies* G or, equivalently, *if F then G*.

The equivalence operator $\equiv$ is equality for Booleans. We can replace $\equiv$ by $=$, but not vice versa. (We can write FALSE $= \neg$TRUE, but not $2 + 2 \equiv 4$.) It's a good idea to write $\equiv$ instead of $=$ to make it clear that the equal expressions are Booleans.[1]

Just like formulas of algebra, formulas of propositional logic are made up of values, operators, and identifiers like x that stand for values. However, propositional-logic formulas use only the two values TRUE and FALSE and the five Boolean operators $\wedge$, $\vee$, $\neg$, $\Rightarrow$, and $\equiv$. In algebraic formulas, $*$ has higher precedence (binds more tightly) than $+$, so $x + y * z$ means $x + (y * z)$. Similarly, $\neg$ has higher precedence than $\wedge$ and $\vee$, which have higher precedence than $\Rightarrow$ and $\equiv$, so $\neg F \wedge G \Rightarrow H$ means $((\neg F) \wedge G) \Rightarrow H$. Other mathematical operators like $+$ and $>$ have higher precedence than the operators of propositional logic, so $n > 0 \Rightarrow n - 1 \geq 0$ means $(n > 0) \Rightarrow (n - 1 \geq 0)$. Redundant parentheses can't hurt and often make a formula easier to read. If you have the slightest doubt about whether parentheses are needed, use them.

The operators $\wedge$ and $\vee$ are associative, just like $+$ and $*$. Associativity of $+$ means that $x + (y + z)$ equals $(x + y) + z$, so we can write $x + y + z$ without parentheses. Similarly, associativity of $\wedge$ and $\vee$ lets us write $F \wedge G \wedge H$ or $F \vee G \vee H$. Like $+$ and $*$, the operators $\wedge$ and $\vee$ are also commutative, so $F \wedge G$ is equivalent to $G \wedge F$, and $F \vee G$ is equivalent to $G \vee F$.

The truth of the formula $(x = 2) \Rightarrow (x + 1 = 3)$ expresses a fact about numbers. To determine that it's true, we have to understand some elementary properties of arithmetic. However, we can tell that $(x = 2) \Rightarrow (x = 2) \vee (y > 7)$ is true even if we know nothing about numbers. This formula is true because $F \Rightarrow F \vee G$ is true, regardless of what the formulas F and G are. In other

---

[1]Section 16.1.3 on page 296 explains a more subtle reason for using $\equiv$ instead of $=$ for equality of Boolean values.

words, $F \Rightarrow F \vee G$ is true for all possible truth values of its identifiers F and G. Such a formula is called a *tautology*.

In general, a tautology of propositional logic is a propositional-logic formula that is true for all possible truth values of its identifiers. Simple tautologies like this should be as obvious as simple algebraic properties of numbers. It should be as obvious that $F \Rightarrow F \vee G$ is a tautology as that $x \leq x + y$ is true for all non-negative numbers x and y. One can derive complicated tautologies from simpler ones by calculations, just as one derives more complicated properties of numbers from simpler ones. However, this takes practice. You've spent years learning how to manipulate number-valued expressions—for example, to deduce that $x \leq -x + y$ holds iff $2 * x \leq y$ does. You probably haven't learned to deduce that $\neg F \vee G$ holds iff $F \Rightarrow G$ does.

If you haven't learned to manipulate Boolean-valued expressions, you will have to do the equivalent of counting on your fingers. You can check if a formula is a tautology by calculating whether it equals TRUE for each possible assignment of Boolean values to its variables. This is best done by constructing a truth table that lists the possible assignments of values to variables and the corresponding values of all subformulas. For example, here is the truth table showing that $(F \Rightarrow G) \equiv (\neg F \vee G)$ is a tautology.

F	G	$F \Rightarrow G$	$\neg F$	$\neg F \vee G$	$(F \Rightarrow G) \equiv \neg F \vee G$
TRUE	TRUE	TRUE	FALSE	TRUE	TRUE
TRUE	FALSE	FALSE	FALSE	FALSE	TRUE
FALSE	TRUE	TRUE	TRUE	TRUE	TRUE
FALSE	FALSE	TRUE	TRUE	TRUE	TRUE

Writing truth tables is a good way to improve your understanding of propositional logic. However, computers are better than people at doing this sort of calculation. Chapter 14 explains, on page 261, how to use the TLC model checker to verify propositional logic tautologies and to perform other TLA$^+$ calculations.

1.2 Sets

Set theory is the foundation of ordinary mathematics. A set is often described as a collection of elements, but saying that a set is a collection doesn't explain very much. The concept of set is so fundamental that we don't try to define it. We take as undefined concepts the notion of a set and the relation $\in$, where $x \in S$ means that x is an element of S. We often say *is in* instead of *is an element of*.

A set can have a finite or infinite number of elements. The set of all natural numbers (0, 1, 2, etc.) is an infinite set. The set of all natural numbers less than

3 is finite, and contains the three elements 0, 1, and 2. We can write this set $\{0, 1, 2\}$.

A set is completely determined by its elements. Two sets are equal iff they have the same elements. Thus, $\{0, 1, 2\}$ and $\{2, 1, 0\}$ and $\{0, 0, 1, 2, 2\}$ are all the same set—the unique set containing the three elements 0, 1, and 2. The empty set, which we write $\{\}$, is the unique set that has no elements.

The most common operations on sets are

$$\cap \text{ intersection} \qquad \cup \text{ union} \qquad \subseteq \text{ subset} \qquad \setminus \text{ set difference}$$

Here are their definitions and examples of their use.

$S \cap T$ The set of elements in both S and T.
$$\{1, -1/2, 3\} \cap \{1, 2, 3, 5, 7\} = \{1, 3\}$$

$S \cup T$ The set of elements in S or T (or both).
$$\{1, -1/2\} \cup \{1, 5, 7\} = \{1, -1/2, 5, 7\}$$

$S \subseteq T$ True iff every element of S is an element of T.
$$\{1, 3\} \subseteq \{3, 2, 1\}$$

$S \setminus T$ The set of elements in S that are not in T.
$$\{1, -1/2, 3\} \setminus \{1, 5, 7\} = \{-1/2, 3\}$$

This is all you need to know about sets before we start looking at how to specify systems. We'll return to set theory in Section 6.1.

1.3   Predicate Logic

Once we have sets, it's natural to say that some formula is true for all the elements of a set, or for some of the elements of a set. Predicate logic extends propositional logic with the two quantifiers

$\forall$ universal quantification (for all)
$\exists$ existential quantification (there exists)

The formula $\forall\, x \in S : F$ asserts that formula F is true for every element x in the set S. For example, $\forall\, n \in Nat : n + 1 > n$ asserts that the formula $n + 1 > n$ is true for all elements n of the set Nat of natural numbers. This formula happens to be true.

The formula $\exists\, x \in S : F$ asserts that formula F is true for at least one element x in S. For example, $\exists\, n \in Nat : n^2 = 2$ asserts that there exists a natural number n whose square equals 2. This formula happens to be false.

Formula F is true for some x in S iff F is not false for all x in S—that is, iff it's not the case that $\neg F$ is true for all x in S. Hence, the formula

$$(1.1) \quad (\exists\, x \in S : F) \;\equiv\; \neg(\forall\, x \in S : \neg F)$$

is a tautology of predicate logic, meaning that it is true for all values of the identifiers S and F.[2]

Since there exists no element in the empty set, the formula $\exists\, x \in \{\} : F$ is false for every formula F. By (1.1), this implies that $\forall\, x \in \{\} : F$ must be true for every F.

The quantification in the formulas $\forall\, x \in S : F$ and $\exists\, x \in S : F$ is said to be *bounded*, since these formulas make an assertion only about elements in the set S. There is also unbounded quantification. The formula $\forall\, x : F$ asserts that F is true for all values x, and $\exists\, x : F$ asserts that F is true for at least one value of x—a value that is not constrained to be in any particular set. Bounded and unbounded quantification are related by the following tautologies:

$$(\forall\, x \in S : F) \;\equiv\; (\forall\, x : (x \in S) \Rightarrow F)$$
$$(\exists\, x \in S : F) \;\equiv\; (\exists\, x : (x \in S) \wedge F)$$

The analog of (1.1) for unbounded quantifiers is also a tautology:

$$(\exists\, x : F) \;\equiv\; \neg(\forall\, x : \neg F)$$

Whenever possible, it is better to use bounded than unbounded quantification in a specification. This makes the specification easier for both people and tools to understand.

Universal quantification generalizes conjunction. If S is a finite set, then $\forall\, x \in S : F$ is the conjunction of the formulas obtained by substituting the different elements of S for x in F. For example,

$$(\forall\, x \in \{2,3,7\} : x < y^x) \;\equiv\; (2 < y^2) \wedge (3 < y^3) \wedge (7 < y^7)$$

We sometimes informally talk about the conjunction of an infinite number of formulas when we formally mean a universally quantified formula. For example, the conjunction of the formulas $x \leq y^x$ for all natural numbers x is the formula $\forall\, x \in Nat : x \leq y^x$. Similarly, existential quantification generalizes disjunction.

Logicians have rules for proving predicate-logic tautologies such as (1.1), but you shouldn't need them. You should become familiar enough with predicate logic that simple tautologies are obvious. Thinking of $\forall$ as conjunction and $\exists$ as disjunction can help. For example, the associativity and commutativity of conjunction and disjunction lead to the tautologies

$$(\forall\, x \in S : F) \wedge (\forall\, x \in S : G) \equiv (\forall\, x \in S : F \wedge G)$$
$$(\exists\, x \in S : F) \vee (\exists\, x \in S : G) \equiv (\exists\, x \in S : F \vee G)$$

for any set S and formulas F and G.

Mathematicians use some obvious abbreviations for nested quantifiers. For example,

---

[2]Strictly speaking, $\in$ isn't an operator of predicate logic, so this isn't really a predicate-logic tautology.

$$\forall\, x \in S, y \in T : F \quad \text{means} \quad \forall\, x \in S : (\forall\, y \in T : F)$$
$$\exists\, w, x, y, z \in S : F \quad \text{means} \quad \exists\, w \in S : (\exists\, x \in S : (\exists\, y \in S : (\exists\, z \in S : F)))$$

In the expression $\exists\, x \in S : F$, logicians say that x is a *bound variable* and that occurrences of x in F are *bound*. For example, n is a bound variable in the formula $\exists\, n \in Nat : n + 1 > n$, and the two occurrences of n in the subexpression $n + 1 > n$ are bound. A variable x that's not bound is said to be *free*, and occurrences of x that are not bound are called *free* occurrences. This terminology is rather misleading. A bound variable doesn't really occur in a formula because replacing it by some new variable doesn't change the formula. The two formulas

$$\exists\, n \in Nat : n + 1 > n \qquad \exists\, x \in Nat : x + 1 > x$$

are equivalent. Calling n a variable of the first formula is a bit like calling a a variable of that formula because it appears in the name *Nat*. Nevertheless, it is convenient to talk about an occurrence of a bound variable in a formula.

1.4   Formulas and Language

When you first studied mathematics, formulas were statements. The formula $2 * x > x$ was just a compact way of writing the statement "2 times x is greater than x." In this book, you are entering the realm of logic, where a formula is a noun. The formula $2 * x > x$ is just a formula; it may be true or false, depending on the value of x. If we want to assert that this formula is true, meaning that $2 * x$ really is greater than x, we should explicitly write "$2 * x > x$ is true."

Using a formula in place of a statement can lead to confusion. On the other hand, formulas are more compact and easier to read than prose. Reading $2 * x > x$ is easier than reading "$2 * x$ is greater than x"; and "$2 * x > x$ is true" may seem redundant. So, like most mathematicians, I will often write sentences like

We know that x is positive, so $2 * x > x$.

If it's not obvious whether a formula is really a formula or is the statement that the formula is true, here's an easy way to tell. Replace the formula with a name and read the sentence. If the sentence is grammatically correct, even though nonsensical, then the formula is a formula; otherwise, it's a statement. The formula $2 * x > x$ in the sentence above is a statement because

We know that x is positive, so Mary.

is ungrammatical. It is a formula in the sentence

To prove $2 * x > x$, we must prove that x is positive.

because the following silly sentence is grammatically correct:

To prove Fred, we must prove that x is positive.

Chapter 2

Specifying a Simple Clock

2.1 Behaviors

Before we try to specify a system, let's look at how scientists do it. For centuries, they have described a system with equations that determine how its state evolves with time, where the state consists of the values of variables. For example, the state of the system comprising the earth and the moon might be described by the values of the four variables e_pos, m_pos, e_vel, and m_vel, representing the positions and velocities of the two bodies. These values are elements in a 3-dimensional space. The earth-moon system is described by equations expressing the variables' values as functions of time and of certain constants—namely, their masses and initial positions and velocities.

A behavior of the earth-moon system consists of a function F from time to states, $F(t)$ representing the state of the system at time t. A computer system differs from the systems traditionally studied by scientists because we can pretend that its state changes in discrete steps. So, we represent the execution of a system as a sequence of states. Formally, we define a *behavior* to be a sequence of states, where a state is an assignment of values to variables. We specify a system by specifying a set of possible behaviors—the ones representing a correct execution of the system.

2.2 An Hour Clock

Let's start with a very trivial system—a digital clock that displays only the hour. To make the system completely trivial, we ignore the relation between the display and the actual time. The hour clock is then just a device whose display cycles through the values 1 through 12. Let the variable hr represent the clock's display. A typical behavior of the clock is the sequence

$$(2.1) \quad [hr = 11] \ \rightarrow \ [hr = 12] \ \rightarrow \ [hr = 1] \ \rightarrow \ [hr = 2] \ \rightarrow \ \cdots$$

of states, where $[hr = 11]$ is a state in which the variable hr has the value 11. A pair of successive states, such as $[hr = 1] \rightarrow [hr = 2]$, is called a *step*.

To specify the hour clock, we describe all its possible behaviors. We write an *initial predicate* that specifies the possible initial values of hr, and a *next-state relation* that specifies how the value of hr can change in any step.

We don't want to specify exactly what the display reads initially; any hour will do. So, we want the initial predicate to assert that hr can have any value from 1 through 12. Let's call the initial predicate $HCini$. We might informally define $HCini$ by

$$HCini \ \triangleq \ hr \in \{1, \ldots, 12\}$$

The symbol $\triangleq$ means *is defined to equal*.

Later, we'll see how to write this definition formally, without the "..." that stands for the informal *and so on*.

The next-state relation $HCnxt$ is a formula expressing the relation between the values of hr in the old (first) state and new (second) state of a step. We let hr represent the value of hr in the old state and hr' represent its value in the new state. (The $'$ in hr' is read *prime*.) We want the next-state relation to assert that hr' equals $hr + 1$ except if hr equals 12, in which case hr' should equal 1. Using an IF/THEN/ELSE construct with the obvious meaning, we can define $HCnxt$ to be the next-state relation by writing

$$HCnxt \ \triangleq \ hr' = \text{IF } hr \neq 12 \text{ THEN } hr + 1 \text{ ELSE } 1$$

$HCnxt$ is an ordinary mathematical formula, except that it contains primed as well as unprimed variables. Such a formula is called an *action*. An action is true or false of a step. A step that satisfies the action $HCnxt$ is called an $HCnxt$ step.

When an $HCnxt$ step occurs, we sometimes say that $HCnxt$ is *executed*. However, it would be a mistake to take this terminology seriously. An action is a formula, and formulas aren't executed.

We want our specification to be a single formula, not the pair of formulas $HCini$ and $HCnxt$. This formula must assert about a behavior that (i) its initial state satisfies $HCini$, and (ii) each of its steps satisfies $HCnxt$. We express (i) as the formula $HCini$, which we interpret as a statement about behaviors to mean that the initial state satisfies $HCini$. To express (ii), we use the temporal-logic operator $\Box$ (pronounced *box*). The temporal formula $\Box F$ asserts that formula F is always true. In particular, $\Box HCnxt$ is the assertion that $HCnxt$ is true for every step in the behavior. So, $HCini \wedge \Box HCnxt$ is true of a behavior iff the initial state satisfies $HCini$ and every step satisfies $HCnxt$. This formula describes all behaviors like the one in (2.1) on this page; it seems to be the specification we're looking for.

If we considered the clock only in isolation and never tried to relate it to another system, then this would be a fine specification. However, suppose the clock is part of a larger system—for example, the hour display of a weather station that displays the current hour and temperature. The state of the station is described by two variables: hr, representing the hour display, and tmp, representing the temperature display. Consider this behavior of the weather station:

$$\begin{bmatrix} hr & = & 11 \\ tmp & = & 23.5 \end{bmatrix} \rightarrow \begin{bmatrix} hr & = & 12 \\ tmp & = & 23.5 \end{bmatrix} \rightarrow \begin{bmatrix} hr & = & 12 \\ tmp & = & 23.4 \end{bmatrix} \rightarrow$$

$$\begin{bmatrix} hr & = & 12 \\ tmp & = & 23.3 \end{bmatrix} \rightarrow \begin{bmatrix} hr & = & 1 \\ tmp & = & 23.3 \end{bmatrix} \rightarrow \cdots$$

In the second and third steps, tmp changes but hr remains the same. These steps are not allowed by $\Box HCnxt$, which asserts that every step must increment hr. The formula $HCini \land \Box HCnxt$ does not describe the hour clock in the weather station.

A formula that describes *any* hour clock must allow steps that leave hr unchanged—in other words, $hr' = hr$ steps. These are called *stuttering steps* of the clock. A specification of the hour clock should allow both $HCnxt$ steps and stuttering steps. So, a step should be allowed iff it is either an $HCnxt$ step or a stuttering step—that is, iff it is a step satisfying $HCnxt \lor (hr' = hr)$. This suggests that we adopt $HCini \land \Box(HCnxt \lor (hr' = hr))$ as our specification. In TLA, we let $[HCnxt]_{hr}$ stand for $HCnxt \lor (hr' = hr)$, so we can write the formula more compactly as $HCini \land \Box[HCnxt]_{hr}$.

I pronounce $[HCnxt]_{hr}$ as *square HCnxt sub hr*.

The formula $HCini \land \Box[HCnxt]_{hr}$ does allow stuttering steps. In fact, it allows the behavior

$$[hr = 10] \rightarrow [hr = 11] \rightarrow [hr = 11] \rightarrow [hr = 11] \rightarrow \cdots$$

that ends with an infinite sequence of stuttering steps. This behavior describes a clock whose display attains the value 11 and then keeps that value forever—in other words, a clock that stops at 11. In a like manner, we can represent a terminating execution of any system by an infinite behavior that ends with a sequence of nothing but stuttering steps. We have no need of finite behaviors (finite sequences of states), so we consider only infinite ones.

It's natural to require that a clock does not stop, so our specification should assert that there are infinitely many nonstuttering steps. Chapter 8 explains how to express this requirement. For now, we content ourselves with clocks that may stop, and we take as our specification of an hour clock the formula HC defined by

$$HC \;\triangleq\; HCini \land \Box[HCnxt]_{hr}$$

2.3   A Closer Look at the Specification

A state is an assignment of values to variables, but what variables? The answer is simple: all variables. In the behavior (2.1) on page 16, $[hr = 1]$ represents some particular state that assigns the value 1 to hr. It might assign the value 23 to the variable tmp and the value $\sqrt{-17}$ to the variable m_pos. We can think of a state as representing a potential state of the entire universe. A state that assigns 1 to hr and a particular point in 3-space to m_pos describes a state of the universe in which the hour clock reads 1 and the moon is in a particular place. A state that assigns $\sqrt{-2}$ to hr doesn't correspond to any state of the universe that we recognize, because the hour clock can't display the value $\sqrt{-2}$. It might represent the state of the universe after a bomb fell on the clock, making its display purely imaginary.

A behavior is an infinite sequence of states—for example:

$$(2.2) \quad [hr = 11] \ \to \ [hr = 77.2] \ \to \ [hr = 78.2] \ \to \ [hr = \sqrt{-2}] \ \to \ \cdots$$

A behavior describes a potential history of the universe. The behavior (2.2) doesn't correspond to a history that we understand, because we don't know how the clock's display can change from 11 to 77.2. Whatever kind of history it represents is not one in which the clock is doing what it's supposed to.

Formula HC is a temporal formula. A temporal formula is an assertion about behaviors. We say that a behavior *satisfies HC* iff HC is a true assertion about the behavior. Behavior (2.1) satisfies formula HC. Behavior (2.2) does not, because HC asserts that every step satisfies $HCnxt$ or leaves hr unchanged, and the first and third steps of (2.2) don't. (The second step, $[hr = 77.2] \to [hr = 78.2]$, does satisfy $HCnxt$.) We regard formula HC to be the specification of an hour clock because it is satisfied by exactly those behaviors that represent histories of the universe in which the clock functions properly.

If the clock is behaving properly, then its display should be an integer from 1 through 12. So, hr should be an integer from 1 through 12 in every state of any behavior satisfying the clock's specification, HC. Formula $HCini$ asserts that hr is an integer from 1 through 12, and $\Box HCini$ asserts that $HCini$ is always true. So, $\Box HCini$ should be true for any behavior satisfying HC. Another way of saying this is that HC implies $\Box HCini$, for any behavior. Thus, the formula $HC \Rightarrow \Box HCini$ should be satisfied by *every* behavior. A temporal formula satisfied by every behavior is called a *theorem*, so $HC \Rightarrow \Box HCini$ should be a theorem.[1] It's easy to see that it is: HC implies that $HCini$ is true initially (in the first state of the behavior), and $\Box[HCnxt]_{hr}$ implies that each step either advances hr to its proper next value or else leaves hr unchanged. We can formalize this reasoning using the proof rules of TLA, but we're not going to delve into proofs and proof rules.

---

[1] Logicians call a formula *valid* if it is satisfied by every behavior; they reserve the term *theorem* for provably valid formulas.

2.4   The Specification in TLA$^+$

Figure 2.1 on the next page shows how the hour-clock specification can be written in TLA$^+$. There are two versions: the ASCII version on the bottom is the actual TLA$^+$ specification, the way you type it; the typeset version on the top is one that the TLATEX program, described in Chapter 13, might produce. Before trying to understand the specification, observe the relation between the two syntaxes.

- Reserved words that appear in small upper-case letters (like EXTENDS) are written in ASCII with ordinary upper-case letters.

- When possible, symbols are represented pictorially in ASCII—for example, □ is typed as [] and $\neq$ as #. (You can also type $\neq$ as /=.)

- When there is no good ASCII representation, TEX notation[2] is used—for example, $\in$ is typed as \in. The major exception is $\triangleq$, which is typed as ==.

A complete list of symbols and their ASCII equivalents appears in Table 8 on page 273. I will usually show the typeset version of a specification; the ASCII versions of all the specifications in this book can be found through the TLA Web page.

Now let's look at what the specification says. It starts with

$$\boxed{\qquad\qquad \text{MODULE } \textit{HourClock} \qquad\qquad}$$

which begins a module named *HourClock*. TLA$^+$ specifications are partitioned into modules; the hour clock's specification consists of this single module.

Arithmetic operators like + are not built into TLA$^+$, but are themselves defined in modules. (You might want to write a specification in which + means addition of matrices rather than numbers.) The usual operators on natural numbers are defined in the *Naturals* module. Their definitions are incorporated into module *HourClock* by the statement

    EXTENDS *Naturals*

Every symbol that appears in a formula must either be a built-in operator of TLA$^+$, or else it must be declared or defined. The statement

    VARIABLE *hr*

declares *hr* to be a variable.

---

[2]The TEX typesetting system is described in *The TEXbook* by Donald E. Knuth, published by Addison-Wesley, Reading, Massachusetts, 1986.

―――――――――――――――――― MODULE *HourClock* ―――――――――――――

EXTENDS *Naturals*

VARIABLE *hr*

$HCini \triangleq hr \in (1 .. 12)$

$HCnxt \triangleq hr' = \text{IF } hr \neq 12 \text{ THEN } hr + 1 \text{ ELSE } 1$

$HC \triangleq HCini \wedge \Box[HCnxt]_{hr}$

THEOREM $HC \Rightarrow \Box HCini$

```
---------------------- MODULE HourClock ----------------------
EXTENDS Naturals
VARIABLE hr
HCini  ==  hr \in (1 .. 12)
HCnxt  ==  hr' = IF hr # 12 THEN hr + 1 ELSE 1
HC     ==  HCini /\ [][HCnxt]_hr
--------------------------------------------------------------
THEOREM  HC => []HCini
==============================================================
```

Figure 2.1: The hour-clock specification—typeset and ASCII versions.

To define $HCini$, we need to express the set $\{1, \ldots, 12\}$ formally, without the ellipsis "...". We can write this set out completely as

$$\{1, 2, 3, 4, 5, 6, 7, 8, 9, 10, 11, 12\}$$

but that's tiresome. Instead, we use the operator "..", defined in the *Naturals* module, to write this set as $1 .. 12$. In general $i .. j$ is the set of integers from i through j, for any integers i and j. (It equals the empty set if $j < i$.) It's now obvious how to write the definition of $HCini$. The definitions of $HCnxt$ and HC are written just as before. (The ordinary mathematical operators of logic and set theory, like $\wedge$ and $\in$, are built into TLA$^+$.)

The line

---

can appear anywhere between statements; it's purely cosmetic and has no meaning. Following it is the statement

THEOREM $HC \Rightarrow \Box HCini$

of the theorem that was discussed above. This statement asserts that the formula $HC \Rightarrow \Box HCini$ is true in the context of the statement. More precisely, it asserts that the formula follows logically from the definitions in this module, the definitions in the *Naturals* module, and the rules of TLA$^+$. If the formula were not true, then the module would be incorrect.

The module is terminated by the symbol

The specification of the hour clock is the definition of *HC*, including the definitions of the formulas *HCnxt* and *HCini* and of the operators .. and + that appear in the definition of *HC*. Formally, nothing in the module tells us that *HC* rather than *HCini* is the clock's specification. TLA$^+$ is a language for writing mathematics—in particular, for writing mathematical definitions and theorems. What those definitions represent, and what significance we attach to those theorems, lies outside the scope of mathematics and therefore outside the scope of TLA$^+$. Engineering requires not just the ability to use mathematics, but the ability to understand what, if anything, the mathematics tells us about an actual system.

2.5 An Alternative Specification

The *Naturals* module also defines the modulus operator, which we write %. The formula $i \% n$, which mathematicians write $i \bmod n$, is the remainder when i is divided by n. More formally, $i \% n$ is the natural number less than n satisfying $i = q * n + (i \% n)$ for some natural number q. Let's express this condition mathematically. The *Naturals* module defines *Nat* to be the set of natural numbers, and the assertion that there exists a q in the set *Nat* satisfying a formula F is written $\exists q \in Nat : F$. Thus, if i and n are elements of *Nat* and $n > 0$, then $i \% n$ is the unique number satisfying

$$(i \% n \in 0 .. (n-1)) \wedge (\exists q \in Nat : i = q * n + (i \% n))$$

We can use % to simplify our hour-clock specification a bit. Observing that $(11 \% 12) + 1$ equals 12 and $(12 \% 12) + 1$ equals 1, we can define a different next-state action *HCnxt2* and a different formula *HC2* to be the clock specification

$$HCnxt2 \;\triangleq\; hr' = (hr \% 12) + 1 \qquad HC2 \;\triangleq\; HCini \wedge \Box[HCnxt2]_{hr}$$

Actions *HCnxt* and *HCnxt2* are not equivalent. The step $[hr = 24] \to [hr = 25]$ satisfies *HCnxt* but not *HCnxt2*, while the step $[hr = 24] \to [hr = 1]$ satisfies *HCnxt2* but not *HCnxt*. However, any step starting in a state with *hr* in 1 .. 12 satisfies *HCnxt* iff it satisfies *HCnxt2*. It's therefore not hard to deduce that any behavior starting in a state satisfying *HCini* satisfies $\Box[HCnxt]_{hr}$ iff it satisfies $\Box[HCnxt2]_{hr}$. Hence, formulas *HC* and *HC2* *are* equivalent. In other words, $HC \equiv HC2$ is a theorem. It doesn't matter which of the two formulas we take to be the specification of an hour clock.

Mathematics provides infinitely many ways of expressing the same thing. The expressions $6 + 6$, $3 * 4$, and $141 - 129$ all have the same meaning; they are just different ways of writing the number 12. We could replace either instance of the number 12 in module *HourClock* by any of these expressions without changing the meaning of any of the module's formulas.

When writing a specification, you will often be faced with a choice of how to express something. When that happens, you should first make sure that the choices yield equivalent specifications. If they do, then you can choose the one that you feel makes the specification easiest to understand. If they don't, then you must decide which one you mean.

Chapter 3

An Asynchronous Interface

We now specify an interface for transmitting data between asynchronous devices. A *sender* and a *receiver* are connected as shown here.

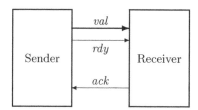

Data is sent on *val*, and the *rdy* and *ack* lines are used for synchronization. The sender must wait for an acknowledgment (an *Ack*) for one data item before it can send the next. The interface uses the standard two-phase handshake protocol, described by the following sample behavior:

$$
\begin{bmatrix} val & = & 26 \\ rdy & = & 0 \\ ack & = & 0 \end{bmatrix} \xrightarrow{Send\ 37} \begin{bmatrix} val & = & 37 \\ rdy & = & 1 \\ ack & = & 0 \end{bmatrix} \xrightarrow{Ack} \begin{bmatrix} val & = & 37 \\ rdy & = & 1 \\ ack & = & 1 \end{bmatrix} \xrightarrow{Send\ 4}
$$

$$
\begin{bmatrix} val & = & 4 \\ rdy & = & 0 \\ ack & = & 1 \end{bmatrix} \xrightarrow{Ack} \begin{bmatrix} val & = & 4 \\ rdy & = & 0 \\ ack & = & 0 \end{bmatrix} \xrightarrow{Send\ 19} \begin{bmatrix} val & = & 19 \\ rdy & = & 1 \\ ack & = & 0 \end{bmatrix} \xrightarrow{Ack} \cdots
$$

(It doesn't matter what value *val* has in the initial state.)

It's easy to see from this sample behavior what the set of all possible behaviors should be—once we decide what the data values are that can be sent. But, before writing the TLA$^+$ specification that describes these behaviors, let's look at what I've just done.

In writing this behavior, I made the decision that *val* and *rdy* should change in a single step. The values of the variables *val* and *rdy* represent voltages on some set of wires in the physical device. Voltages on different wires don't change at precisely the same instant. I decided to ignore this aspect of the physical system and pretend that the values of *val* and *rdy* represented by those voltages change instantaneously. This simplifies the specification, but at the price of ignoring what may be an important detail of the system. In an actual implementation of the protocol, the voltage on the *rdy* line shouldn't change until the voltages on the *val* lines have stabilized; but you won't learn that from my specification. Had I wanted the specification to convey this requirement, I would have written a behavior in which the value of *val* and the value of *rdy* change in separate steps.

A specification is an abstraction. It describes some aspects of the system and ignores others. We want the specification to be as simple as possible, so we want to ignore as many details as we can. But, whenever we omit some aspect of the system from the specification, we admit a potential source of error. With my specification, we can verify the correctness of a system that uses this interface, and the system could still fail because the implementer didn't know that the *val* line should stabilize before the *rdy* line is changed.

The hardest part of writing a specification is choosing the proper abstraction. I can teach you about TLA$^+$, so expressing an abstract view of a system as a TLA$^+$ specification becomes a straightforward task. But I don't know how to teach you about abstraction. A good engineer knows how to abstract the essence of a system and suppress the unimportant details when specifying and designing it. The art of abstraction is learned only through experience.

When writing a specification, you must first choose the abstraction. In a TLA$^+$ specification, this means choosing the variables that represent the system's state and the granularity of the steps that change those variables' values. Should the *rdy* and *ack* lines be represented as separate variables or as a single variable? Should *val* and *rdy* change in one step, two steps, or an arbitrary number of steps? To help make these choices, I recommend that you start by writing the first few steps of one or two sample behaviors, just as I did at the beginning of this section. Chapter 7 has more to say about these choices.

3.1   The First Specification

Let's specify the asynchronous interface with a module *AsynchInterface*. The specification uses subtraction of natural numbers, so our module EXTENDS the *Naturals* module to incorporate the definition of the subtraction operator "−". We next decide what the possible values of *val* should be—that is, what data values may be sent. We could write a specification that places no restriction on the data values. The specification could allow the sender first to send 37, then to send $\sqrt{-15}$, and then to send *Nat* (the entire set of natural numbers). However, any real device can send only a restricted set of values. We could pick some specific set—for example, 32-bit numbers. However, the protocol is the same regardless of whether it's used to send 32-bit numbers or 128-bit numbers. So, we compromise between the two extremes of allowing anything to be sent and allowing only 32-bit numbers to be sent by assuming only that there is some set *Data* of data values that may be sent. The constant *Data* is a parameter of the specification. It's declared by the statement

> CONSTANT *Data*

Our three variables are declared by

> VARIABLES *val, rdy, ack*

The keywords VARIABLE and VARIABLES are synonymous, as are CONSTANT and CONSTANTS.

The variable *rdy* can assume any value—for example, $-1/2$. That is, there exist states that assign the value $-1/2$ to *rdy*. When discussing the specification, we usually say that *rdy* can assume only the values 0 and 1. What we really mean is that the value of *rdy* equals 0 or 1 in every state of any behavior satisfying the specification. But a reader of the specification shouldn't have to understand the complete specification to figure this out. We can make the specification easier to understand by telling the reader what values the variables can assume in a behavior that satisfies the specification. We could do this with comments, but I prefer to use a definition like this one:

$$TypeInvariant \triangleq (val \in Data) \wedge (rdy \in \{0,1\}) \wedge (ack \in \{0,1\})$$

I call the set $\{0,1\}$ the *type* of *rdy*, and I call *TypeInvariant* a *type invariant*. Let's define *type* and some other terms more precisely.

- A *state function* is an ordinary expression (one with no prime or □) that can contain variables and constants.

- A *state predicate* is a Boolean-valued state function.

- An *invariant Inv* of a specification *Spec* is a state predicate such that $Spec \Rightarrow \Box Inv$ is a theorem.

- A variable v has *type T* in a specification *Spec* iff $v \in T$ is an invariant of *Spec*.

We can make the definition of *TypeInvariant* easier to read by writing it as follows.

$$TypeInvariant \triangleq \begin{array}{l} \wedge\ val \in Data \\ \wedge\ rdy \in \{0,1\} \\ \wedge\ ack \in \{0,1\} \end{array}$$

Each conjunct begins with a ∧ and must lie completely to the right of that ∧. (The conjunct may occupy multiple lines). We use a similar notation for disjunctions. When using this bulleted-list notation, the ∧'s or ∨'s must line up precisely (even in the ASCII input). Because the indentation is significant, we can eliminate parentheses, making this notation especially useful when conjunctions and disjunctions are nested.

The formula *TypeInvariant* will not appear as part of the specification. We do not assume that *TypeInvariant* is an invariant; the specification should imply that it is. In fact, its invariance will be asserted as a theorem.

The initial predicate is straightforward. Initially, *val* can equal any element of *Data*. We can start with *rdy* and *ack* either both 0 or both 1.

$$
\begin{aligned}
Init \;\triangleq\; &\land\; val \in Data \\
&\land\; rdy \in \{0,1\} \\
&\land\; ack = rdy
\end{aligned}
$$

Now for the next-state action *Next*. A step of the protocol either sends a value or receives a value. We define separately the two actions *Send* and *Rcv* that describe the sending and receiving of a value. A *Next* step (one satisfying action *Next*) is either a *Send* step or a *Rcv* step, so it is a *Send* ∨ *Rcv* step. Therefore, *Next* is defined to equal *Send* ∨ *Rcv*. Let's now define *Send* and *Rcv*.

We say that action *Send* is *enabled* in a state from which it is possible to take a *Send* step. From the sample behavior above, we see that *Send* is enabled iff *rdy* equals *ack*. Usually, the first question we ask about an action is, when is it enabled? So, the definition of an action usually begins with its enabling condition. The first conjunct in the definition of *Send* is therefore *rdy* = *ack*. The next conjuncts tell us what the new values of the variables *val*, *rdy*, and *ack* are. The new value *val'* of *val* can be any element of *Data*—that is, any value satisfying *val'* ∈ *Data*. The value of *rdy* changes from 0 to 1 or from 1 to 0, so *rdy'* equals 1 − *rdy* (because 1 = 1 − 0 and 0 = 1 − 1). The value of *ack* is left unchanged.

TLA$^+$ defines UNCHANGED *v* to mean that the expression *v* has the same value in the old and new states. More precisely, UNCHANGED *v* equals *v'* = *v*, where *v'* is the expression obtained from *v* by priming all its variables. So, we define *Send* by

$$
\begin{aligned}
Send \;\triangleq\; &\land\; rdy = ack \\
&\land\; val' \in Data \\
&\land\; rdy' = 1 - rdy \\
&\land\; \text{UNCHANGED } ack
\end{aligned}
$$

(I could have written *ack'* = *ack* instead of UNCHANGED *ack*, but I prefer to use the UNCHANGED construct in specifications.)

A *Rcv* step is enabled iff *rdy* is different from *ack*; it complements the value of *ack* and leaves *val* and *rdy* unchanged. Both *val* and *rdy* are left unchanged iff

─────────────────────── MODULE *AsynchInterface* ───────────────────────

EXTENDS *Naturals*
CONSTANT *Data*
VARIABLES *val*, *rdy*, *ack*

$TypeInvariant \overset{\Delta}{=} \wedge val \in Data$
$\qquad\qquad\qquad\quad \wedge rdy \in \{0, 1\}$
$\qquad\qquad\qquad\quad \wedge ack \in \{0, 1\}$

───

$Init \overset{\Delta}{=} \wedge val \in Data$
$\qquad\quad \wedge rdy \in \{0, 1\}$
$\qquad\quad \wedge ack = rdy$

$Send \overset{\Delta}{=} \wedge rdy = ack$
$\qquad\quad \wedge val' \in Data$
$\qquad\quad \wedge rdy' = 1 - rdy$
$\qquad\quad \wedge$ UNCHANGED *ack*

$Rcv \overset{\Delta}{=} \wedge rdy \neq ack$
$\qquad\quad \wedge ack' = 1 - ack$
$\qquad\quad \wedge$ UNCHANGED $\langle val, rdy \rangle$

$Next \overset{\Delta}{=} Send \vee Rcv$
$Spec \overset{\Delta}{=} Init \wedge \Box[Next]_{\langle val, rdy, ack \rangle}$

───

THEOREM $Spec \Rightarrow \Box TypeInvariant$

───

Figure 3.1: Our first specification of an asynchronous interface.

the pair of values *val*, *rdy* is left unchanged. TLA$^+$ uses angle brackets $\langle$ and $\rangle$ to enclose ordered tuples, so *Rcv* asserts that $\langle val, rdy \rangle$ is left unchanged. (Angle brackets are typed in ASCII as << and >>.) The definition of *Rcv* is therefore

$Rcv \overset{\Delta}{=} \wedge rdy \neq ack$
$\qquad\quad \wedge ack' = 1 - ack$
$\qquad\quad \wedge$ UNCHANGED $\langle val, rdy \rangle$

As in our clock example, the complete specification *Spec* should allow stuttering steps—in this case, ones that leave all three variables unchanged. So, *Spec* allows steps that leave $\langle val, rdy, ack \rangle$ unchanged. Its definition is

$Spec \overset{\Delta}{=} Init \wedge \Box[Next]_{\langle val, rdy, ack \rangle}$

Module *AsynchInterface* also asserts the invariance of *TypeInvariant*. It appears in full in Figure 3.1 on this page.

3.2   Another Specification

Module *AsynchInterface* is a fine description of the interface and its handshake protocol. However, it's not well suited for helping to specify systems that use the interface. Let's rewrite the interface specification in a form that makes it more convenient to use as part of a larger specification.

The first problem with the original specification is that it uses three variables to describe a single interface. A system might use several different instances of the interface. To avoid a proliferation of variables, we replace the three variables *val*, *rdy*, *ack* with a single variable *chan* (short for *channel*). A mathematician would do this by letting the value of *chan* be an ordered triple—for example, a state $[chan = \langle -1/2,\, 0,\, 1 \rangle]$ might replace the state with $val = -1/2$, $rdy = 0$, and $ack = 1$. But programmers have learned that using tuples like this leads to mistakes; it's easy to forget if the *ack* line is represented by the second or third component. TLA[+] therefore provides records in addition to more conventional mathematical notation.

Let's represent the state of the channel as a record with *val*, *rdy*, and *ack* fields. If r is such a record, then $r.val$ is its *val* field. The type invariant asserts that the value of *chan* is an element of the set of all such records r in which $r.val$ is an element of the set *Data* and $r.rdy$ and $r.ack$ are elements of the set $\{0, 1\}$. This set of records is written

$$[val : Data,\ rdy : \{0,1\},\ ack : \{0,1\}]$$

The fields of a record are not ordered, so it doesn't matter in what order we write them. This same set of records can also be written as

$$[ack : \{0,1\},\ val : Data,\ rdy : \{0,1\}]$$

Initially, *chan* can equal any element of this set whose *ack* and *rdy* fields are equal, so the initial predicate is the conjunction of the type invariant and the condition $chan.ack = chan.rdy$.

A system that uses the interface may perform an operation that sends some data value d and performs some other changes that depend on the value d. We'd like to represent such an operation as an action that is the conjunction of two separate actions: one that describes the sending of d and the other that describes the other changes. Thus, instead of defining an action *Send* that sends some unspecified data value, we define the action $Send(d)$ that sends data value d. The next-state action is satisfied by a $Send(d)$ step, for some d in *Data*, or a *Rcv* step. (The value received by a *Rcv* step equals $chan.val$.) Saying that a step is a $Send(d)$ step for some d in *Data* means that there exists a d in *Data* such that the step satisfies $Send(d)$—in other words, that the step is an $\exists\, d \in Data : Send(d)$ step. So we define

$$Next\ \triangleq\ (\exists\, d \in Data\ :\ Send(d)) \vee Rcv$$

The $Send(d)$ action asserts that $chan'$ equals the record r such that

$$r.val = d \qquad r.rdy = 1 - chan.rdy \qquad r.ack = chan.ack$$

This record is written in TLA$^+$ as

$$[val \mapsto d, \ \ rdy \mapsto 1 - chan.rdy, \ \ ack \mapsto chan.ack]$$

(The symbol $\mapsto$ is typed in ASCII as |->.) Since the fields of records are not ordered, this record can just as well be written

$$[ack \mapsto chan.ack, \ \ val \mapsto d, \ \ rdy \mapsto 1 - chan.rdy]$$

The enabling condition of $Send(d)$ is that the rdy and ack lines are equal, so we can define

$$
\begin{aligned}
Send(d) \ \ \triangleq \ & \\
\wedge \ & chan.rdy = chan.ack \\
\wedge \ & chan' = [val \mapsto d, \ rdy \mapsto 1 - chan.rdy, \ ack \mapsto chan.ack]
\end{aligned}
$$

This is a perfectly good definition of $Send(d)$. However, I prefer a slightly different one. We can describe the value of $chan'$ by saying that it is the same as the value of $chan$ except that its val field equals d and its rdy field equals $1 - chan.rdy$. In TLA$^+$, we can write this value as

$$[chan \ \text{EXCEPT} \ !.val = d, \ !.rdy = 1 - chan.rdy]$$

Think of the ! as standing for the new record that the EXCEPT expression forms by modifying $chan$. So, the expression can be read as the record ! that is the same as $chan$ except !.val equals d and !.rdy equals $1 - chan.rdy$. In the expression that !.rdy equals, the symbol @ stands for $chan.rdy$, so we can write this EXCEPT expression as

$$[chan \ \text{EXCEPT} \ !.val = d, \ !.rdy = 1 - @]$$

In general, for any record r, the expression

$$[r \ \text{EXCEPT} \ !.c_1 = e_1, \ \dots, \ !.c_n = e_n]$$

is the record obtained from r by replacing $r.c_i$ with e_i, for each i in $1 \dots n$. An @ in the expression e_i stands for $r.c_i$. Using this notation, we define

$$
\begin{aligned}
Send(d) \ \ \triangleq \ \ & \wedge \ chan.rdy = chan.ack \\
& \wedge \ chan' = [chan \ \text{EXCEPT} \ !.val = d, \ !.rdy = 1 - @]
\end{aligned}
$$

The definition of Rcv is straightforward. A value can be received when $chan.rdy$ does not equal $chan.ack$, and receiving the value complements $chan.ack$:

$$
\begin{aligned}
Rcv \ \ \triangleq \ \ & \wedge \ chan.rdy \neq chan.ack \\
& \wedge \ chan' = [chan \ \text{EXCEPT} \ !.ack = 1 - @]
\end{aligned}
$$

The complete specification appears in Figure 3.2 on the next page.

─────────────── MODULE *Channel* ───────────────

EXTENDS *Naturals*
CONSTANT *Data*
VARIABLE *chan*
$TypeInvariant \stackrel{\Delta}{=} chan \in [val : Data, \ rdy : \{0, 1\}, \ ack : \{0, 1\}]$

───

$Init \stackrel{\Delta}{=} \wedge TypeInvariant$
$\qquad\qquad \wedge chan.ack = chan.rdy$

$Send(d) \stackrel{\Delta}{=} \wedge chan.rdy = chan.ack$
$\qquad\qquad\quad \wedge chan' = [chan \text{ EXCEPT } !.val = d, \ !.rdy = 1 - @]$

$Rcv \qquad \stackrel{\Delta}{=} \wedge chan.rdy \neq chan.ack$
$\qquad\qquad\quad \wedge chan' = [chan \text{ EXCEPT } !.ack = 1 - @]$

$Next \stackrel{\Delta}{=} (\exists\, d \in Data \ : \ Send(d)) \vee Rcv$

$Spec \stackrel{\Delta}{=} Init \wedge \Box[Next]_{chan}$

───

THEOREM $Spec \Rightarrow \Box TypeInvariant$

───

Figure 3.2: Our second specification of an asynchronous interface.

3.3   Types: A Reminder

As defined in Section 3.1, a variable v has type T in specification *Spec* iff $v \in T$ is an invariant of *Spec*. Thus, hr has type $1 .. 12$ in the specification *HC* of the hour clock. This assertion does *not* mean that the variable hr can assume only values in the set $1 .. 12$. A state is an arbitrary assignment of values to variables, so there exist states in which the value of hr is $\sqrt{-2}$. The assertion does mean that, in every behavior satisfying formula *HC*, the value of hr is an element of $1 .. 12$.

If you are used to types in programming languages, it may seem strange that TLA$^+$ allows a variable to assume any value. Why not restrict our states to ones in which variables have the values of the right type? In other words, why not add a formal type system to TLA$^+$? A complete answer would take us too far afield. The question is addressed further in Section 6.2. For now, remember that TLA$^+$ is an untyped language. Type correctness is just a name for a certain invariance property. Assigning the name *TypeInvariant* to a formula gives it no special status.

3.4 Definitions

Let's examine what a definition means. If *Id* is a simple identifier like *Init* or *Spec*, then the definition $Id \triangleq exp$ defines *Id* to be synonymous with the expression *exp*. Replacing *Id* by *exp*, or vice-versa, in any expression does not change the meaning of that expression. This replacement must be done after the expression is parsed, not in the "raw input". For example, the definition $x \triangleq a + b$ makes $x * c$ equal to $(a + b) * c$, not to $a + b * c$, which equals $a + (b * c)$.

The definition of *Send* has the form $Id(p) \triangleq exp$, where *Id* and *p* are identifiers. For any expression *e*, this defines $Id(e)$ to be the expression obtained by substituting *e* for *p* in *exp*. For example, the definition of *Send* in the *Channel* module defines $Send(-5)$ to equal

$$\wedge\ chan.rdy = chan.ack$$
$$\wedge\ chan' = [chan \text{ EXCEPT } !.val = -5,\ !.rdy = 1 - @]$$

$Send(e)$ is an expression, for any expression *e*. Thus, we can write the formula $Send(-5) \wedge (chan.ack = 1)$. The identifier *Send* by itself is not an expression, and $Send \wedge (chan.ack = 1)$ is not a grammatically well-formed string. It's non-syntactic nonsense, like $a + * b +$.

We say that *Send* is an *operator* that takes a single argument. We define operators that take more than one argument in the obvious way, the general form being

$$(3.1) \quad Id(p_1, \ldots, p_n) \triangleq exp$$

where the p_i are distinct identifiers and *exp* is an expression. We can consider defined identifiers like *Init* and *Spec* to be operators that take no argument, but we generally use *operator* to mean an operator that takes one or more arguments.

I will use the term *symbol* to mean an identifier like *Send* or an operator symbol like $+$. Every symbol that is used in a specification must either be a built-in operator of TLA^+ (like $\in$) or it must be declared or defined. Every symbol declaration or definition has a *scope* within which the symbol may be used. The scope of a VARIABLE or CONSTANT declaration, and of a definition, is the part of the module that follows it. Thus, we can use *Init* in any expression that follows its definition in module *Channel*. The statement EXTENDS *Naturals* extends the scope of symbols like $+$ defined in the *Naturals* module to the *Channel* module.

The operator definition (3.1) implicitly includes a declaration of the identifiers $p_1, \ldots, p_n$ whose scope is the expression *exp*. An expression of the form

$$\exists\, v \in S\ :\ exp$$

has a declaration of *v* whose scope is the expression *exp*. Thus the identifier *v* has a meaning within the expression *exp* (but not within the expression *S*).

A symbol cannot be declared or defined if it already has a meaning. The expression

$$(\exists\, v \in S \,:\, exp1) \,\wedge\, (\exists\, v \in T \,:\, exp2)$$

is all right, because neither declaration of v lies within the scope of the other. Similarly, the two declarations of the symbol d in the *Channel* module (in the definition of *Send* and in the expression $\exists\, d$ in the definition of *Next*) have disjoint scopes. However, the expression

$$(\exists\, v \in S \,:\, (exp1 \,\wedge\, \exists\, v \in T \,:\, exp2))$$

is illegal because the declaration of v in the second $\exists\, v$ lies inside the scope of its declaration in the first $\exists\, v$. Although conventional mathematics and programming languages allow such redeclarations, TLA$^+$ forbids them because they can lead to confusion and errors.

3.5   Comments

Even simple specifications like the ones in modules *AsynchInterface* and *Channel* can be hard to understand from the mathematics alone. That's why I began with an intuitive explanation of the interface. That explanation made it easier for you to understand formula *Spec* in the module, which is the actual specification. Every specification should be accompanied by an informal prose explanation. The explanation may be in an accompanying document, or it may be included as comments in the specification.

Figure 3.3 on the next page shows how the hour clock's specification in module *HourClock* might be explained by comments. In the typeset version, comments are distinguished from the specification itself by the use of a different font. As shown in the figure, TLA$^+$ provides two ways of writing comments in the ASCII version. A comment may appear anywhere enclosed between (* and *). An end-of-line comment is preceded by *. Comments may be nested, so you can comment out a section of a specification by enclosing it between (* and *), even if the section contains comments.

A comment almost always appears on a line by itself or at the end of a line. I put a comment between *HCnxt* and $\stackrel{\Delta}{=}$ just to show that it can be done.

To save space, I will write few comments in the example specifications. But specifications should have lots of comments. Even if there is an accompanying document describing the system, comments are needed to help the reader understand how the specification formalizes that description.

Comments can help solve a problem posed by the logical structure of a specification. A symbol has to be declared or defined before it can be used. In module *Channel*, the definition of *Spec* has to follow the definition of *Next*, which has to follow the definitions of *Send* and *Rcv*. But it's usually easiest to

---

────────────────── MODULE *HourClock* ──────────────────

This module specifies a digital clock that displays the current hour. It ignores real time, not specifying when the display can change.

EXTENDS *Naturals*

VARIABLE *hr*   Variable *hr* represents the display.

$HCini \triangleq hr \in (1 \,..\, 12)$ Initially, *hr* can have any value from 1 through 12.

HCnxt This is a weird place for a comment. $\triangleq$

The value of *hr* cycles from 1 through 12.

$hr' = \text{IF } hr \neq 12 \text{ THEN } hr + 1 \text{ ELSE } 1$

$HC \triangleq HCini \land \Box[HCnxt]_{hr}$

The complete spec. It permits the clock to stop.

THEOREM $HC \Rightarrow \Box HCini$ Type-correctness of the spec.

```
--------------------- MODULE HourClock ---------------------
(***********************************************************)
(* This module specifies a digital clock that displays  *)
(* the current hour.  It ignores real time, not         *)
(* specifying when the display can change.               *)
(***********************************************************)
EXTENDS Naturals
VARIABLE hr      \* Variable hr represents the display.
HCini == hr \in (1 .. 12)   \* Initially, hr can have any
                            \* value from 1 through 12.
HCnxt (* This is a weird place for a comment. *)  ==
  (**************************************************)
  (* The value of hr cycles from 1 through 12.   *)
  (**************************************************)
  hr' = IF hr # 12 THEN hr + 1 ELSE 1
HC  ==  HCini /\ [][HCnxt]_hr
  (* The complete spec.  It permits the clock to stop. *)
------------------------------------------------------------
THEOREM  HC => []HCini  \* Type-correctness of the spec.
============================================================
```

Figure 3.3: The hour-clock specification with comments.

understand a top-down description of a system. We would probably first want to read the declarations of *Data* and *chan*, then the definition of *Spec*, then the definitions of *Init* and *Next*, and then the definitions of *Send* and *Rcv*. In other words, we want to read the specification more or less from bottom to top. This is easy enough to do for a module as short as *Channel*; it's inconvenient for longer specifications. We can use comments to guide the reader through a longer specification. For example, we could precede the definition of *Send* in the *Channel* module with the comment

> Actions *Send* and *Rcv* below are the disjuncts of the next-state action *Next*.

The module structure also allows us to choose the order in which a specification is read. For example, we can rewrite the hour-clock specification by splitting the *HourClock* module into three separate modules:

HCVar   A module that declares the variable *hr*.

HCActions A module that EXTENDS modules *Naturals* and *HCVar* and defines *HCini* and *HCnxt*.

HCSpec   A module that EXTENDS module *HCActions*, defines formula *HC*, and asserts the type-correctness theorem.

The EXTENDS relation implies a logical ordering of the modules: *HCVar* precedes *HCActions*, which precedes *HCSpec*. But the modules don't have to be read in that order. The reader can be told to read *HCVar* first, then *HCSpec*, and finally *HCActions*. The INSTANCE construct introduced below in Chapter 4 provides another tool for modularizing specifications.

Splitting a tiny specification like *HourClock* in this way would be ludicrous. But the proper splitting of modules can help make a large specification easier to read. When writing a specification, you should decide in what order it should be read. You can then design the module structure to permit reading it in that order, when each individual module is read from beginning to end. Finally, you should ensure that the comments within each module make sense when the different modules are read in the appropriate order.

Chapter 4

A FIFO

Our next example is a FIFO buffer, called a FIFO for short—a device with which a sender process transmits a sequence of values to a receiver. The sender and receiver use two channels, *in* and *out*, to communicate with the buffer:

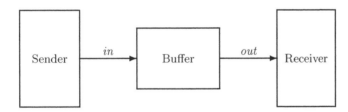

Values are sent over *in* and *out* using the asynchronous protocol specified by the *Channel* module of Figure 3.2 on page 30. The system's specification will allow behaviors with four kinds of nonstuttering steps: *Send* and *Rcv* steps on both the *in* channel and the *out* channel.

4.1 The Inner Specification

The specification of the FIFO first EXTENDS modules *Naturals* and *Sequences*. The *Sequences* module defines operations on finite sequences. We represent a finite sequence as a tuple, so the sequence of three numbers 3, 2, 1 is the triple $\langle 3, 2, 1 \rangle$. The *Sequences* module defines the following operators on sequences:

$Seq(S)$ The set of all sequences of elements of the set S. For example, $\langle 3, 7 \rangle$ is an element of $Seq(Nat)$.

$Head(s)$ The first element of sequence s. For example, $Head(\langle 3, 7 \rangle)$ equals 3.

$Tail(s)$ The tail of sequence s, which consists of s with its head removed. For example, $Tail(\langle 3, 7 \rangle)$ equals $\langle 7 \rangle$.

$Append(s, e)$ The sequence obtained by appending element e to the tail of sequence s. For example, $Append(\langle 3, 7 \rangle, 3)$ equals $\langle 3, 7, 3 \rangle$.

$s \circ t$ The sequence obtained by concatenating the sequences s and t. For example, $\langle 3, 7 \rangle \circ \langle 3 \rangle$ equals $\langle 3, 7, 3 \rangle$. (We type $\circ$ in ASCII as \o.)

$Len(s)$ The length of sequence s. For example, $Len(\langle 3, 7 \rangle)$ equals 2.

The FIFO's specification continues by declaring the constant *Message*, which represents the set of all messages that can be sent.[1] It then declares the variables. There are three variables: *in* and *out*, representing the channels, and a third variable q that represents the queue of buffered messages. The value of q is the sequence of messages that have been sent by the sender but not yet received by the receiver. (Section 4.3 has more to say about this additional variable q.)

We want to use the definitions in the *Channel* module to specify operations on the channels *in* and *out*. This requires two instances of that module—one in which the variable *chan* of the *Channel* module is replaced with the variable *in* of our current module, and the other in which *chan* is replaced with *out*. In both instances, the constant *Data* of the *Channel* module is replaced with *Message*. We obtain the first of these instances with the statement

$$InChan \ \triangleq \ \textsc{instance} \ Channel \ \textsc{with} \ Data \leftarrow Message, \ chan \leftarrow in$$

For every symbol σ defined in module *Channel*, this defines $InChan!\sigma$ to have the same meaning in the current module as σ had in module *Channel*, except with *Message* substituted for *Data* and *in* substituted for *chan*. For example, this statement defines $InChan!TypeInvariant$ to equal

$$in \ \in \ [val : Message, \ rdy : \{0, 1\}, \ ack : \{0, 1\}]$$

(The statement does *not* define $InChan!Data$ because *Data* is declared, not defined, in module *Channel*.) We introduce our second instance of the *Channel* module with the analogous statement

$$OutChan \ \triangleq \ \textsc{instance} \ Channel \ \textsc{with} \ Data \leftarrow Message, \ chan \leftarrow out$$

The initial states of the *in* and *out* channels are specified by $InChan!Init$ and $OutChan!Init$. Initially, no messages have been sent or received, so q should

---

[1] I like to use a singular noun like *Message* rather than a plural like *Messages* for the name of a set. That way, the $\in$ in the expression $m \in Message$ can be read *is a*. This is the same convention that most programmers use for naming types.

equal the empty sequence. The empty sequence is the 0-tuple (there's only one, and it's written $\langle \rangle$), so we define the initial predicate to be

$$
Init \quad \triangleq \quad \wedge\ InChan!Init \\
\wedge\ OutChan!Init \\
\wedge\ q = \langle \rangle
$$

We next define the type invariant. The type invariants for *in* and *out* come from the *Channel* module, and the type of q is the set of finite sequences of messages. The type invariant for the FIFO specification is therefore

$$
TypeInvariant \quad \triangleq \quad \wedge\ InChan!TypeInvariant \\
\wedge\ OutChan!TypeInvariant \\
\wedge\ q \in Seq(Message)
$$

The four kinds of nonstuttering steps allowed by the next-state action are described by four actions:

$SSend(msg)$ The sender sends message *msg* on the *in* channel.

$BufRcv$ The buffer receives the message from the *in* channel and appends it to the tail of q.

$BufSend$ The buffer removes the message from the head of q and sends it on channel *out*.

$RRcv$ The receiver receives the message from the *out* channel.

The definitions of these actions, along with the rest of the specification, are in module *InnerFIFO* of Figure 4.1 on the next page. The reason for the adjective *Inner* is explained in Section 4.3 below.

4.2 Instantiation Examined

The INSTANCE statement is seldom used except in one idiom for hiding variables, which is described in Section 4.3. So, most readers can skip this section and go directly to page 41.

4.2.1 Instantiation Is Substitution

Consider the definition of *Next* in module *Channel* (page 30). We can remove every defined symbol that appears in that definition by using the symbol's definition. For example, we can eliminate the expression $Send(d)$ by expanding the definition of *Send*. We can repeat this process. For example, the "$-$" that appears in the expression $1 - @$ (obtained by expanding the definition of *Send*)

───────────────────── MODULE *InnerFIFO* ─────────────────────

EXTENDS *Naturals*, *Sequences*
CONSTANT *Message*
VARIABLES *in*, *out*, *q*
InChan $\triangleq$ INSTANCE *Channel* WITH *Data* ← *Message*, *chan* ← *in*
OutChan $\triangleq$ INSTANCE *Channel* WITH *Data* ← *Message*, *chan* ← *out*

───

Init $\triangleq$ ∧ *InChan*!*Init*
             ∧ *OutChan*!*Init*
             ∧ *q* = ⟨ ⟩

TypeInvariant $\triangleq$ ∧ *InChan*!*TypeInvariant*
                    ∧ *OutChan*!*TypeInvariant*
                    ∧ *q* ∈ *Seq*(*Message*)

SSend(*msg*) $\triangleq$ ∧ *InChan*!*Send*(*msg*)   Send *msg* on channel *in*.
                 ∧ UNCHANGED ⟨*out*, *q*⟩

BufRcv $\triangleq$ ∧ *InChan*!*Rcv*   Receive message from channel *in*
             ∧ *q'* = *Append*(*q*, *in*.*val*)   and append it to tail of *q*.
             ∧ UNCHANGED *out*

BufSend $\triangleq$ ∧ *q* ≠ ⟨ ⟩   Enabled only if *q* is nonempty.
              ∧ *OutChan*!*Send*(*Head*(*q*))   Send *Head*(*q*) on channel *out*
              ∧ *q'* = *Tail*(*q*)   and remove it from *q*.
              ∧ UNCHANGED *in*

RRcv $\triangleq$ ∧ *OutChan*!*Rcv*   Receive message from channel *out*.
             ∧ UNCHANGED ⟨*in*, *q*⟩

Next $\triangleq$ ∨ ∃ *msg* ∈ *Message* : *SSend*(*msg*)
             ∨ *BufRcv*
             ∨ *BufSend*
             ∨ *RRcv*

Spec $\triangleq$ *Init* ∧ □[*Next*]$_{⟨in,\,out,\,q⟩}$

───

THEOREM *Spec* ⇒ □*TypeInvariant*

───

Figure 4.1: The specification of a FIFO, with the internal variable *q* visible.

can be eliminated by using the definition of "$-$" from the *Naturals* module. Continuing in this way, we eventually obtain a definition for *Next* in terms of only the built-in operators of TLA$^+$ and the parameters *Data* and *chan* of the *Channel* module. We consider this to be the "real" definition of *Next* in module *Channel*. The statement

$$InChan \triangleq \text{INSTANCE } Channel \text{ WITH } Data \leftarrow Message, \ chan \leftarrow in$$

in module *InnerFIFO* defines *InChan!Next* to be the formula obtained from this real definition of *Next* by substituting *Message* for *Data* and *in* for *chan*. This defines *InChan!Next* in terms of only the built-in operators of TLA$^+$ and the parameters *Message* and *in* of module *InnerFIFO*.

Let's now consider an arbitrary INSTANCE statement

$$IM \triangleq \text{INSTANCE } M \text{ WITH } p_1 \leftarrow e_1, \ \ldots, \ p_n \leftarrow e_n$$

Let Σ be a symbol defined in module M and let d be its "real" definition. The INSTANCE statement defines $IM!\Sigma$ to have as its real definition the expression obtained from d by replacing all instances of p_i by the expression e_i, for each i. The definition of $IM!\Sigma$ must contain only the parameters (declared constants and variables) of the current module, not the ones of module M. Hence, the p_i must consist of all the parameters of module M. The e_i must be expressions that are meaningful in the current module.

4.2.2 Parametrized Instantiation

The FIFO specification uses two instances of module *Channel*—one with *in* substituted for *chan* and the other with *out* substituted for *chan*. We could instead use a single parametrized instance by putting the following statement in module *InnerFIFO*:

$$Chan(ch) \triangleq \text{INSTANCE } Channel \text{ WITH } Data \leftarrow Message, chan \leftarrow ch$$

For any symbol Σ defined in module *Channel* and any expression *exp*, this defines *Chan(exp)!Σ* to equal formula Σ with *Message* substituted for *Data* and *exp* substituted for *chan*. The *Rcv* action on channel *in* could then be written *Chan(in)!Rcv*, and the *Send(msg)* action on channel *out* could be written *Chan(out)!Send(msg)*.

The instantiation above defines *Chan!Send* to be an operator with two arguments. Writing *Chan(out)!Send(msg)* instead of *Chan!Send(out, msg)* is just an idiosyncrasy of the syntax. It is no stranger than the syntax for infix operators, which has us write $a + b$ instead of $+(a, b)$.

Parametrized instantiation is used almost exclusively in the TLA$^+$ idiom for variable hiding, described in Section 4.3. You can use that idiom without understanding it, so you probably don't need to know anything about parametrized instantiation.

4.2.3   Implicit Substitutions

The use of *Message* as the name for the set of transmitted values in the FIFO specification is a bit strange, since we had just used the name *Data* for the analogous set in the asynchronous channel specifications. Suppose we had used *Data* in place of *Message* as the constant parameter of module *InnerFIFO*. The first instantiation statement would then have been

$$InChan \;\triangleq\; \text{INSTANCE } Channel \text{ WITH } Data \leftarrow Data, chan \leftarrow in$$

The substitution $Data \leftarrow Data$ indicates that the constant parameter *Data* of the instantiated module *Channel* is replaced with the expression *Data* of the current module. TLA$^+$ allows us to drop any substitution of the form $\Sigma \leftarrow \Sigma$, for a symbol Σ. So, the statement above can be written as

$$InChan \;\triangleq\; \text{INSTANCE } Channel \text{ WITH } chan \leftarrow in$$

We know there is an implied $Data \leftarrow Data$ substitution because an INSTANCE statement must have a substitution for every parameter of the instantiated module. If some parameter p has no explicit substitution, then there is an implicit substitution $p \leftarrow p$. This means that the INSTANCE statement must lie within the scope of a declaration or definition of the symbol p.

It is quite common to instantiate a module with this kind of implicit substitution. Often, every parameter has an implicit substitution, in which case the list of explicit substitutions is empty. The WITH is then omitted.

4.2.4   Instantiation Without Renaming

So far, all the instantiations we've used have been with renaming. For example, the first instantiation of module *Channel* renames the defined symbol *Send* as *InChan!Send*. This kind of renaming is necessary if we are using multiple instances of the module, or a single parametrized instance. The two instances *InChan!Init* and *OutChan!Init* of *Init* in module *InnerFIFO* are different formulas, so they need different names.

Sometimes we need only a single instance of a module. For example, suppose we are specifying a system with only a single asynchronous channel. We then need only one instance of *Channel*, so we don't have to rename the instantiated symbols. In that case, we can write something like

$$\text{INSTANCE } Channel \text{ WITH } Data \leftarrow D, chan \leftarrow x$$

This instantiates *Channel* with no renaming, but with substitution. Thus, it defines *Rcv* to be the formula of the same name from the *Channel* module, except with D substituted for *Data* and x substituted for *chan*. The expressions substituted for an instantiated module's parameters must be defined. So, this INSTANCE statement must be within the scope of the definitions or declarations of D and x.

4.3   Hiding the Queue

Module *InnerFIFO* of Figure 4.1 defines *Spec* to be *Init* $\wedge\ \Box[\textit{Next}]_{\ldots}$, the sort of formula we've become accustomed to as a system specification. However, formula *Spec* describes the value of variable q, as well as of the variables *in* and *out*. The picture of the FIFO system I drew on page 35 shows only channels *in* and *out*; it doesn't show anything inside the boxes. A specification of the FIFO should describe only the values sent and received on the channels. The variable q, which represents what's going on inside the box labeled *Buffer*, is used to specify what values are sent and received. It is an *internal* variable and, in the final specification, it should be hidden.

In TLA, we hide a variable with the existential quantifier $\exists$ of temporal logic. The formula $\exists\, x : F$ is true of a behavior iff there exists some sequence of values—one in each state of the behavior—that can be assigned to the variable x that will make formula F true. (The meaning of $\exists$ is defined more precisely in Section 8.8.)

The obvious way to write a FIFO specification in which q is hidden is with the formula $\exists\, q : Spec$. However, we can't put this definition in module *InnerFIFO* because q is already declared there, and a formula $\exists\, q : \ldots$ would redeclare it. Instead, we use a new module with a parametrized instantiation of the *InnerFIFO* module (see Section 4.2.2 on page 39):

$$\overline{\qquad\qquad\qquad \text{MODULE } \textit{FIFO} \qquad\qquad\qquad}$$

CONSTANT *Message*
VARIABLES *in, out*

$Inner(q) \;\triangleq\;$ INSTANCE *InnerFIFO*
$Spec \;\triangleq\; \exists\, q \,:\, Inner(q)!\,Spec$

Observe that the INSTANCE statement is an abbreviation for

$Inner(q) \;\triangleq\;$ INSTANCE *InnerFIFO*
                   WITH $q \leftarrow q,\ in \leftarrow in,\ out \leftarrow out,\ Message \leftarrow Message$

The variable parameter q of module *InnerFIFO* is instantiated with the parameter q of the definition of *Inner*. The other parameters of the *InnerFIFO* module are instantiated with the parameters of module *FIFO*.

If this seems confusing, don't worry about it. Just learn the TLA$^+$ idiom for hiding variables used here and be content with its intuitive meaning. In fact, for most applications, there's no need to hide variables in the specification. You can just write the inner specification and note in the comments which variables should be regarded as visible and which as internal (hidden).

4.4   A Bounded FIFO

We have specified an unbounded FIFO—a buffer that can hold an unbounded number of messages. Any real system has a finite amount of resources, so it can contain only a bounded number of in-transit messages. In many situations, we wish to abstract away the bound on resources and describe a system in terms of unbounded FIFOs. In other situations, we may care about that bound. We then want to strengthen our specification by placing a bound N on the number of outstanding messages.

A specification of a bounded FIFO differs from our specification of the unbounded FIFO only in that action *BufRcv* should not be enabled unless there are fewer than N messages in the buffer—that is, unless $Len(q)$ is less than N. It would be easy to write a complete new specification of a bounded FIFO by copying module *InnerFIFO* and just adding the conjunct $Len(q) < N$ to the definition of *BufRcv*. But let's use module *InnerFIFO* as it is, rather than copying it.

The next-state action *BNext* for the bounded FIFO is the same as the FIFO's next-state action *Next* except that it allows a *BufRcv* step only if $Len(q)$ is less than N. In other words, *BNext* should allow a step only if (i) it's a *Next* step and (ii) if it's a *BufRcv* step, then $Len(q) < N$ is true in the first state. In other words, *BNext* should equal

$$Next \ \land \ (BufRcv \Rightarrow (Len(q) < N))$$

Module *BoundedFIFO* in Figure 4.2 on the next page contains the specification. It introduces the new constant parameter N. It also contains the statement

ASSUME $(N \in Nat) \land (N > 0)$

which asserts that, in this module, we are assuming that N is a positive natural number. Such an assumption has no effect on any definitions made in the module. However, it may be taken as a hypothesis when proving any theorems asserted in the module. In other words, a module asserts that its assumptions imply its theorems. It's a good idea to state this kind of simple assumption about constants.

An ASSUME statement should be used only to make assumptions about constants. The formula being assumed should not contain any variables. It might be tempting to assert type declarations as assumptions—for example, to add to module *InnerFIFO* the assumption $q \in Seq(Message)$. However, that would be wrong because it asserts that, in any state, q is a sequence of messages. As we observed in Section 3.3, a state is a completely arbitrary assignment of values to variables, so there are states in which q has the value $\sqrt{-17}$. Assuming that such a state doesn't exist would lead to a logical contradiction.

You may wonder why module *BoundedFIFO* assumes that N is a positive natural, but doesn't assume that *Message* is a set. Similarly, why didn't we

4.5. WHAT WE'RE SPECIFYING43

MODULE *BoundedFIFO*

EXTENDS *Naturals, Sequences*
VARIABLES *in, out*
CONSTANT *Message, N*
ASSUME $(N \in Nat) \wedge (N > 0)$

$Inner(q) \triangleq$ INSTANCE *InnerFIFO*

$BNext(q) \triangleq \wedge Inner(q)!Next$
$\qquad\qquad\quad \wedge Inner(q)!BufRcv \Rightarrow (Len(q) < N)$

$Spec \triangleq \exists q : Inner(q)!Init \wedge \Box[BNext(q)]_{\langle in,out,q \rangle}$

Figure 4.2: A specification of a FIFO buffer of length *N*.

assume that the constant parameter *Data* in our asynchronous interface specifications is a set? The answer is that, in TLA$^+$, every value is a set.[2] A value like the number 3, which we don't think of as a set, is formally a set. We just don't know what its elements are. The formula $2 \in 3$ is a perfectly reasonable one, but TLA$^+$ does not specify whether it's true or false. So, we don't have to assume that *Message* is a set because we know that it is one.

Although *Message* is automatically a set, it isn't necessarily a finite set. For example, *Message* could be instantiated with the set *Nat* of natural numbers. If you want to assume that a constant parameter is a finite set, then you need to state this as an assumption. (You can do this with the *IsFiniteSet* operator from the *FiniteSets* module, described in Section 6.1.) However, most specifications make perfect sense for infinite sets of messages or processors, so there is no reason to assume these sets to be finite.

4.5 What We're Specifying

I wrote at the beginning of this chapter that we were going to specify a FIFO buffer. Formula *Spec* of the *FIFO* module actually specifies a set of behaviors, each representing a sequence of sending and receiving operations on the channels *in* and *out*. The sending operations on *in* are performed by the sender, and the receiving operations on *out* are performed by the receiver. The sender and receiver are not part of the FIFO buffer; they form its *environment*.

Our specification describes a system consisting of the FIFO buffer and its environment. The behaviors satisfying formula *Spec* of module *FIFO* represent those histories of the universe in which both the system and its environment

---

[2]TLA$^+$ is based on the mathematical formalism known as Zermelo-Fränkel set theory, also called ZF.

behave correctly. It's often helpful in understanding a specification to indicate explicitly which steps are system steps and which are environment steps. We can do this by defining the next-state action to be

$$Next \;\triangleq\; SysNext \vee EnvNext$$

where *SysNext* describes system steps and *EnvNext* describes environment steps. For the FIFO, we have

$$SysNext \;\triangleq\; BufRcv \vee BufSend$$
$$EnvNext \;\triangleq\; (\exists\, msg \in Message : SSend(msg)) \vee RRcv$$

While suggestive, this way of defining the next-state action has no formal significance. The specification *Spec* equals $Init \wedge \Box[Next]_{...}$; changing the way we structure the definition of *Next* doesn't change its meaning. If a behavior fails to satisfy *Spec*, nothing tells us if the system or its environment is to blame.

A formula like *Spec*, which describes the correct behavior of both the system and its environment, is called a *closed-system* or *complete-system* specification. An *open-system* specification is one that describes only the correct behavior of the system. A behavior satisfies an open-system specification if it represents a history in which either the system operates correctly, or it failed to operate correctly only because its environment did something wrong. Section 10.7 explains how to write open-system specifications.

Open-system specifications are philosophically more satisfying. However, closed-system specifications are a little easier to write, and the mathematics underlying them is simpler. So, we almost always write closed-system specifications. It's usually quite easy to turn a closed-system specification into an open-system specification. But in practice, there's seldom any reason to do so.

Chapter 5

A Caching Memory

A memory system consists of a set of processors connected to a memory by some abstract interface, which we label *memInt*.

In this section we specify what the memory is supposed to do, then we specify a particular implementation of the memory using caches. We begin by specifying the memory interface, which is common to both specifications.

5.1 The Memory Interface

The asynchronous interface described in Chapter 3 uses a handshake protocol. Receipt of a data value must be acknowledged before the next data value can be sent. In the memory interface, we abstract away this kind of detail and represent both the sending of a data value and its receipt as a single step. We call it a *Send* step if a processor is sending the value to the memory; it's a *Reply* step if the memory is sending to a processor. Processors do not send values to one another, and the memory sends to only one processor at a time.

We represent the state of the memory interface by the value of the variable *memInt*. A *Send* step changes *memInt* in some way, but we don't want to specify exactly how. The way to leave something unspecified in a specification is to make it a parameter. For example, in the bounded FIFO of Section 4.4, we left the size of the buffer unspecified by making it a parameter N. We'd therefore like to declare a parameter *Send* so that *Send*(p, d) describes how *memInt* is changed by a step that represents processor p sending data value d to the memory. However, TLA$^+$ provides only CONSTANT and VARIABLE parameters, not action parameters.[1] So, we declare *Send* to be a constant operator and write *Send*$(p, d, memInt, memInt')$ instead of *Send*(p, d).

In TLA$^+$, we declare *Send* to be a constant operator that takes four arguments by writing

CONSTANT *Send*$(\_, \_, \_, \_)$

This means that *Send*$(p, d, miOld, miNew)$ is an expression, for any expressions p, d, *miOld*, and *miNew*, but it says nothing about what the value of that expression is. We want it to be a Boolean value that is true iff a step in which *memInt* equals *miOld* in the first state and *miNew* in the second state represents the sending by p of value d to the memory.[2] We can assert that the value is a Boolean by the assumption

ASSUME $\forall\ p, d, miOld, miNew$:
$\qquad$ *Send*$(p, d, miOld, miNew) \in$ BOOLEAN

This asserts that the formula

Send$(p, d, miOld, miNew) \in$ BOOLEAN

is true for all values of p, d, *miOld*, and *miNew*. The built-in symbol BOOLEAN denotes the set {TRUE, FALSE}, whose elements are the two Boolean values TRUE and FALSE.

This ASSUME statement asserts formally that the value of

Send$(p, d, miOld, miNew)$

is a Boolean. But the only way to assert formally what that value signifies would be to say what it actually equals—that is, to define *Send* rather than making it a parameter. We don't want to do that, so we just state informally what the value means. This statement is part of the intrinsically informal description of the relation between our mathematical abstraction and a physical memory system.

To allow the reader to understand the specification, we have to describe informally what *Send* means. The ASSUME statement asserting that *Send*$(\ldots)$ is a Boolean is then superfluous as an explanation. But it's a good idea to include it anyway.

---

[1] Even if TLA$^+$ allowed us to declare an action parameter, we would have no way to specify that a *Send*(p, d) action constrains only *memInt* and not other variables.

[2] We expect *Send*$(p, d, miOld, miNew)$ to have this meaning only when p is a processor and d a value that p is allowed to send, but we simplify the specification a bit by requiring it to be a Boolean for all values of p and d.

A specification that uses the memory interface can use the operators *Send* and *Reply* to specify how the variable *memInt* changes. The specification must also describe *memInt*'s initial value. We therefore declare a constant parameter *InitMemInt* that is the set of possible initial values of *memInt*.

We also introduce three constant parameters that are needed to describe the interface:

Proc The set of processor identifiers. (We usually shorten *processor identifier* to *processor* when referring to an element of *Proc*.)

Adr The set of memory addresses.

Val The set of possible memory values that can be assigned to an address.

Finally, we define the values that the processors and memory send to one another over the interface. A processor sends a request to the memory. We represent a request as a record with an *op* field that specifies the type of request and additional fields that specify its arguments. Our simple memory allows only read and write requests. A read request has *op* field "Rd" and an *adr* field specifying the address to be read. The set of all read requests is therefore the set

$$[op : \{\text{"Rd"}\},\ adr : Adr]$$

of all records whose *op* field equals "Rd" (is an element of the set $\{\text{"Rd"}\}$ whose only element is the string "Rd") and whose *adr* field is an element of *Adr*. A write request must specify the address to be written and the value to write. It is represented by a record with *op* field equal to "Wr", and with *adr* and *val* fields specifying the address and value. We define *MReq*, the set of all requests, to equal the union of these two sets. (Set operations, including union, are described in Section 1.2 on page 11.)

The memory responds to a read request with the memory value it read. We will also have it respond to a write request, and it seems nice to let the response be different from the response to any read request. We therefore require the memory to respond to a write request by returning a value *NoVal* that is different from any memory value. We could declare *NoVal* to be a constant parameter and add the assumption $NoVal \notin Val$. (The symbol $\notin$ is typed in ASCII as \notin.) But it's best, when possible, to avoid introducing parameters. Instead, we define *NoVal* by

$$NoVal \;\triangleq\; \text{CHOOSE } v \,:\, v \notin Val$$

The expression CHOOSE $x : F$ equals an arbitrarily chosen value x that satisfies the formula F. (If no such x exists, the expression has a completely arbitrary value.) This statement defines *NoVal* to be some value that is not an element of

```
┌──────────────────────── MODULE MemoryInterface ──────────────────────────┐
│                                                                           │
│ VARIABLE memInt                                                           │
│ CONSTANTS  Send(_, _, _, _),    A Send(p, d, memInt, memInt′) step represents processor p │
│                                 sending value d to the memory.            │
│                                                                           │
│            Reply(_, _, _, _),   A Reply(p, d, memInt, memInt′) step represents the memory │
│                                 sending value d to processor p.           │
│                                                                           │
│            InitMemInt,   The set of possible initial values of memInt.    │
│            Proc,         The set of processor identifiers.                │
│            Adr,          The set of memory addresses.                     │
│            Val           The set of memory values.                        │
│                                                                           │
│ ASSUME ∀ p, d, miOld, miNew :  ∧ Send(p, d, miOld, miNew) ∈ BOOLEAN       │
│                                ∧ Reply(p, d, miOld, miNew) ∈ BOOLEAN      │
├───────────────────────────────────────────────────────────────────────────┤
│                                                                           │
│ MReq  ≜  [op : {"Rd"}, adr : Adr]  ∪  [op : {"Wr"}, adr : Adr, val : Val] │
│      The set of all requests; a read specifies an address, a write specifies an address and a value. │
│                                                                           │
│ NoVal  ≜  CHOOSE v : v ∉ Val    An arbitrary value not in Val.            │
│                                                                           │
└───────────────────────────────────────────────────────────────────────────┘
```

Figure 5.1: The specification of a memory interface.

Val. We have no idea what the value of *NoVal* is; we just know what it isn't—namely, that it isn't an element of *Val*. The CHOOSE operator is discussed in Section 6.6 on page 73.

The complete memory interface specification is module *MemoryInterface* in Figure 5.1 on this page.

5.2   Functions

A memory assigns values to addresses. The state of the memory is therefore an assignment of elements of *Val* (memory values) to elements of *Adr* (memory addresses). In a programming language, such an assignment is called an array of type *Val* indexed by *Adr*. In mathematics, it's called a function from *Adr* to *Val*. Before writing the memory specification, let's look at the mathematics of functions, and how it is described in TLA$^+$.

A function f has a domain, written DOMAIN f, and it assigns to each element x of its domain the value $f[x]$. (Mathematicians write this as $f(x)$, but TLA$^+$ uses the array notation of programming languages, with square brackets.) Two functions f and g are equal iff they have the same domain and $f[x] = g[x]$ for all x in their domain.

The *range* of a function f is the set of all values of the form $f[x]$ with x in DOMAIN f. For any sets S and T, the set of all functions whose domain equals S and whose range is any subset of T is written $[S \to T]$.

Ordinary mathematics does not have a convenient notation for writing an expression whose value is a function. TLA$^+$ defines $[x \in S \mapsto e]$ to be the function f with domain S such that $f[x] = e$ for every $x \in S$.[3] For example,

$$succ \;\triangleq\; [n \in Nat \mapsto n + 1]$$

defines *succ* to be the successor function on the natural numbers—the function with domain *Nat* such that $succ[n] = n + 1$ for all $n \in Nat$.

A record is a function whose domain is a finite set of strings. For example, a record with *val*, *ack*, and *rdy* fields is a function whose domain is the set $\{$ "val", "ack", "rdy"$\}$ consisting of the three strings "val", "ack", and "rdy". The expression $r.ack$, the *ack* field of a record r, is an abbreviation for $r[$"ack"$]$. The record

$$[val \mapsto 42, \; ack \mapsto 1, \; rdy \mapsto 0]$$

can be written

$$[i \in \{ \text{"val"}, \text{"ack"}, \text{"rdy"}\} \mapsto$$
$$\text{IF } i = \text{"val"} \text{ THEN } 42 \text{ ELSE IF } i = \text{"ack"} \text{ THEN } 1 \text{ ELSE } 0]$$

The EXCEPT construct for records, explained in Section 3.2, is a special case of a general EXCEPT construct for functions, where $!.c$ is an abbreviation for $![$"c"$]$. For any function f, the expression $[f \text{ EXCEPT } ![c] = e]$ is the function $\hat{f}$ that is the same as f except with $\hat{f}[c] = e$. This function can also be written

$$[x \in \text{DOMAIN } f \mapsto \text{ IF } x = c \text{ THEN } e \text{ ELSE } f[x]]$$

assuming that the symbol x does not occur in any of the expressions f, c, and e. For example, $[succ \text{ EXCEPT } ![42] = 86]$ is the function g that is the same as *succ* except that $g[42]$ equals 86 instead of 43.

As in the EXCEPT construct for records, the expression e in

$$[f \text{ EXCEPT } ![c] = e]$$

can contain the symbol @, where it means $f[c]$. For example,

$$[succ \text{ EXCEPT } ![42] = 2 * @] \;\; = \;\; [succ \text{ EXCEPT } ![42] = 2 * succ[42]]$$

In general,

$$[f \text{ EXCEPT } ![c_1] = e_1, \; \ldots, \; ![c_n] = e_n]$$

---

[3]The $\in$ in $[x \in S \mapsto e]$ is just part of the syntax; TLA$^+$ uses that particular symbol to help you remember what the construct means. Computer scientists write $\lambda x : S.e$ to represent something similar to $[x \in S \mapsto e]$, except that their λ expressions aren't quite the same as the functions of ordinary mathematics that are used in TLA$^+$.

is the function $\hat{f}$ that is the same as f except with $\hat{f}[c_i] = e_i$ for each i. More precisely, this expression equals

$$[\ldots[[f \text{ EXCEPT } ![c_1] = e_1] \text{ EXCEPT } ![c_2] = e_2] \ldots \text{ EXCEPT } ![c_n] = e_n]$$

Functions correspond to the arrays of programming languages. The domain of a function corresponds to the index set of an array. Function $[f \text{ EXCEPT } ![c] = e]$ corresponds to the array obtained from f by assigning e to $f[c]$. A function whose range is a set of functions corresponds to an array of arrays. TLA$^+$ defines $[f \text{ EXCEPT } ![c][d] = e]$ to be the function corresponding to the array obtained by assigning e to $f[c][d]$. It can be written as

$$[f \text{ EXCEPT } ![c] = [@ \text{ EXCEPT } ![d] = e]]$$

The generalization to $[f \text{ EXCEPT } ![c_1] \ldots [c_n] = e]$ for any n should be obvious. Since a record is a function, this notation can be used for records as well. TLA$^+$ uniformly maintains the notation that $\sigma.c$ is an abbreviation for $\sigma[\text{``c''}]$. For example, this implies

$$[f \text{ EXCEPT } ![c].d = e] = [f \text{ EXCEPT } ![c][\text{``d''}] = e]$$
$$= [f \text{ EXCEPT } ![c] = [@ \text{ EXCEPT } !.d = e]]$$

The TLA$^+$ definition of records as functions makes it possible to manipulate them in ways that have no counterparts in programming languages. For example, we can define an operator R such that $R(r, s)$ is the record obtained from r by replacing the value of each field c that is also a field of the record s with $s.c$. In other words, for every field c of r, if c is a field of s then $R(r, s).c = s.c$; otherwise $R(r, s).c = r.c$. The definition is

$$R(r, s) \;\triangleq\; [c \in \text{DOMAIN } r \mapsto \text{IF } c \in \text{DOMAIN } s \text{ THEN } s[c] \text{ ELSE } r[c]]$$

So far, we have seen only functions of a single argument, which are the mathematical analog of the one-dimensional arrays of programming languages. Mathematicians also use functions of multiple arguments, which are the analog of multi-dimensional arrays. In TLA$^+$, as in ordinary mathematics, a function of multiple arguments is one whose domain is a set of tuples. For example, $f[5, 3, 1]$ is an abbreviation for $f[\langle 5, 3, 1 \rangle]$, the value of the function f applied to the triple $\langle 5, 3, 1 \rangle$.

The function constructs of TLA$^+$ have extensions for functions of multiple arguments. For example, $[g \text{ EXCEPT } ![a, b] = e]$ is the function $\hat{g}$ that is the same as g except with $\hat{g}[a, b]$ equal to e. The expression

(5.1) $[n \in Nat, \, r \in Real \mapsto n * r]$

equals the function f such that $f[n, r]$ equals $n * r$, for all $n \in Nat$ and $r \in Real$. Just as $\forall i \in S : \forall j \in S : P$ can be written as $\forall i, j \in S : P$, we can write the function $[i \in S, \, j \in S \mapsto e]$ as $[i, j \in S \mapsto e]$.

Section 16.1.7 on page 301 describes the general versions of the TLA$^+$ function constructs for functions with any number of arguments. However, functions of a single argument are all you're likely to need. You can almost always replace a function of multiple arguments with a function-valued function—for example, writing $f[a][b]$ instead of $f[a, b]$.

5.3 A Linearizable Memory

We now specify a very simple memory system in which a processor p issues a memory request and then waits for a response before issuing the next request. In our specification, the request is executed by accessing (reading or modifying) a variable *mem*, which represents the current state of the memory. Because the memory can receive requests from other processors before responding to processor p, it matters when *mem* is accessed. We let the access of *mem* occur any time between the request and the response. This specifies what is called a *linearizable* memory. Less restrictive, more practical memory specifications are described in Section 11.2.

In addition to *mem*, the specification has the internal variables *ctl* and *buf*, where *ctl*[p] describes the status of processor p's request, and *buf*[p] contains either the request or the response. Consider the request *req* that equals

$$[op \mapsto \text{``Wr''}, \ adr \mapsto a, \ val \mapsto v]$$

It is a request to write v to memory address a, and it generates the response *NoVal*. The processing of this request is represented by the following three steps:

$$\begin{bmatrix} ctl[p] & = & \text{``rdy''} \\ buf[p] & = & \cdots \\ mem[a] & = & \cdots \end{bmatrix} \xrightarrow{Req(p)} \begin{bmatrix} ctl[p] & = & \text{``busy''} \\ buf[p] & = & req \\ mem[a] & = & \cdots \end{bmatrix}$$

$$\xrightarrow{Do(p)} \begin{bmatrix} ctl[p] & = & \text{``done''} \\ buf[p] & = & NoVal \\ mem[a] & = & v \end{bmatrix} \xrightarrow{Rsp(p)} \begin{bmatrix} ctl[p] & = & \text{``rdy''} \\ buf[p] & = & NoVal \\ mem[a] & = & v \end{bmatrix}$$

A *Req(p)* step represents the issuing of a request by processor p. It is enabled when *ctl*[p] = "rdy"; it sets *ctl*[p] to "busy" and sets *buf*[p] to the request. A *Do(p)* step represents the memory access; it is enabled when *ctl*[p] = "busy" and it sets *ctl*[p] to "done" and *buf*[p] to the response. A *Rsp(p)* step represents the memory's response to p; it is enabled when *ctl*[p] = "done" and it sets *ctl*[p] to "rdy".

Writing the specification is a straightforward exercise in representing these changes to the variables in TLA$^+$ notation. The internal specification, with *mem*, *ctl*, and *buf* visible (free variables), appears in module *InternalMemory* on the following two pages. The memory specification, which hides the three internal variables, is module *Memory* in Figure 5.3 on page 53.

─────────────── MODULE *InternalMemory* ───────────────

EXTENDS *MemoryInterface*
VARIABLES *mem*, *ctl*, *buf*

───

IInit ≜   The initial predicate
    ∧ *mem* ∈ [*Adr* → *Val*]   Initially, memory locations have any values in *Val*,
    ∧ *ctl* = [*p* ∈ *Proc* ↦ "rdy"]   each processor is ready to issue requests,
    ∧ *buf* = [*p* ∈ *Proc* ↦ *NoVal*]   each *buf*[*p*] is arbitrarily initialized to *NoVal*,
    ∧ *memInt* ∈ *InitMemInt*   and *memInt* is any element of *InitMemInt*.

TypeInvariant ≜   The type-correctness invariant.
    ∧ *mem* ∈ [*Adr* → *Val*]   *mem* is a function from *Adr* to *Val*.
    ∧ *ctl* ∈ [*Proc* → {"rdy", "busy", "done"}]   *ctl*[*p*] equals "rdy", "busy", or "done".
    ∧ *buf* ∈ [*Proc* → *MReq* ∪ *Val* ∪ {*NoVal*}]   *buf*[*p*] is a request or a response.

Req(*p*) ≜   Processor *p* issues a request.
    ∧ *ctl*[*p*] = "rdy"   Enabled iff *p* is ready to issue a request.
    ∧ ∃ *req* ∈ *MReq* :   For some request *req*:
         ∧ *Send*(*p*, *req*, *memInt*, *memInt*′)   Send *req* on the interface.
         ∧ *buf*′ = [*buf* EXCEPT ![*p*] = *req*]   Set *buf*[*p*] to the request.
         ∧ *ctl*′ = [*ctl* EXCEPT ![*p*] = "busy"]   Set *ctl*[*p*] to "busy".
    ∧ UNCHANGED *mem*

Do(*p*) ≜   Perform *p*'s request to memory.
    ∧ *ctl*[*p*] = "busy"   Enabled iff *p*'s request is pending.
    ∧ *mem*′ = IF *buf*[*p*].*op* = "Wr"
                  THEN [*mem* EXCEPT   Write to memory on a
                            ![*buf*[*p*].*adr*] = *buf*[*p*].*val*]   "Wr" request.
                  ELSE *mem*   Leave *mem* unchanged on a "Rd" request.
    ∧ *buf*′ = [*buf* EXCEPT
                  ![*p*] = IF *buf*[*p*].*op* = "Wr"   Set *buf*[*p*] to the response:
                           THEN *NoVal*   *NoVal* for a write;
                           ELSE *mem*[*buf*[*p*].*adr*]]   the memory value for a read.
    ∧ *ctl*′ = [*ctl* EXCEPT ![*p*] = "done"]   Set *ctl*[*p*] to "done".
    ∧ UNCHANGED *memInt*

Figure 5.2a: The internal memory specification (beginning).

$Rsp(p) \triangleq$ Return the response to p's request.

   $\wedge\ ctl[p] =$ "done"   Enabled iff req. is done but resp. not sent.

   $\wedge\ Reply(p,\ buf[p],\ memInt,\ memInt')$ Send the response on the interface.

   $\wedge\ ctl' = [ctl$ EXCEPT $![p] =$ "rdy"$]$ Set $ctl[p]$ to "rdy".

   $\wedge$ UNCHANGED $\langle mem, buf \rangle$

$INext \triangleq \exists\, p \in Proc\ :\ Req(p) \vee Do(p) \vee Rsp(p)$ The next-state action.

$ISpec \triangleq IInit \wedge \square[INext]_{\langle memInt,\, mem,\, ctl,\, buf \rangle}$ The specification.

THEOREM $ISpec \Rightarrow \square TypeInvariant$

Figure 5.2b: The internal memory specification (end).

5.4 Tuples as Functions

Before writing our caching memory specification, let's take a closer look at tuples. Recall that $\langle a, b, c \rangle$ is the 3-tuple with components a, b, and c. In TLA$^+$, this 3-tuple is actually the function with domain $\{1, 2, 3\}$ that maps 1 to a, 2 to b, and 3 to c. Thus, $\langle a, b, c \rangle[2]$ equals b.

TLA$^+$ provides the Cartesian product operator $\times$ of ordinary mathematics, where $A \times B \times C$ is the set of all 3-tuples $\langle a, b, c \rangle$ such that $a \in A$, $b \in B$, and $c \in C$. Note that $A \times B \times C$ is different from $A \times (B \times C)$, which is the set of pairs $\langle a, p \rangle$ with a in A and p in the set of pairs $B \times C$.

The *Sequences* module defines finite sequences to be tuples. Hence, a sequence of length n is a function with domain $1 .. n$. In fact, s is a sequence iff it equals $[i \in 1 .. Len(s) \mapsto s[i]]$. Below are a few operator definitions from the *Sequences* module. (The meanings of the operators are described in Section 4.1.)

$$Head(s) \triangleq s[1]$$
$$Tail(s) \triangleq [i \in 1 .. (Len(s) - 1) \mapsto s[i+1]]$$
$$s \circ t \triangleq [i \in 1 .. (Len(s) + Len(t)) \mapsto$$
$$\text{IF } i \le Len(s) \text{ THEN } s[i] \text{ ELSE } t[i - Len(s)]]$$

— MODULE *Memory* —

EXTENDS *MemoryInterface*

$Inner(mem, ctl, buf) \triangleq$ INSTANCE *InternalMemory*

$Spec \triangleq \exists\, mem, ctl, buf\ :\ Inner(mem, ctl, buf)!ISpec$

Figure 5.3: The memory specification.

5.5   Recursive Function Definitions

We need one more tool to write the caching memory specification: recursive function definitions. Recursively defined functions are familiar to programmers. The classic example is the factorial function, which I'll call *fact*. It's usually defined by writing

$$fact[n] \;=\; \text{IF } n = 0 \text{ THEN } 1 \text{ ELSE } n * fact[n-1]$$

for all $n \in Nat$. The TLA$^+$ notation for writing functions suggests trying to define *fact* by

$$fact \;\triangleq\; [n \in Nat \mapsto \text{IF } n = 0 \text{ THEN } 1 \text{ ELSE } n * fact[n-1]]$$

This definition is illegal because the occurrence of *fact* to the right of the $\triangleq$ is undefined—*fact* is defined only after its definition.

TLA$^+$ does allow the apparent circularity of recursive function definitions. We can define the factorial function *fact* by

$$fact[n \in Nat] \;\triangleq\; \text{IF } n = 0 \text{ THEN } 1 \text{ ELSE } n * fact[n-1]$$

In general, a definition of the form $f[x \in S] \triangleq e$ can be used to define recursively a function f with domain S.

The function definition notation has a straightforward generalization to definitions of functions of multiple arguments. For example,

$$\begin{aligned}
Acker[m, \, n \in Nat] \;&\triangleq\; \\
\text{IF } m = 0 \text{ THEN } &\; n + 1 \\
\text{ELSE IF } n = 0 \text{ THEN } &\; Acker[m-1, \, 0] \\
\text{ELSE } &\; Acker[m-1, \, Acker[m, \, n-1]]
\end{aligned}$$

defines $Acker[m, n]$ for all natural numbers m and n.

Section 6.3 explains exactly what recursive definitions mean. For now, we will just write recursive definitions without worrying about their meaning.

5.6   A Write-Through Cache

We now specify a simple write-through cache that implements the memory specification. The system is described by the picture of Figure 5.4 on the next page. Each processor p communicates with a local controller, which maintains three state components: $buf[p]$, $ctl[p]$, and $cache[p]$. The value of $cache[p]$ represents the processor's cache; $buf[p]$ and $ctl[p]$ play the same role as in the internal memory specification (module *InternalMemory*). (However, as we will see below, $ctl[p]$ can assume an additional value "waiting".) These local controllers

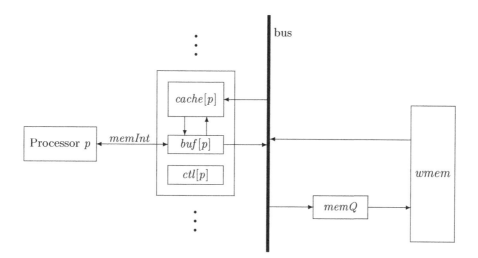

Figure 5.4: The write-through cache.

communicate with the main memory *wmem*,[4] and with one another, over a bus. Requests from the processors to the main memory are in the queue *memQ* of maximum length *QLen*.

A write request by processor p is performed by the action *DoWr*(p). This is a write-through cache, meaning that every write request updates main memory. So, the *DoWr*(p) action writes the value into *cache*[p] and adds the write request to the tail of *memQ*. When the request reaches the head of *memQ*, the action *MemQWr* stores the value in *wmem*. The *DoWr*(p) action also updates *cache*[q] for every other processor q that has a copy of the address in its cache.

A read request by processor p is performed by the action *DoRd*(p), which obtains the value from the cache. If the value is not in the cache, the action *RdMiss*(p) adds the request to the tail of *memQ* and sets *ctl*[p] to "waiting". When the enqueued request reaches the head of *memQ*, the action *MemQRd* reads the value and puts it in *cache*[p], enabling the *DoRd*(p) action.

We might expect the *MemQRd* action to read the value from *wmem*. However, this could cause an error if there is a write to that address enqueued in *memQ* behind the read request. In that case, reading the value from memory could lead to two processors having different values for the address in their caches: the one that issued the read request, and the one that issued the write request that followed the read in *memQ*. So, the *MemQRd* action must read the value from the last write to that address in *memQ*, if there is such a write; otherwise, it reads the value from *wmem*.

---

[4]We use the name *wmem* to distinguish this variable from variable *mem* of module *InternalMemory*. We don't have to, since *mem* is not a free (visible) variable of the actual memory specification in module *Memory*, but it helps us avoid getting confused.

Eviction of an address from processor p's cache is represented by a separate *Evict(p)* action. Since all cached values have been written to memory, eviction does nothing but remove the address from the cache. There is no reason to evict an address until the space is needed, so in an implementation, this action would be executed only when a request for an uncached address is received from p and p's cache is full. But that's a performance optimization; it doesn't affect the correctness of the algorithm, so it doesn't appear in the specification. We allow a cached address to be evicted from p's cache at any time—except if the address was just put there by a *MemQRd* action for a read request whose *DoRd(p)* action has not yet been performed. This is the case when *ctl[p]* equals "waiting" and *buf[p].adr* equals the cached address.

The actions *Req(p)* and *Rsp(p)*, which represent processor p issuing a request and the memory issuing a reply to p, are the same as the corresponding actions of the memory specification, except that they also leave the new variables *cache* and *memQ* unchanged, and they leave unchanged *vmem* instead of *mem*.

To specify all these actions, we must decide how the processor caches and the queue of requests to memory are represented by the variables *memQ* and *cache*. We let *memQ* be a sequence of pairs of the form $\langle p, req \rangle$, where *req* is a request and p is the processor that issued it. For any memory address a, we let *cache[p][a]* be the value in p's cache for address a (the "copy" of a in p's cache). If p's cache does not have a copy of a, we let *cache[p][a]* equal *NoVal*.

The specification appears in module *WriteThroughCache* on pages 57–59. I'll now go through this specification, explaining some of the finer points and some notation that we haven't encountered before.

The EXTENDS, declaration statements, and ASSUME are familiar. We can reuse some of the definitions from the *InternalMemory* module, so an INSTANCE statement instantiates a copy of that module with *wmem* substituted for *mem*. (The other parameters of module *InternalMemory* are instantiated by the parameters of the same name in module *WriteThroughCache*.)

The initial predicate *Init* contains the conjunct *M!IInit*, which asserts that *ctl* and *buf* have the same initial values as in the internal memory specification, and that *wmem* has the same initial value as *mem* does in that specification. The write-through cache allows *ctl[p]* to have the value "waiting" that it didn't in the internal memory specification, so we can't reuse the internal memory's type invariant *M!TypeInvariant*. Formula *TypeInvariant* therefore explicitly describes the types of *wmem*, *ctl*, and *buf*. The type of *memQ* is the set of sequences of $\langle$processor, request$\rangle$ pairs.

The module next defines the predicate *Coherence*, which asserts the basic cache coherence property of the write-through cache: for any processors p and q and any address a, if p and q both have copies of address a in their caches, then those copies are equal. Note the trick of writing $x \notin \{y, z\}$ instead of the equivalent but longer formula $(x \neq y) \wedge (x \neq z)$.

─────────────── MODULE *WriteThroughCache* ───────────────

EXTENDS *Naturals, Sequences, MemoryInterface*
VARIABLES *wmem, ctl, buf, cache, memQ*
CONSTANT *QLen*
ASSUME $(QLen \in Nat) \wedge (QLen > 0)$
$M \triangleq$ INSTANCE *InternalMemory* WITH *mem* ← *wmem*

───

$Init \triangleq$ The initial predicate
   $\wedge$ *M!IInit*   *wmem, buf,* and *ctl* are initialized as in the internal memory spec.
   $\wedge$ *cache* =   All caches are initially empty ($cache[p][a] = NoVal$ for all p, a).
      $[p \in Proc \mapsto [a \in Adr \mapsto NoVal]]$
   $\wedge$ $memQ = \langle \rangle$ The queue *memQ* is initially empty.

$TypeInvariant \triangleq$ The type invariant.
   $\wedge$ *wmem* $\in [Adr \rightarrow Val]$
   $\wedge$ *ctl* $\in [Proc \rightarrow \{\text{"rdy"}, \text{"busy"}, \text{"waiting"}, \text{"done"}\}]$
   $\wedge$ *buf* $\in [Proc \rightarrow MReq \cup Val \cup \{NoVal\}]$
   $\wedge$ *cache* $\in [Proc \rightarrow [Adr \rightarrow Val \cup \{NoVal\}]]$
   $\wedge$ $memQ \in Seq(Proc \times MReq)$ *memQ* is a sequence of ⟨proc., request⟩ pairs.

$Coherence \triangleq$ Asserts that if two processors' caches both have copies
  $\forall p, q \in Proc, a \in Adr :$ of an address, then those copies have equal values.
   $(NoVal \notin \{cache[p][a], cache[q][a]\}) \Rightarrow (cache[p][a] = cache[q][a])$

───

$Req(p) \triangleq$ Processor p issues a request.
  $M!Req(p) \wedge$ UNCHANGED $\langle cache, memQ \rangle$

$Rsp(p) \triangleq$ The system issues a response to processor p.
  $M!Rsp(p) \wedge$ UNCHANGED $\langle cache, memQ \rangle$

$RdMiss(p) \triangleq$ Enqueue a request to write value from memory to p's cache.
   $\wedge$ $(ctl[p] = \text{"busy"}) \wedge (buf[p].op = \text{"Rd"})$ Enabled on a read request when
   $\wedge$ $cache[p][buf[p].adr] = NoVal$ the address is not in p's cache
   $\wedge$ $Len(memQ) < QLen$ and *memQ* is not full.
   $\wedge$ $memQ' = Append(memQ, \langle p, buf[p] \rangle)$ Append ⟨p, request⟩ to *memQ*.
   $\wedge$ $ctl' = [ctl$ EXCEPT $![p] = \text{"waiting"}]$ Set $ctl[p]$ to "waiting".
   $\wedge$ UNCHANGED $\langle memInt, wmem, buf, cache \rangle$

Figure 5.5a: The write-through cache specification (beginning).

$DoRd(p)$ $\triangleq$ Perform a read by p of a value in its cache.
  $\quad \wedge \; ctl[p] \in \{ \text{``busy''}, \text{``waiting''} \}$ Enabled if a read
  $\quad \wedge \; buf[p].op = \text{``Rd''}$ request is pending and
  $\quad \wedge \; cache[p][buf[p].adr] \neq NoVal$ address is in cache.
  $\quad \wedge \; buf' = [buf \text{ EXCEPT } ![p] = cache[p][buf[p].adr]]$ Get result from cache.
  $\quad \wedge \; ctl' \; = [ctl \text{ EXCEPT } ![p] = \text{``done''}]$ Set $ctl[p]$ to "done".
  $\quad \wedge \; \text{UNCHANGED} \; \langle memInt, wmem, cache, memQ \rangle$

$DoWr(p)$ $\triangleq$ Write to p's cache, update other caches, and enqueue memory update.
  $\quad \text{LET} \;\; r \;\; \triangleq \;\; buf[p]$ Processor p's request.
  $\quad \text{IN} \quad \wedge \; (ctl[p] = \text{``busy''}) \wedge (r.op = \text{``Wr''})$ Enabled if write request pending
  $\qquad\qquad \wedge \; Len(memQ) < QLen$ and $memQ$ is not full.
  $\qquad\qquad \wedge \; cache' = $ Update p's cache and any other cache that has a copy.
  $\qquad\qquad\quad [q \in Proc \mapsto \text{IF } (p = q) \vee (cache[q][r.adr] \neq NoVal)$
  $\qquad\qquad\qquad\qquad\qquad \text{THEN } [cache[q] \text{ EXCEPT } ![r.adr] = r.val]$
  $\qquad\qquad\qquad\qquad\qquad \text{ELSE } \;\; cache[q]]$
  $\qquad\qquad \wedge \; memQ' = Append(memQ, \langle p, r \rangle)$ Enqueue write at tail of $memQ$.
  $\qquad\qquad \wedge \; buf' = [buf \text{ EXCEPT } ![p] = NoVal]$ Generate response.
  $\qquad\qquad \wedge \; ctl' \; = [ctl \text{ EXCEPT } ![p] = \text{``done''}]$ Set ctl to indicate request is done.
  $\qquad\qquad \wedge \; \text{UNCHANGED} \; \langle memInt, wmem \rangle$

$vmem$ $\triangleq$ The value $wmem$ will have after all the writes in $memQ$ are performed.
  $\quad \text{LET} \; f[i \in 0 .. Len(memQ)] \;\; \triangleq$ The value $wmem$ will have after the first
  $\qquad\qquad \text{IF } \; i = 0 \; \text{THEN } \; wmem$ i writes in $memQ$ are performed.
  $\qquad\qquad\qquad \text{ELSE } \; \text{IF } \; memQ[i][2].op = \text{``Rd''}$
  $\qquad\qquad\qquad\qquad \text{THEN } \; f[i-1]$
  $\qquad\qquad\qquad\qquad \text{ELSE } \; [f[i-1] \text{ EXCEPT } ![memQ[i][2].adr] =$
  $\qquad\qquad\qquad\qquad\qquad\qquad\qquad\qquad memQ[i][2].val]$
  $\quad \text{IN} \quad f[Len(memQ)]$

$MemQWr$ $\triangleq$ Perform write at head of $memQ$ to memory.
  $\quad \text{LET} \; r \;\; \triangleq \;\; Head(memQ)[2]$ The request at the head of $memQ$.
  $\quad \text{IN} \quad \wedge \; (memQ \neq \langle \rangle) \wedge (r.op = \text{``Wr''})$ Enabled if $Head(memQ)$ a write.
  $\qquad\qquad \wedge \; wmem' =$ Perform the write to memory.
  $\qquad\qquad\quad [wmem \text{ EXCEPT } ![r.adr] = r.val]$
  $\qquad\qquad \wedge \; memQ' = Tail(memQ)$ Remove the write from $memQ$.
  $\qquad\qquad \wedge \; \text{UNCHANGED} \; \langle memInt, buf, ctl, cache \rangle$

Figure 5.5b: The write-through cache specification (middle).

$MemQRd \stackrel{\Delta}{=}$ Perform an enqueued read to memory.

  LET $p \stackrel{\Delta}{=} Head(memQ)[1]$ The requesting processor.

   $r \stackrel{\Delta}{=} Head(memQ)[2]$ The request at the head of $memQ$.

  IN $\wedge\ (memQ \neq \langle\rangle) \wedge (r.op = \text{``Rd''})$ Enabled if $Head(memQ)$ is a read.

   $\wedge\ memQ' = Tail(memQ)$ Remove the head of $memQ$.

   $\wedge\ cache' =$ Put value from memory or $memQ$ in p's cache.

    $[cache \text{ EXCEPT } ![p][r.adr] = vmem[r.adr]]$

   $\wedge\ \text{UNCHANGED } \langle memInt, wmem, buf, ctl \rangle$

$Evict(p, a) \stackrel{\Delta}{=}$ Remove address a from p's cache.

   $\wedge\ (ctl[p] = \text{``waiting''}) \Rightarrow (buf[p].adr \neq a)$ Can't evict a if it was just read

   $\wedge\ cache' = [cache \text{ EXCEPT } ![p][a] = NoVal]$ into cache from memory.

   $\wedge\ \text{UNCHANGED } \langle memInt, wmem, buf, ctl, memQ \rangle$

$Next \stackrel{\Delta}{=}\ \vee\ \exists\, p \in Proc :\ \vee\ Req(p) \vee Rsp(p)$

     $\vee\ RdMiss(p) \vee DoRd(p) \vee DoWr(p)$

     $\vee\ \exists\, a \in Adr :\ Evict(p, a)$

    $\vee\ MemQWr \vee MemQRd$

$Spec \stackrel{\Delta}{=}\ Init \wedge \square[Next]_{\langle memInt, wmem, buf, ctl, cache, memQ \rangle}$

THEOREM $Spec \Rightarrow \square(TypeInvariant \wedge Coherence)$

$LM \stackrel{\Delta}{=}$ INSTANCE $Memory$ The memory spec. with internal variables hidden.

THEOREM $Spec \Rightarrow LM\,!\,Spec$ Formula $Spec$ implements the memory spec.

Figure 5.5c: The write-through cache specification (end).

The actions $Req(p)$ and $Rsp(p)$, which represent a processor sending a request and receiving a reply, are essentially the same as the corresponding actions in module $InternalMemory$. However, they must also specify that the variables $cache$ and $memQ$, not present in module $InternalMemory$, are left unchanged.

In the definition of $RdMiss$, the expression $Append(memQ, \langle p, buf[p]\rangle)$ is the sequence obtained by appending the element $\langle p, buf[p]\rangle$ to the end of $memQ$.

The $DoRd(p)$ action represents the performing of the read from p's cache. If $ctl[p] = \text{``busy''}$, then the address was originally in the cache. If $ctl[p] = \text{``waiting''}$, then the address was just read into the cache from memory.

The $DoWr(p)$ action writes the value to p's cache and updates the value in any other caches that have copies. It also enqueues a write request in $memQ$. In an implementation, the request is put on the bus, which transmits it to the other caches and to the $memQ$ queue. In our high-level view of the system, we represent all this as a single step.

The definition of *DoWr* introduces the TLA$^+$ LET/IN construct. The LET clause consists of a sequence of definitions whose scope extends until the end of the IN clause. In the definition of *DoWr*, the LET clause defines r to equal $buf[p]$ within the IN clause. Observe that the definition of r contains the parameter p of the definition of *DoWr*. Hence, we could not move the definition of r outside the definition of *DoWr*.

A definition in a LET is just like an ordinary definition in a module; in particular, it can have parameters. These local definitions can be used to shorten an expression by replacing common subexpressions with an operator. In the definition of *DoWr*, I replaced five instances of $buf[p]$ by the single symbol r. This was a silly thing to do, because it makes almost no difference in the length of the definition and it requires the reader to remember the definition of the new symbol r. But using a LET to eliminate common subexpressions can often greatly shorten and simplify an expression.

A LET can also be used to make an expression easier to read, even if the operators it defines appear only once in the IN expression. We write a specification with a sequence of definitions, instead of just defining a single monolithic formula, because a formula is easier to understand when presented in smaller chunks. The LET construct allows the process of splitting a formula into smaller parts to be done hierarchically. A LET can appear as a subexpression of an IN expression. Nested LETs are common in large, complicated specifications.

Next comes the definition of the state function *vmem*, which is used in defining action *MemQRd* below. It equals the value that the main memory *wmem* will have after all the write operations currently in *memQ* have been performed. Recall that the value read by *MemQRd* must be the most recent one written to that address—a value that may still be in *memQ*. That value is the one in *vmem*. The function *vmem* is defined in terms of the recursively defined function f, where $f[i]$ is the value *wmem* will have after the first i operations in *memQ* have been performed. Note that $memQ[i][2]$ is the second component (the request) of $memQ[i]$, the i-th element in the sequence *memQ*.

The next two actions, *MemQWr* and *MemQRd*, represent the processing of the request at the head of the *memQ* queue—*MemQWr* for a write request, and *MemQRd* for a read request. These actions also use a LET to make local definitions. Here, the definitions of p and r could be moved before the definition of *MemQWr*. In fact, we could save space by replacing the two local definitions of r with one global (within the module) definition. However, making the definition of r global in this way would be somewhat distracting, since r is used only in the definitions of *MemQWr* and *MemQRd*. It might be better instead to combine these two actions into one. Whether you put a definition into a LET or make it more global should depend on what makes the specification easier to read.

The *Evict*(p, a) action represents the operation of removing address a from processor p's cache. As explained above, we allow an address to be evicted at any time—unless the address was just written to satisfy a pending read request, which is the case iff $ctl[p] =$ "waiting" and $buf[p].adr = a$. Note the use of the "double subscript" in the EXCEPT expression of the action's second conjunct. This conjunct "assigns $NoVal$ to $cache[p][a]$". If address a is not in p's cache, then $cache[p][a]$ already equals $NoVal$ and an $Evict(p, a)$ step is a stuttering step.

The definitions of the next-state action $Next$ and of the complete specification $Spec$ are straightforward. The module closes with two theorems that are discussed next.

5.7 Invariance

Module $WriteThroughCache$ contains the theorem

THEOREM $Spec \Rightarrow \Box(TypeInvariant \wedge Coherence)$

which asserts that $TypeInvariant \wedge Coherence$ is an invariant of $Spec$. A state predicate $P \wedge Q$ is always true iff both P and Q are always true, so $\Box(P \wedge Q)$ is equivalent to $\Box P \wedge \Box Q$. This implies that the theorem above is equivalent to the two theorems

THEOREM $Spec \Rightarrow \Box TypeInvariant$

THEOREM $Spec \Rightarrow \Box Coherence$

The first theorem is the usual type-invariance assertion. The second, which asserts that $Coherence$ is an invariant of $Spec$, expresses an important property of the algorithm.

Although $TypeInvariant$ and $Coherence$ are both invariants of the temporal formula $Spec$, they differ in a fundamental way. If s is any state satisfying $TypeInvariant$, then any state t such that $s \rightarrow t$ is a $Next$ step also satisfies $TypeInvariant$. This property is expressed by

THEOREM $TypeInvariant \wedge Next \Rightarrow TypeInvariant'$

(Recall that $TypeInvariant'$ is the formula obtained by priming all the variables in formula $TypeInvariant$.) In general, when $P \wedge N \Rightarrow P'$ holds, we say that predicate P is an invariant of action N. Predicate $TypeInvariant$ is an invariant of $Spec$ because it is an invariant of $Next$ and it is implied by the initial predicate $Init$.

An invariant of a specification S that is also an invariant of its next-state action is sometimes called an *inductive* invariant of S.

Predicate $Coherence$ is not an invariant of the next-state action $Next$. For example, suppose s is a state in which

- $cache[p1][a] = 1$

- $cache[q][b] = NoVal$, for all $\langle q, b \rangle$ different from $\langle p1, a \rangle$

- $wmem[a] = 2$

- $memQ$ contains the single element $\langle p2, [op \mapsto \text{"Rd"}, adr \mapsto a] \rangle$

for two different processors $p1$ and $p2$ and some address a. Such a state s (an assignment of values to variables) exists, assuming that there are at least two processors and at least one address. Then *Coherence* is true in state s. Let t be the state obtained from s by taking a *MemQRd* step. In state t, we have $cache[p2][a] = 2$ and $cache[p1][a] = 1$, so *Coherence* is false. Hence *Coherence* is not an invariant of the next-state action.

Coherence is an invariant of formula *Spec* because states like s cannot occur in a behavior satisfying *Spec*. Proving its invariance is not so easy. We must find a predicate *Inv* that is an invariant of *Next* such that *Inv* implies *Coherence* and is implied by the initial predicate *Init*.

Important properties of a specification can often be expressed as invariants. Proving that a state predicate P is an invariant of a specification means proving a formula of the form

$$Init \wedge \Box[Next]_v \Rightarrow \Box P$$

This is done by finding an appropriate state predicate *Inv* and proving

$$Init \Rightarrow Inv, \qquad Inv \wedge [Next]_v \Rightarrow Inv', \qquad Inv \Rightarrow P$$

Since our subject is specification, not proof, I won't discuss how to find *Inv*.

5.8   Proving Implementation

Module *WriteThroughCache* ends with the theorem

THEOREM $Spec \Rightarrow LM!Spec$

where $LM!Spec$ is formula *Spec* of module *Memory*. This theorem asserts that every behavior satisfying specification *Spec* of the write-through cache also satisfies $LM!Spec$, the specification of a linearizable memory. In other words, it asserts that the write-through cache implements a linearizable memory. In TLA, implementation is implication. A system described by a formula *Sys* implements a specification *Spec* iff *Sys* implies *Spec*—that is, iff $Sys \Rightarrow Spec$ is a theorem. TLA makes no distinction between system descriptions and specifications; they are both just formulas.

By definition of formula *Spec* of the *Memory* module (page 53), we can restate the theorem as

THEOREM $Spec \Rightarrow \exists\, mem, ctl, buf : LM!Inner(mem, ctl, buf)!ISpec$

where $LM!Inner(mem, ctl, buf)!ISpec$ is formula *ISpec* of the *InternalMemory* module. The rules of logic tell us that to prove such a theorem, we must find "witnesses" for the quantified variables *mem*, *ctl*, and *buf*. These witnesses are state functions (ordinary expressions with no primes), which I'll call *omem*, *octl*, and *obuf*, that satisfy

(5.2) $Spec \Rightarrow LM\,!\,Inner(omem,\ octl,\ obuf)\,!\,ISpec$

Formula $LM\,!\,Inner(omem,\ octl,\ obuf)\,!\,ISpec$ is formula *ISpec* with the substitutions

$$mem \leftarrow omem, \quad ctl \leftarrow octl, \quad buf \leftarrow obuf$$

The tuple $\langle\, omem,\ octl,\ obuf\, \rangle$ of witness functions is called a *refinement mapping*, and we describe (5.2) as the assertion that *Spec* implements formula *ISpec* under this refinement mapping. Intuitively, this means *Spec* implies that the value of the tuple $\langle\, memInt,\ omem,\ octl,\ obuf\, \rangle$ of state functions changes the way *ISpec* asserts that the tuple $\langle\, memInt,\ mem,\ ctl,\ buf\, \rangle$ of variables should change.

I will now briefly describe how we prove (5.2); for details, see the technical papers about TLA, available through the TLA Web page. Let me first introduce a bit of non-TLA$^+$ notation. For any formula F of module *InternalMemory*, let $\overline{F}$ equal $LM\,!\,Inner(omem,\ octl,\ obuf)\,!\,F$, which is formula F with *omem*, *octl*, and *obuf* substituted for *mem*, *ctl*, and *buf*. In particular, $\overline{mem}$, $\overline{ctl}$, and $\overline{buf}$ equal *omem*, *octl*, and *obuf*, respectively.

With this notation, we can write (5.2) as $Spec \Rightarrow \overline{ISpec}$. Replacing *Spec* and *ISpec* by their definitions, this formula becomes

(5.3) $Init \wedge \Box[Next]_{\langle memInt,\, wmem,\, buf,\, ctl,\, cache,\, memQ \rangle}$

$\Rightarrow\ \overline{IInit} \wedge \Box[\overline{INext}]_{\langle memInt,\, \overline{mem},\, \overline{ctl},\, \overline{buf} \rangle}$

> $\overline{memInt}$ equals *memInt*, since *memInt* is a variable distinct from *mem*, *ctl*, and *buf*.

Formula (5.3) is then proved by finding an invariant *Inv* of *Spec* such that

$\wedge\ Init \Rightarrow \overline{IInit}$

$\wedge\ Inv \wedge Next\ \Rightarrow\ \vee\ \overline{INext}$

$\qquad\qquad\qquad \vee\ \text{UNCHANGED}\ \langle\, memInt,\ \overline{mem},\ \overline{ctl},\ \overline{buf}\, \rangle$

The second conjunct is called *step simulation*. It asserts that a *Next* step starting in a state satisfying the invariant *Inv* is either an $\overline{INext}$ step—a step that changes the 4-tuple $\langle\, memInt,\ omem,\ octl,\ obuf\, \rangle$ the way an *INext* step changes $\langle\, memInt,\ mem,\ ctl,\ buf\, \rangle$—or else it leaves that 4-tuple unchanged. For our memory specifications, the state functions *omem*, *octl*, and *obuf* are defined by

$omem \ \triangleq\ vmem$

$octl \quad \triangleq\ [p \in Proc \mapsto \text{IF}\ \ ctl[p] = \text{``waiting''}\ \text{THEN}\ \text{``busy''}\ \text{ELSE}\ ctl[p]]$

$obuf \ \triangleq\ buf$

The mathematics of an implementation proof is simple, so the proof is straightforward—in theory. For specifications of real systems, such proofs can be quite difficult. Going from theory to practice requires turning the mathematics

of proofs into an engineering discipline. This is a subject that deserves a book
to itself, and I won't try to discuss it here.

You will probably never prove that one specification implements another.
However, you should understand refinement mappings and step simulation. You
will then be able to use TLC to check that one specification implements another;
Chapter 14 explains how.

Chapter 6

Some More Math

The mathematics we use to write specifications is built on a small, simple collection of concepts. You've already seen most of what's needed to describe almost any kind of mathematics. All you lack is a handful of operators on sets that are described below in Section 6.1. After learning about them, you will be able to define all the data structures and operations that occur in specifications.

While our mathematics is simple, its foundations are nonobvious—for example, the meanings of recursive function definitions and the CHOOSE operator are subtle. This section discusses some of those foundations. Understanding them will help you use TLA$^+$ more effectively.

6.1   Sets

The simple operations on sets described in Section 1.2 are all you need to write most system specifications. However, you may occasionally have to use more sophisticated operators—especially if you need to define data structures beyond tuples, records, and simple functions.

Two powerful operators of set theory are the unary operators UNION and SUBSET, defined as follows:

UNION S The union of the elements of S. In other words, a value e is an element of UNION S iff it is an element of an element of S. For example:

$$\text{UNION } \{\{1,2\},\{2,3\},\{3,4\}\} \ = \ \{1,2,3,4\}$$

> Mathematicians write UNION S as $\bigcup S$.

SUBSET S The set of all subsets of S. In other words, $T \in$ SUBSET S iff $T \subseteq S$. For example:

$$\text{SUBSET } \{1,2\} \ = \ \{\{\},\{1\},\{2\},\{1,2\}\}$$

> Mathematicians call SUBSET S the *power set* of S and write it $\mathcal{P}(S)$ or 2^S.

65

Mathematicians often describe a set as "the set of all ... such that ...". TLA$^+$ has two constructs that formalize such a description:

$\{x \in S : p\}$ The subset of S consisting of all elements x satisfying property p. For example, the set of odd natural numbers can be written $\{n \in Nat : n \% 2 = 1\}$. The identifier x is bound in p; it may not occur in S.

$\{e : x \in S\}$ The set of elements of the form e, for all x in the set S. For example, $\{2 * n + 1 : n \in Nat\}$ is the set of all odd natural numbers. The identifier x is bound in e; it may not occur in S.

The modulus operator % is described in Section 2.5 on page 21.

The construct $\{e : x \in S\}$ has the same generalizations as $\exists\, x \in S : F$. For example, $\{e : x \in S,\, y \in T\}$ is the set of all elements of the form e, for x in S and y in T. In the construct $\{x \in S : P\}$, we can let x be a tuple. For example, $\{\langle y, z \rangle \in S : P\}$ is the set of all pairs $\langle y, z \rangle$ in the set S that satisfy P. The grammar of TLA$^+$ in Chapter 15 specifies precisely what set expressions you can write.

All the set operators we've seen so far are built-in operators of TLA$^+$. There is also a standard module *FiniteSets* that defines two operators:

Cardinality(S) The number of elements in set S, if S is a finite set.

IsFiniteSet(S) True iff S is a finite set.

The *FiniteSets* module appears on 341. The definition of *Cardinality* is discussed below on page 70.

Careless reasoning about sets can lead to problems. The classic example of this is Russell's paradox:

Let $\mathcal{R}$ be the set of all sets S such that $S \notin S$. The definition of $\mathcal{R}$ implies that $\mathcal{R} \in \mathcal{R}$ is true iff $\mathcal{R} \notin \mathcal{R}$ is true.

The formula $\mathcal{R} \notin \mathcal{R}$ is the negation of $\mathcal{R} \in \mathcal{R}$, and a formula and its negation can neither both be true nor both be false. The source of the paradox is that $\mathcal{R}$ isn't a set. There's no way to write it in TLA$^+$. Intuitively, $\mathcal{R}$ is too big to be a set. A collection $\mathcal{C}$ is too big to be a set if it is as big as the collection of all sets—meaning that we can assign to every set a different element of $\mathcal{C}$. That is, $\mathcal{C}$ is too big to be a set if we can define an operator *SMap* such that

- *SMap*(S) is in $\mathcal{C}$, for any set S.

- If S and T are two different sets, then *SMap*$(S) \neq$ *SMap*(T).

For example, the collection of all sequences of length 2 is too big to be a set; we can define the operator *SMap* by

$$SMap(S) \;\triangleq\; \langle 1, S \rangle$$

This operator assigns to every set S a different sequence of length 2.

6.2 Silly Expressions

Most modern programming languages introduce some form of type checking to prevent you from writing silly expressions like 3/"abc". TLA$^+$ is based on the usual formalization of mathematics by mathematicians, which doesn't have types. In an untyped formalism, every syntactically well-formed expression has a meaning—even a silly expression like 3/"abc". Mathematically, the expression 3/"abc" is no sillier than the expression 3/0, and mathematicians implicitly write that silly expression all the time. For example, consider the true formula

$$\forall\, x \in Real\ :\ (x \neq 0) \Rightarrow (x * (3/x) = 3)$$

where *Real* is the set of all real numbers. This asserts that $(x \neq 0) \Rightarrow (x*(3/x) = 3)$ is true for all real numbers x. Substituting 0 for x yields the true formula $(0 \neq 0) \Rightarrow (0*(3/0) = 3)$ that contains the silly expression 3/0. It's true because $0 \neq 0$ equals FALSE, and FALSE $\Rightarrow P$ is true for any formula P.

A correct formula can contain silly expressions. For example, $3/0 = 3/0$ is a correct formula because any value equals itself. However, the truth of a correct formula cannot depend on the meaning of a silly expression. If an expression is silly, then its meaning is probably unspecified. The definitions of / and * (which are in the standard module *Reals*) don't specify the value of $0 * (3/0)$, so there's no way of knowing whether that value equals 3.

No sensible syntactic rules can prevent you from writing 3/0 without also preventing you from writing perfectly reasonable expressions. The typing rules of programming languages introduce complexity and limitations on what you can write that don't exist in ordinary mathematics. In a well-designed programming language, the costs of types are balanced by benefits: types allow a compiler to produce more efficient code, and type checking catches errors. For programming languages, the benefits seem to outweigh the costs. For writing specifications, I have found that the costs outweigh the benefits.

If you're used to the constraints of programming languages, it may be a while before you start taking advantage of the freedom afforded by mathematics. At first, you won't think of defining anything like the operator R defined on page 50 of Section 5.2, which couldn't be written in a typed programming language.

6.3 Recursion Revisited

Section 5.5 introduced recursive function definitions. Let's now examine what such definitions mean mathematically. Mathematicians usually define the factorial function *fact* by writing

$$fact[n]\ =\ \text{IF}\ n = 0\ \text{THEN}\ 1\ \text{ELSE}\ n * fact[n-1],\ \text{for all }n \in Nat$$

This definition can be justified by proving that it defines a unique function *fact* with domain *Nat*. In other words, *fact* is the unique value satisfying

(6.1) $fact = [n \in Nat \mapsto \text{IF } n = 0 \text{ THEN } 1 \text{ ELSE } n * fact[n-1]]$

The CHOOSE operator, introduced on pages 47–48 of Section 5.1, allows us to express "the value x satisfying property p" as CHOOSE $x : p$. We can therefore define *fact* as follows to be the value satisfying (6.1):

(6.2) $fact \triangleq \text{CHOOSE } fact :$
$$fact = [n \in Nat \mapsto \text{IF } n = 0 \text{ THEN } 1$$
$$\text{ELSE } n * fact[n-1]]$$

(Since the symbol *fact* is not yet defined in the expression to the right of the "$\triangleq$", we can use it as the bound identifier in the CHOOSE expression.) The TLA$^+$ definition

$$fact[n \in Nat] \triangleq \text{IF } n = 0 \text{ THEN } 1 \text{ ELSE } n * fact[n-1]$$

is simply an abbreviation for (6.2). In general, $f[x \in S] \triangleq e$ is an abbreviation for

(6.3) $f \triangleq \text{CHOOSE } f : f = [x \in S \mapsto e]$

TLA$^+$ allows you to write silly definitions. For example, you can write

(6.4) $circ[n \in Nat] \triangleq \text{CHOOSE } y : y \neq circ[n]$

This appears to define *circ* to be a function such that $circ[n] \neq circ[n]$ for any natural number n. There obviously is no such function, so *circ* can't be defined to equal it. A recursive function definition doesn't necessarily define a function. If there is no f that equals $[x \in S \mapsto e]$, then (6.3) defines f to be some unspecified value. Thus, the nonsensical definition (6.4) defines *circ* to be some unknown value.

Although TLA$^+$ allows the apparent circularity of a recursive function definition, it does not allow circular definitions in which two or more functions are defined in terms of one another. Mathematicians occasionally write such mutually recursive definitions. For example, they might try to define functions f and g, with domains equal to the set *Nat*, by writing

$$f[n \in Nat] \triangleq \text{IF } n = 0 \text{ THEN } 17 \text{ ELSE } f[n-1] * g[n]$$
$$g[n \in Nat] \triangleq \text{IF } n = 0 \text{ THEN } 42 \text{ ELSE } f[n-1] + g[n-1]$$

This pair of definitions is not allowed in TLA$^+$.

TLA$^+$ does not allow mutually recursive definitions. However, we can define these functions f and g in TLA$^+$ as follows. We first define a function *mr* such that $mr[n]$ is a record whose f and g fields equal $f[n]$ and $g[n]$, respectively:

$$mr[n \in Nat] \triangleq$$
$$[f \mapsto \text{IF } n = 0 \text{ THEN } 17 \text{ ELSE } mr[n-1].f * mr[n].g,$$
$$g \mapsto \text{IF } n = 0 \text{ THEN } 42 \text{ ELSE } mr[n-1].f + mr[n-1].g]$$

We can then define f and g in terms of mr:

$$f[n \in Nat] \triangleq mr[n].f$$
$$g[n \in Nat] \triangleq mr[n].g$$

This trick can be used to convert any mutually recursive definitions into a single recursive definition of a record-valued function whose fields are the desired functions.

If we want to reason about a function f defined by $f[x \in S] \triangleq e$, we need to prove that there exists an f that equals $[x \in S \mapsto e]$. The existence of f is obvious if f does not occur in e. If it does, so this is a recursive definition, then there is something to prove. Since I'm not discussing proofs, I won't describe how to prove it. Intuitively, you have to check that, as in the case of the factorial function, the definition uniquely determines the value of $f[x]$ for every x in S.

Recursion is a common programming technique because programs must compute values using a small repertoire of simple elementary operations. It's not used as often in mathematical definitions, where we needn't worry about how to compute the value and can use the powerful operators of logic and set theory. For example, the operators *Head*, *Tail*, and $\circ$ are defined in Section 5.4 without recursion, even though computer scientists usually define them recursively. Still, there are some things that are best defined inductively, using a recursive function definition.

6.4 Functions versus Operators

Consider these definitions, which we've seen before:

$$Tail(s) \triangleq [i \in 1 \mathrel{..} (Len(s) - 1) \mapsto s[i+1]]$$
$$fact[n \in Nat] \triangleq \text{IF } n = 0 \text{ THEN } 1 \text{ ELSE } n * fact[n-1]$$

They define two very different kinds of objects: *fact* is a function, and *Tail* is an operator. Functions and operators differ in a few basic ways.

Their most obvious difference is that a function like *fact* by itself is a complete expression that denotes a value, but an operator like *Tail* is not. Both $fact[n] \in S$ and $fact \in S$ are syntactically correct expressions. But, while $Tail(n) \in S$ is syntactically correct, $Tail \in S$ is not. It is gibberish—a meaningless string of symbols, like $x+ > 0$.

Unlike an operator, a function must have a domain, which is a set. We cannot define a function *Tail* so that $Tail[s]$ is the tail of any nonempty sequence s; the domain of such a function would have to include all nonempty sequences, and the collection of all such sequences is too big to be a set. (As explained on page 66, a collection $\mathcal{C}$ is too big to be a set if we can assign to each set a different member of $\mathcal{C}$. The operator *SMap* defined by $SMap(S) \triangleq \langle S \rangle$ assigns to every set a different nonempty sequence.) Hence, we can't define *Tail* to be a function.

Unlike a function, an operator cannot be defined recursively in TLA⁺. However, we can usually transform an illegal recursive operator definition into a nonrecursive one using a recursive function definition. For example, let's try to define the *Cardinality* operator on finite sets. (Recall that the cardinality of a finite set S is the number of elements in S.) The collection of all finite sets is too big to be a set. (The operator $SMap(S) \triangleq \{S\}$ assigns to each set a different set of cardinality 1.) The *Cardinality* operator has a simple intuitive definition:

- $Cardinality(\{\}) = 0$.

- If S is a nonempty finite set, then

$$Cardinality(S) = 1 + Cardinality(S \setminus \{x\})$$

  where x is an arbitrary element of S.

$S \setminus \{x\}$ is the set of all elements in S except x.

Using the CHOOSE operator to describe an arbitrary element of S, we can write this as the more formal-looking, but still illegal, definition

$$Cardinality(S) \;\triangleq\; \qquad \text{This is not a legal TLA}^{+}\text{ definition.}$$
$$\text{IF } \; S = \{\} \; \text{THEN } \; 0$$
$$\text{ELSE } \; 1 + Cardinality(S \setminus \{\text{CHOOSE } x : x \in S\})$$

This definition is illegal because it's circular—only in a recursive function definition can the symbol being defined appear to the right of the $\triangleq$.

To turn this into a legal definition, observe that, for a given finite set S, we can define a function CS such that $CS[T]$ equals the cardinality of T for every subset T of S. The definition is

$$CS[T \in \text{SUBSET } S] \;\triangleq\;$$
$$\text{IF } \; T = \{\} \; \text{THEN } \; 0$$
$$\text{ELSE } \; 1 + CS[T \setminus \{\text{CHOOSE } x : x \in T\}]$$

Since S is a subset of itself, this defines $CS[S]$ to equal $Cardinality(S)$, if S is a finite set. (We don't know or care what $CS[S]$ equals if S is not finite.) So, we can define the *Cardinality* operator by

$$Cardinality(S) \;\triangleq\;$$
$$\text{LET } \; CS[T \in \text{SUBSET } S] \;\triangleq\;$$
$$\text{IF } \; T = \{\} \; \text{THEN } \; 0$$
$$\text{ELSE } \; 1 + CS[T \setminus \{\text{CHOOSE } x : x \in T\}]$$
$$\text{IN } \quad CS[S]$$

Operators also differ from functions in that an operator can take an operator as an argument. For example, we can define an operator *IsPartialOrder* so that

IsPartialOrder(R, S) equals true iff the operator R defines an irreflexive partial order on S. The definition is

$$IsPartialOrder(R(\_, \_), S) \;\triangleq$$
$$\wedge\; \forall\, x, y, z \in S \;:\; R(x, y) \wedge R(y, z) \Rightarrow R(x, z)$$
$$\wedge\; \forall\, x \in S \;:\; \neg R(x, x)$$

If you don't know what an irreflexive partial order is, read this definition of *IsPartialOrder* to find out.

We could also use an infix-operator symbol like $\prec$ instead of R as the parameter of the definition, writing

$$IsPartialOrder(\_\prec\_, S) \;\triangleq$$
$$\wedge\; \forall\, x, y, z \in S \;:\; (x \prec y) \wedge (y \prec z) \Rightarrow (x \prec z)$$
$$\wedge\; \forall\, x \in S \;:\; \neg(x \prec x)$$

The first argument of *IsPartialOrder* is an operator that takes two arguments; its second argument is an expression. Since $>$ is an operator that takes two arguments, the expression *IsPartialOrder*($>, Nat$) is syntactically correct. In fact, it equals TRUE, if $>$ is defined to be the usual operator on numbers. The expression *IsPartialOrder*($+, 3$) is also syntactically correct, but it's silly and we have no idea whether or not it equals TRUE.

There is one difference between functions and operators that is subtle and not very important, but I will mention it anyway for completeness. The definition of *Tail* defines *Tail*(s) for all values of s. For example, it defines *Tail*($1/2$) to equal

$$(6.5) \quad [i \in 1\; ..\; (Len(1/2) - 1) \mapsto (1/2)[i + 1]]$$

We have no idea what this expression means, because we don't know what *Len*($1/2$) or $(1/2)[i + 1]$ mean. But, whatever (6.5) means, it equals *Tail*($1/2$). The definition of *fact* defines *fact*$[n]$ only for $n \in Nat$. It tells us nothing about the value of *fact*$[1/2]$. The expression *fact*$[1/2]$ is syntactically well-formed, so it too denotes some value. However, the definition of *fact* tells us nothing about what that value is.

The last difference between operators and functions has nothing to do with mathematics and is an idiosyncrasy of TLA$^+$: the language doesn't permit us to define infix functions. Mathematicians often define / to be a function of two arguments, but we can't do that in TLA$^+$. If we want to define /, we have no choice but to make it an operator.

One can write equally nonsensical things using functions or operators. However, whether you use functions or operators may determine whether the nonsense you write is nonsyntactic gibberish or syntactically correct but semantically silly. The string of symbols 2("a") is not a syntactically correct formula because 2 is not an operator. However, 2["a"], which can also be written 2.*a*, is a syntactically correct expression. It's nonsensical because 2 isn't a function,[1] so

---

[1] More precisely, we don't know whether or not 2 is a function.

we don't know what 2["a"] means. Similarly, $Tail(s, t)$ is syntactically incorrect because $Tail$ is an operator that takes a single argument. However, as explained in Section 16.1.7 on page 301, $fact[m, n]$ is syntactic sugar for $fact[\langle m, n \rangle]$, so it is a syntactically correct, semantically silly formula. Whether an error is syntactic or semantic determines what kind of tool can catch it. In particular, the parser described in Chapter 12 catches syntactic errors, but not semantic silliness. The TLC model checker, described in Chapter 14, will report an error if it tries to evaluate a semantically silly expression.

The distinction between functions and operators seems to confuse some people. One reason is that, although this distinction exists in ordinary math, it usually goes unnoticed by mathematicians. If you ask a mathematician whether SUBSET is a function, she's likely to say yes. But if you point out to her that SUBSET can't be a function because its domain can't be a set, she will probably realize for the first time that mathematicians use operators like SUBSET and $\in$ without noticing that they form a class of objects different from functions. Logicians will observe that the distinction between operators and values, including functions, arises because TLA$^+$ is a first-order logic rather than a higher-order logic.

When defining an object V, you may have to decide whether to make V an operator that takes an argument or a function. The differences between operators and functions will often determine the decision. For example, if a variable may have V as its value, then V must be a function. Thus, in the memory specification of Section 5.3, we had to represent the state of the memory by a function rather than an operator, since the variable mem couldn't equal an operator. If these differences don't determine whether to use an operator or a function, then the choice is a matter of taste. I usually prefer operators.

6.5   Using Functions

Consider the following two formulas:

(6.6) $f' = [i \in Nat \mapsto i + 1]$

(6.7) $\forall i \in Nat : f'[i] = i + 1$

Both formulas imply that $f'[i] = i + 1$ for every natural number i, but they are not equivalent. Formula (6.6) uniquely determines f', asserting that it's a function with domain Nat. Formula (6.7) is satisfied by lots of different values of f'. For example, it is satisfied if f' is the function

$$[i \in Real \mapsto \text{IF} \ i \in Nat \ \text{THEN} \ i + 1 \ \text{ELSE} \ i^2]$$

In fact, from (6.7), we can't even deduce that f' is a function. Formula (6.6) implies formula (6.7), but not vice-versa.

When writing specifications, we almost always want to specify the new value of a variable f rather than the new values of $f[i]$ for all i in some set. We therefore usually write (6.6) rather than (6.7).

6.6 Choose

The CHOOSE operator was introduced in the memory interface of Section 5.1 in the simple idiom CHOOSE $v : v \notin S$, which is an expression whose value is not an element of S. In Section 6.3 above, we saw that it is a powerful tool that can be used in rather subtle ways.

The CHOOSE operator is known to logicians as Hilbert's ε.

The most common use for the CHOOSE operator is to "name" a uniquely specified value. For example, a/b is the unique real number that satisfies the formula $a = b * (a/b)$, if a and b are real numbers and $b \neq 0$. So, the standard module *Reals* defines division on the set *Real* of real numbers by

$$a/b \;\triangleq\; \text{CHOOSE } c \in Real \,:\, a = b * c$$

(The expression CHOOSE $x \in S : p$ means CHOOSE $x : (x \in S) \wedge p$.) If a is a nonzero real number, then there is no real number c such that $a = 0 * c$. Therefore, $a/0$ has an unspecified value. We don't know what a real number times a string equals, so we cannot say whether or not there is a real number c such that a equals "xyz" $* c$. Hence, we don't know what the value of $a/$"xyz" is.

People who do a lot of programming and not much mathematics often think that CHOOSE must be a nondeterministic operator. In mathematics, there is no such thing as a nondeterministic operator or a nondeterministic function. If some expression equals 42 today, then it will equal 42 tomorrow, and it will still equal 42 a million years from tomorrow. The specification

$$(x = \text{CHOOSE } n \,:\, n \in Nat) \;\wedge\; \Box[x' = \text{CHOOSE } n \,:\, n \in Nat]_x$$

allows only a single behavior—one in which x always equals CHOOSE $n : n \in Nat$, which is some particular, unspecified natural number. It is very different from the specification

$$(x \in Nat) \;\wedge\; \Box[x' \in Nat]_x$$

that allows all behaviors in which x is always a natural number—possibly a different number in each state. This specification is highly nondeterministic, allowing lots of different behaviors.

Chapter 7

Writing a Specification: Some Advice

You have now learned all you need to know about TLA$^+$ to write your own specifications. Here are a few additional hints to help you get started.

7.1 Why Specify

Writing a specification requires effort; the benefit it provides must justify that effort. The purpose of writing a specification is to help avoid errors. Here are some ways it can do that.

- Writing a TLA$^+$ specification can help the design process. Having to describe a design precisely often reveals problems—subtle interactions and "corner cases" that are easily overlooked. These problems are easier to correct when discovered in the design phase rather than after implementation has begun.

- A TLA$^+$ specification can provide a clear, concise way of communicating a design. It helps ensure that the designers agree on what they have designed, and it provides a valuable guide to the engineers who implement and test the system. It may also help users understand the system.

- A TLA$^+$ specification is a formal description to which tools can be applied to help find errors in the design and to help in testing the system. The most useful tool written so far for this purpose is the TLC model checker, described in Chapter 14.

Whether the benefit justifies the effort of writing the specification depends on the nature of the project. Specification is not an end in itself; it is just a tool that an engineer should be able to use when appropriate.

7.2   What to Specify

Although we talk about specifying a system, that's not what we do. A specification is a mathematical model of a particular view of some part of a system. When writing a specification, the first thing you must choose is exactly what part of the system you want to model. Sometimes the choice is obvious; often it isn't. The cache-coherence protocol of a real multiprocessor computer may be intimately connected with how the processors execute instructions. Finding an abstraction that describes the coherence protocol while suppressing the details of instruction execution may be difficult. It may require defining an interface between the processor and the memory that doesn't exist in the actual system design.

The primary purpose of a specification is to help avoid errors. You should specify those parts of the system for which a specification is most likely to reveal errors. TLA$^+$ is particularly effective at revealing concurrency errors—ones that arise through the interaction of asynchronous components. So, when writing a TLA$^+$ specification, you will probably concentrate your efforts on the parts of the system that are most likely to have such errors. If that's not where you should be concentrating your efforts, then you probably shouldn't be using TLA$^+$.

7.3   The Grain of Atomicity

After choosing what part of the system to specify, you must choose the specification's level of abstraction. The most important aspect of the level of abstraction is the grain of atomicity, the choice of what system changes are represented as a single step of a behavior. Sending a message in an actual system involves multiple suboperations, but we usually represent it as a single step. On the other hand, the sending of a message and its receipt are usually represented as separate steps when specifying a distributed system.

The same sequence of system operations is represented by a shorter sequence of steps in a coarser-grained representation than in a finer-grained one. This almost always makes the coarser-grained specification simpler than the finergrained one. However, the finer-grained specification more accurately describes the behavior of the actual system. A coarser-grained specification may fail to reveal important details of the system.

There is no simple rule for deciding on the grain of atomicity. However, there is one way of thinking about granularity that can help. To describe it, we need the TLA$^+$ action-composition operator "$\cdot$". If A and B are actions, then the action $A \cdot B$ is executed by executing first A then B as a single step. More precisely, $A \cdot B$ is the action defined by letting $s \to t$ be an $A \cdot B$ step iff there exists a state u such that $s \to u$ is an A step and $u \to t$ is a B step.

When determining the grain of atomicity, we must decide whether to represent the execution of an operation as a single step or as a sequence of steps, each corresponding to the execution of a suboperation. Let's consider the simple case of an operation consisting of two suboperations that are executed sequentially, where those suboperations are described by the two actions R and L. (Executing R enables L and disables R.) When the operation's execution is represented by two steps, each of those steps is an R step or an L step. The operation is then described with the action $R \vee L$. When its execution is represented by a single step, the operation is described with the action $R \cdot L$.[1] Let $S2$ be the finer-grained specification in which the operation is executed in two steps, and let $S1$ be the coarser-grained specification in which it is executed as a single $R \cdot L$ step. To choose the grain of atomicity, we must choose whether to take $S1$ or $S2$ as the specification. Let's examine the relation between the two specifications.

We can transform any behavior σ satisfying $S1$ into a behavior $\hat{\sigma}$ satisfying $S2$ by replacing each step $s \xrightarrow{R \cdot L} t$ with the pair of steps $s \xrightarrow{R} u \xrightarrow{L} t$, for some state u. If we regard σ as being equivalent to $\hat{\sigma}$, then we can regard $S1$ as being a strengthened version of $S2$—one that allows fewer behaviors. Specification $S1$ requires that each R step be followed immediately by an L step, while $S2$ allows behaviors in which other steps come between the R and L steps. To choose the appropriate grain of atomicity, we must decide whether those additional behaviors allowed by $S2$ are important.

The additional behaviors allowed by $S2$ are not important if the actual system executions they describe are also described by behaviors allowed by $S1$. So, we can ask whether each behavior τ satisfying $S2$ has a corresponding behavior $\tilde{\tau}$ satisfying $S1$ that is, in some sense, equivalent to τ. One way to construct $\tilde{\tau}$ from τ is to transform a sequence of steps

$$(7.1) \qquad s \xrightarrow{R} u_1 \xrightarrow{A_1} u_2 \xrightarrow{A_2} u_3 \ldots u_n \xrightarrow{A_n} u_{n+1} \xrightarrow{L} t$$

into the sequence

$$(7.2) \qquad s \xrightarrow{A_1} v_1 \ldots v_{k-2} \xrightarrow{A_k} v_{k-1} \xrightarrow{R} v_k \xrightarrow{L} v_{k+1} \xrightarrow{A_{k+1}} v_{k+2} \ldots v_{n+1} \xrightarrow{A_n} t$$

where the A_i are other system actions that can be executed between the R and L steps. Both sequences start in state s and end in state t, but the intermediate states may be different.

---

[1]We actually describe the operation with an ordinary action, like the ones we've been writing, that is equivalent to $R \cdot L$. The operator "$\cdot$" rarely appears in an actual specification. If you're ever tempted to use it, look for a better way to write the specification; you can probably find one.

When is such a transformation possible? An answer can be given in terms of commutativity relations. We say that actions A and B commute if performing them in either order produces the same result. Formally, A and B commute iff $A \cdot B$ is equivalent to $B \cdot A$. A simple sufficient condition for commutativity is that two actions commute if (i) each one leaves unchanged any variable whose value may be changed by the other, and (ii) neither enables or disables the other. It's not hard to see that we can transform (7.1) to (7.2) in the following two cases:

- R commutes with each A_i. (In this case, $k = n$.)

- L commutes with each A_i. (In this case, $k = 0$.)

In general, if an operation consists of a sequence of m subactions, we must decide whether to choose the finer-grained representation $O_1 \vee O_2 \vee \ldots \vee O_m$ or the coarser-grained one $O_1 \cdot O_2 \cdots O_m$. The generalization of the transformation from (7.1) to (7.2) is one that transforms an arbitrary behavior satisfying the finer-grained specification into one in which the sequence of O_1, O_2, ..., O_m steps come one right after the other. Such a transformation is possible if all but one of the actions O_i commute with every other system action. Commutativity can be replaced by weaker conditions, but it is the most common case.

By commuting actions and replacing a sequence $s \xrightarrow{O_1} \cdots \xrightarrow{O_m} t$ of steps by a single $O_1 \cdots O_m$ step, you may be able to transform any behavior of a finer-grained specification into a corresponding behavior of a coarser-grained one. But that doesn't mean that the coarser-grained specification is just as good as the finer-grained one. The sequences (7.1) and (7.2) are not the same, and a sequence of O_i steps is not the same as a single $O_1 \cdots O_m$ step. Whether you can consider the transformed behavior to be equivalent to the original one, and use the coarser-grained specification, depends on the particular system you are specifying and on the purpose of the specification. Understanding the relation between finer- and coarser-grained specifications can help you choose between them; it won't make the choice for you.

7.4   The Data Structures

Another aspect of a specification's level of abstraction is the accuracy with which it describes the system's data structures. For example, should the specification of a program interface describe the actual layout of a procedure's arguments in memory, or should the arguments be represented more abstractly?

To answer such a question, you must remember that the purpose of the specification is to help catch errors. A precise description of the layout of procedure arguments will help prevent errors caused by misunderstandings about that layout, but at the cost of complicating the program interface's specification. The cost is justified only if such errors are likely to be a real problem and the TLA$^+$ specification provides the best way to avoid them.

If the purpose of the specification is to catch errors caused by the asynchronous interaction of concurrently executing components, then detailed descriptions of data structures will be a needless complication. So, you will probably want to use high-level, abstract descriptions of the system's data structures in the specification. For example, to specify a program interface, you might introduce constant parameters to represent the actions of calling and returning from a procedure—parameters analogous to *Send* and *Reply* of the memory interface described in Section 5.1 (page 45).

7.5 Writing the Specification

Once you've chosen the part of the system to specify and the level of abstraction, you're ready to start writing the TLA$^+$ specification. We've already seen how this is done; let's review the steps.

First, pick the variables and define the type invariant and initial predicate. In the course of doing this, you will determine the constant parameters and assumptions about them that you need. You may also have to define some additional constants.

Next, write the next-state action, which forms the bulk of the specification. Sketching a few sample behaviors may help you get started. You must first decide how to decompose the next-state action as the disjunction of actions describing the different kinds of system operations. You then define those actions. The goal is to make the action definitions as compact and easy to read as possible, which requires carefully structuring them. One way to reduce the size of a specification is to define state predicates and state functions that are used in several different action definitions. When writing the action definitions, you will determine which of the standard modules you need and will add the appropriate EXTENDS statement. You may also have to define some constant operators for the data structures that you are using.

You must now write the temporal part of the specification. If you want to specify liveness properties, you have to choose the fairness conditions, as described below in Chapter 8. You then combine the initial predicate, next-state action, and any fairness conditions you've chosen into the definition of a single temporal formula that is the specification.

Finally, you can assert theorems about the specification. If nothing else, you probably want to add a type-correctness theorem.

7.6   Some Further Hints

Here are a few miscellaneous suggestions that may help you write better specifications.

Don't be too clever.

Cleverness can make a specification hard to read—and even wrong. The formula $q = \langle h' \rangle \circ q'$ may look like a nice, short way of writing

$$(7.3) \quad (h' = Head(q)) \wedge (q' = Tail(q))$$

But not only is $q = \langle h' \rangle \circ q'$ harder to understand than (7.3), it's also wrong. We don't know what $a \circ b$ equals if a and b are not both sequences, so we don't know whether $h' = Head(q)$ and $q' = Tail(q)$ are the only values of h' and q' that satisfy $q = \langle h' \rangle \circ q'$. There could be other values of h' and q', which are not sequences, that satisfy the formula.

In general, the best way to specify the new value of a variable v is with a conjunct of the form $v' = exp$ or $v' \in exp$, where exp is a state function—an expression with no primes.

A type invariant is not an assumption.

Type invariance is a property of a specification, not an assumption. When writing a specification, we usually define a type invariant. But that's just a definition; a definition is not an assumption. Suppose you define a type invariant that asserts that a variable n is of type Nat. You may be tempted then to think that a conjunct $n' > 7$ in an action asserts that n' is a natural number greater than 7. It doesn't. The formula $n' > 7$ asserts only that $n' > 7$. It is satisfied if $n' = \sqrt{96}$ as well as if $n' = 8$. Since we don't know whether or not "abc" > 7 is true, it might be satisfied even if $n' =$ "abc". The meaning of the formula is not changed just because you've defined a type invariant that asserts $n \in Nat$.

In general, you may want to describe the new value of a variable x by asserting some property of x'. However, the next-state action should imply that x' is an element of some suitable set. For example, a specification might define[2]

$$
\begin{aligned}
Action1 &\stackrel{\Delta}{=} (n' > 7) \wedge \dots \\
Action2 &\stackrel{\Delta}{=} (n' \leq 6) \wedge \dots \\
Next &\stackrel{\Delta}{=} (n' \in Nat) \wedge (Action1 \vee Action2)
\end{aligned}
$$

---

[2]An alternative approach is to define $Next$ to equal $Action1 \vee Action2$ and to let the specification be $Init \wedge \Box[Next]_{\dots} \wedge \Box(n \in Nat)$. But it's usually better to stick to the simple form $Init \wedge \Box[Next]_{\dots}$ for specifications.

Don't be too abstract.

Suppose a user interacts with the system by typing on a keyboard. We could describe the interaction abstractly with a variable *typ* and an operator parameter *KeyStroke*, where the action *KeyStroke*("a", *typ*, *typ'*) represents the user typing an "a". This is the approach we took in describing the communication between the processors and the memory in the *MemoryInterface* module on page 48.

A more concrete description would be to let *kbd* represent the state of the keyboard, perhaps letting *kbd* = {} mean that no key is depressed, and *kbd* = {"a"} mean that the *a* key is depressed. The typing of an *a* is represented by two steps, a [*kbd* = {}] → [*kbd* = {"a"}] step represents the pressing of the *a* key, and a [*kbd* = {"a"}] → [*kbd* = {}] step represents its release. This is the approach we took in the asynchronous interface specifications of Chapter 3.

The abstract interface is simpler; typing an *a* is represented by a single *KeyStroke*("a", *typ*, *typ'*) step instead of a pair of steps. However, using the concrete representation leads us naturally to ask: what if the user presses the *a* key and, before releasing it, presses the *b* key? That's easy to describe with the concrete representation. The state with both keys depressed is *kbd* = {"a", "b"}. Pressing and releasing a key are represented simply by the two actions

$$Press(k) \triangleq kbd' = kbd \cup \{k\} \qquad Release(k) \triangleq kbd' = kbd \setminus \{k\}$$

The possibility of having two keys depressed cannot be expressed with the simple abstract interface. To express it abstractly, we would have to replace the parameter *KeyStroke* with two parameters *PressKey* and *ReleaseKey*, and we would have to express explicitly the property that a key can't be released until it has been depressed, and vice-versa. The more concrete representation is then simpler.

We might decide that we don't want to consider the possibility of two keys being depressed, and that we prefer the abstract representation. But that should be a conscious decision. Our abstraction should not blind us to what can happen in the actual system. When in doubt, it's safer to use a concrete representation that more accurately describes the real system. That way, you are less likely to overlook real problems.

Don't assume values that look different are unequal.

The rules of TLA$^+$ do not imply that $1 \neq$ "a". If the system can send a message that is either a string or a number, represent the message as a record with a *type* and *value* field—for example,

$$[type \mapsto \text{"String"}, value \mapsto \text{"a"}] \quad \text{or} \quad [type \mapsto \text{"Nat"}, value \mapsto 1]$$

We know that these two values are different because they have different *type* fields.

Move quantification to the outside.

Specifications are usually easier to read if $\exists$ is moved outside disjunctions and $\forall$ is moved outside conjunctions. For example, instead of

$$
\begin{aligned}
Up &\triangleq \exists\, e \in Elevator : \ldots \\
Down &\triangleq \exists\, e \in Elevator : \ldots \\
Move &\triangleq Up \vee Down
\end{aligned}
$$

it's usually better to write

$$
\begin{aligned}
Up(e) &\triangleq \ldots \\
Down(e) &\triangleq \ldots \\
Move &\triangleq \exists\, e \in Elevator : Up(e) \vee Down(e)
\end{aligned}
$$

Prime only what you mean to prime.

When writing an action, be careful where you put your primes. The expression $f[e]'$ equals $f'[e']$; it equals $f'[e]$ only if $e' = e$, which need not be true if the expression e contains variables. Be especially careful when priming an operator whose definition contains a variable. For example, suppose x is a variable and op is defined by

$$
op(a) \triangleq x + a
$$

Then $op(y)'$ equals $(x+y)'$, which equals $x'+y'$, while $op(y')$ equals $x+y'$. There is no way to use op and $'$ to write the expression $x' + y$. (Writing $op'(y)$ doesn't work because it's illegal—you can prime only an expression, not an operator.)

Write comments as comments.

Don't put comments into the specification itself. I have seen people write things like the following action definition:

$$
\begin{aligned}
A \triangleq\ & \vee \wedge x \geq 0 \\
& \quad\ \wedge \ldots \\
& \vee \wedge x < 0 \\
& \quad\ \wedge \text{FALSE}
\end{aligned}
$$

The second disjunct is meant to indicate that the writer intended A not to be enabled when $x < 0$. But that disjunct is completely redundant, since $F \wedge \text{FALSE}$ equals FALSE, and $F \vee \text{FALSE}$ equals F, for any formula F. So the second disjunct of the definition serves only as a form of comment. It's better to write

$$
\begin{aligned}
A \triangleq\ & \wedge x \geq 0 \quad \text{\small\fbox{A is not enabled if $x < 0$}} \\
& \wedge \ldots
\end{aligned}
$$

7.7 When and How to Specify

Specifications are often written later than they should be. Engineers are usually under severe time constraints, and they may feel that writing a specification will slow them down. Only after a design has become so complex that they need help understanding it do most engineers think about writing a precise specification.

Writing a specification helps you think clearly. Thinking clearly is hard; we can use all the help we can get. Making specification part of the design process can improve the design.

I have described how to write a specification assuming that the system design already exists. But it's better to write the specification as the system is being designed. The specification will start out being incomplete and probably incorrect. For example, an initial specification of the write-through cache of Section 5.6 (page 54) might include the definition

$RdMiss(p) \triangleq$ Enqueue a request to write value from memory to p's cache.
> Some enabling condition must be conjoined here.

$\wedge\ memQ' = Append(memQ, buf[p])$ Append request to $memQ$.

$\wedge\ ctl' = [ctl\ \text{EXCEPT}\ ![p] = \text{``?''}]$ Set $ctl[p]$ to value to be determined later.

$\wedge\ \text{UNCHANGED}\ \langle memInt, wmem, buf, cache \rangle$

Some system functionality will at first be omitted; it can be included later by adding new disjuncts to the next-state action. Tools can be applied to these preliminary specifications to help find design errors.

Part II

More Advanced Topics

Chapter 8

Liveness and Fairness

The specifications we have written so far say what a system must *not* do. The clock must not advance from 11 to 9; the receiver must not receive a message if the FIFO is empty. They don't require that the system ever actually do anything. The clock need never tick; the sender need never send any messages. Our specifications have described what are called *safety properties*. If a safety property is violated, it is violated at some particular point in the behavior—by a step that advances the clock from 11 to 9, or that reads the wrong value from memory. Therefore, we can talk about a safety property being satisfied by a finite behavior, which means that it has not been violated by any step so far.

We now learn how to specify that something *does* happen—that the clock keeps ticking, or that a value is eventually read from memory. We specify *liveness properties*—ones that cannot be violated at any particular instant. Only by examining an entire infinite behavior can we tell that the clock has stopped ticking, or that a message is never sent.

We express liveness properties as temporal formulas. This means that, to add liveness conditions to your specifications, you have to understand temporal logic—the logic of temporal formulas. The chapter begins, in Section 8.1, with a more rigorous look at what a temporal formula means. To understand a logic, you have to understand what its true formulas are. Section 8.2 is about temporal tautologies, the true formulas of temporal logic. Sections 8.4–8.7 describe how to use temporal formulas to specify liveness properties. Section 8.8 completes our study of temporal logic by examining the temporal quantifier ∃. Finally, Section 8.9 reviews what we've done and explains why the undisciplined use of temporal logic is dangerous.

This chapter is the only one that contains proofs. It would be nice if you learned to write similar proofs yourself, but it doesn't matter if you don't. The proofs are here because studying them can help you develop the intuitive understanding of temporal formulas that you need to write specifications—

an understanding that makes the truth of a simple temporal tautology like $\Box\Box F \equiv \Box F$ as obvious as the truth of a simple theorem about numbers like $\forall n \in Nat : 2 * n \geq n$.

Many readers will find that this chapter taxes their mathematical ability. Don't worry if you have trouble understanding it. Treat this chapter as an exercise to stretch your mind and prepare you to add liveness properties to your specifications. And remember that liveness properties are likely to be the least important part of your specification. You will probably not lose much if you simply omit them.

8.1   Temporal Formulas

Recall that a state assigns a value to every variable, and a behavior is an infinite sequence of states. A temporal formula is true or false of a behavior. Formally, a temporal formula F assigns a Boolean value, which we write $\sigma \models F$, to a behavior σ. We say that F is true of σ, or that σ satisfies F, iff $\sigma \models F$ equals TRUE. To define the meaning of a temporal formula F, we have to explain how to determine the value of $\sigma \models F$ for any behavior σ. For now, we consider only temporal formulas that don't contain the temporal existential quantifier $\exists$.

It's easy to define the meaning of a Boolean combination of temporal formulas in terms of the meanings of those formulas. The formula $F \wedge G$ is true of a behavior σ iff both F and G are true of σ, and $\neg F$ is true of σ iff F is not true of σ. These definitions are written more formally as

$$\sigma \models (F \wedge G) \;\triangleq\; (\sigma \models F) \wedge (\sigma \models G) \qquad \sigma \models \neg F \;\triangleq\; \neg(\sigma \models F)$$

These are the definitions of the meaning of $\wedge$ and of $\neg$ as operators on temporal formulas. The meanings of the other Boolean operators are similarly defined. We can also define in this way the ordinary predicate-logic quantifiers $\forall$ and $\exists$ as operators on temporal formulas—for example:

$$\sigma \models (\exists r : F) \;\triangleq\; \exists r : (\sigma \models F)$$

Ordinary quantification over constant sets is defined the same way. For example, if S is an ordinary constant expression—that is, one containing no variables—then

$$\sigma \models (\forall r \in S : F) \;\triangleq\; \forall r \in S : (\sigma \models F)$$

Quantifiers are discussed further in Section 8.8 below.

All the unquantified temporal formulas that we've seen have been Boolean combinations of three simple kinds of formulas, which have the following meanings:

- A state predicate, viewed as a temporal formula, is true of a behavior iff it is true in the first state of the behavior.

State function and *state predicate* are defined on page 25.

- A formula $\Box P$, where P is a state predicate, is true of a behavior iff P is true in every state of the behavior.

- A formula $\Box [N]_v$, where N is an action and v is a state function, is true of a behavior iff every successive pair of steps in the behavior is a $[N]_v$ step.

Since a state predicate is an action that contains no primed variables, we can both combine and generalize these three kinds of temporal formulas into the two kinds of formulas A and $\Box A$, where A is an action. I'll first explain the meanings of these two kinds of formulas, and then define the operator $\Box$ in general. To do this, I will use the notation that σ_i is the $(i + 1)^{\text{st}}$ state of the behavior σ, for any natural number i, so σ is the behavior $\sigma_0 \to \sigma_1 \to \sigma_2 \to \cdots$.

We interpret an arbitrary action A as a temporal formula by defining $\sigma \models A$ to be true iff the first two states of σ are an A step. That is, we define $\sigma \models A$ to be true iff $\sigma_0 \to \sigma_1$ is an A step. In the special case when A is a state predicate, $\sigma_0 \to \sigma_1$ is an A step iff A is true in state σ_0, so this definition of $\sigma \models A$ generalizes our interpretation of a state predicate as a temporal formula.

We have already seen that $\Box [N]_v$ is true of a behavior iff each step is a $[N]_v$ step. This leads us to define $\sigma \models \Box A$ to be true iff $\sigma_n \to \sigma_{n+1}$ is an A step, for all natural numbers n.

We now generalize from the definition of $\sigma \models \Box A$ for an action A to the definition of $\sigma \models \Box F$ for an arbitrary temporal formula F. We defined $\sigma \models \Box A$ to be true iff $\sigma_n \to \sigma_{n+1}$ is an A step for all n. This is true iff A, interpreted as a temporal formula, is true of a behavior whose first step is $\sigma_n \to \sigma_{n+1}$, for all n. Let's define σ^{+n} to be the suffix of σ obtained by deleting its first n states:

$$\sigma^{+n} \quad \triangleq \quad \sigma_n \to \sigma_{n+1} \to \sigma_{n+2} \to \cdots$$

Then $\sigma_n \to \sigma_{n+1}$ is the first step of σ^{+n}, so $\sigma \models \Box A$ is true iff $\sigma^{+n} \models A$ is true for all n. In other words

$$\sigma \models \Box A \quad \equiv \quad \forall\, n \in \mathit{Nat} \,:\, \sigma^{+n} \models A$$

The obvious generalization is

$$\sigma \models \Box F \quad \triangleq \quad \forall\, n \in \mathit{Nat} \,:\, \sigma^{+n} \models F$$

for any temporal formula F. In other words, σ satisfies $\Box F$ iff every suffix σ^{+n} of σ satisfies F. This defines the meaning of the temporal operator $\Box$.

We have now defined the meaning of any temporal formula built from actions (including state predicates), Boolean operators, and the $\Box$ operator. For example:

$$\sigma \models \Box((x = 1) \Rightarrow \Box(y > 0))$$
$$\equiv \ \forall\, n \in \mathit{Nat} \,:\, \sigma^{+n} \models ((x = 1) \Rightarrow \Box(y > 0)) \qquad \text{By the meaning of } \Box.$$
$$\equiv \ \forall\, n \in \mathit{Nat} \,:\, (\sigma^{+n} \models (x = 1)) \Rightarrow (\sigma^{+n} \models \Box(y > 0)) \qquad \text{By the meaning of } \Rightarrow.$$
$$\equiv \ \forall\, n \in \mathit{Nat} \,:\, (\sigma^{+n} \models (x = 1)) \Rightarrow \qquad \text{By the meaning of } \Box.$$
$$(\forall\, m \in \mathit{Nat} \,:\, (\sigma^{+n})^{+m} \models (y > 0))$$

Thus, $\sigma \models \Box((x=1) \Rightarrow \Box(y>0))$ is true iff, for all $n \in Nat$, if $x=1$ is true in state σ_n, then $y>0$ is true in all states σ_{n+m} with $m \geq 0$.

To understand temporal formulas intuitively, think of σ_n as the state of the universe at time instant n during the behavior σ.[1] For any state predicate P, the expression $\sigma^{+n} \models P$ asserts that P is true at time n. Thus, $\Box((x=1) \Rightarrow \Box(y>0))$ asserts that, any time $x=1$ is true, $y>0$ is true from then on. For an arbitrary temporal formula F, we also interpret $\sigma^{+n} \models F$ as the assertion that F is true at time instant n. The formula $\Box F$ then asserts that F is true at all times. We can therefore read $\Box$ as *always* or *henceforth* or *from then on*.

We saw in Section 2.2 that a specification should allow stuttering steps—ones that leave unchanged all the variables appearing in the formula. A stuttering step represents a change only to some part of the system not described by the formula; adding it to the behavior should not affect the truth of the formula. We say that a formula F is *invariant under stuttering*[2] iff adding or deleting a stuttering step to a behavior σ does not affect whether σ satisfies F. A sensible formula should be invariant under stuttering. There's no point writing formulas that aren't sensible, so TLA allows you to write only temporal formulas that are invariant under stuttering.

A state predicate (viewed as a temporal formula) is invariant under stuttering, since its truth depends only on the first state of a behavior, and adding a stuttering step doesn't change the first state. An arbitrary action is not invariant under stuttering. For example, the action $[x' = x+1]_x$ is satisfied by a behavior σ in which x is left unchanged in the first step and incremented by 2 in the second step; it isn't satisfied by the behavior obtained by removing the initial stuttering step from σ. However, the formula $\Box[x' = x+1]_x$ is invariant under stuttering, since it is satisfied by a behavior iff every step that changes x is an $x' = x+1$ step—a condition not affected by adding or deleting stuttering steps.

In general, the formula $\Box[A]_v$ is invariant under stuttering, for any action A and state function v. However, $\Box A$ is not invariant under stuttering for an arbitrary action A. For example, $\Box(x' = x+1)$ can be made false by adding a step that does not change x. So, even though we have assigned a meaning to $\Box(x' = x+1)$, it isn't a legal TLA formula.

Invariance under stuttering is preserved by $\Box$ and by the Boolean operators— that is, if F and G are invariant under stuttering, then so are $\Box F$, $\neg F$, $F \wedge G$, $\forall x \in S : F$, etc. So, state predicates, formulas of the form $\Box[N]_v$, and all formulas obtainable from them by applying $\Box$ and Boolean operators are invariant under stuttering.

---

[1]It is because we think of σ_n as the state at time n, and because we usually measure time starting from 0, that I number the states of a behavior starting with 0 rather than 1.

[2]This is a completely new sense of the word *invariant*; it has nothing to do with the concept of invariance discussed already.

We now examine five especially important classes of formulas that are constructed from arbitrary temporal formulas F and G. We introduce new operators for expressing the first three.

$\Diamond F$ is defined to equal $\neg\Box\neg F$. It asserts that F is not always false, which means that F is true at some time:

$$
\begin{aligned}
\sigma &\models \Diamond F \\
&\equiv \sigma \models \neg\Box\neg F && \text{By definition of } \Diamond. \\
&\equiv \neg\,(\sigma \models \Box\neg F) && \text{By the meaning of } \neg. \\
&\equiv \neg\,(\forall\, n \in Nat \;:\; \sigma^{+n} \models \neg F) && \text{By the meaning of } \Box. \\
&\equiv \neg\,(\forall\, n \in Nat \;:\; \neg\,(\sigma^{+n} \models F)) && \text{By the meaning of } \neg. \\
&\equiv \exists\, n \in Nat \;:\; \sigma^{+n} \models F && \text{Because } \neg\forall\neg \text{ is equivalent to } \exists.
\end{aligned}
$$

We usually read $\Diamond$ as *eventually*, taking eventually to include now.

$F \rightsquigarrow G$ is defined to equal $\Box(F \Rightarrow \Diamond G)$. The same kind of calculation we just did for $\sigma \models \Diamond F$ shows

$$
\begin{aligned}
\sigma \models (F \rightsquigarrow G) \;\equiv\; \\
\forall\, n \in Nat \;:\; (\sigma^{+n} \models F) \Rightarrow (\exists\, m \in Nat \;:\; (\sigma^{+(n+m)} \models G))
\end{aligned}
$$

The formula $F \rightsquigarrow G$ asserts that whenever F is true, G is eventually true—that is, G is true then or at some later time. We read $\rightsquigarrow$ as *leads to*.

$\Diamond\langle A\rangle_v$ is defined to equal $\neg\Box[\neg A]_v$, where A is an action and v a state function. It asserts that not every step is a $(\neg A) \vee (v' = v)$ step, so some step is a $\neg((\neg A) \vee (v' = v))$ step. Since $\neg(P \vee Q)$ is equivalent to $(\neg P) \wedge (\neg Q)$, for any P and Q, action $\neg((\neg A) \vee (v' = v))$ is equivalent to $A \wedge (v' \neq v)$. Hence, $\Diamond\langle A\rangle_v$ asserts that some step is an $A \wedge (v' \neq v)$ step—that is, an A step that changes v. We define the action $\langle A\rangle_v$ by

$$
\langle A\rangle_v \;\triangleq\; A \wedge (v' \neq v)
$$

so $\Diamond\langle A\rangle_v$ asserts that eventually an $\langle A\rangle_v$ step occurs. We think of $\Diamond\langle A\rangle_v$ as the formula obtained by applying the operator $\Diamond$ to $\langle A\rangle_v$, although technically it's not because $\langle A\rangle_v$ isn't a temporal formula.

> I pronounce $\langle A\rangle_v$ as *angle A sub v*.

$\Box\Diamond F$ asserts that at all times, F is true then or at some later time. For time 0, this implies that F is true at some time $n_0 \geq 0$. For time $n_0 + 1$, it implies that F is true at some time $n_1 \geq n_0 + 1$. For time $n_1 + 1$, it implies that F is true at some time $n_2 \geq n_1 + 1$. Continuing the process, we see that F is true at an infinite sequence of time instants $n_0, n_1, n_2, \ldots$. So, $\Box\Diamond F$ implies that F is true at infinitely many instants. Conversely, if F is true at infinitely many instants, then, at every instant, F must be true at some later instant, so $\Box\Diamond F$ is true. Therefore, $\Box\Diamond F$ asserts that F is *infinitely often* true. In particular, $\Box\Diamond\langle A\rangle_v$ asserts that infinitely many $\langle A\rangle_v$ steps occur.

$\Diamond\Box F$ asserts that eventually (at some time), F becomes true and remains true thereafter. In other words, $\Diamond\Box F$ asserts that F is *eventually always* true. In particular, $\Diamond\Box[N]_v$ asserts that, eventually, every step is a $[N]_v$ step.

The operators $\Box$ and $\Diamond$ have higher precedence (bind more tightly) than the Boolean operators, so $\Diamond F \vee \Box G$ means $(\Diamond F) \vee (\Box G)$. The operator $\rightsquigarrow$ has lower precedence than $\wedge$ and $\vee$.

8.2 Temporal Tautologies

A temporal theorem is a temporal formula that is satisfied by all behaviors. In other words, F is a theorem iff $\sigma \models F$ equals TRUE for all behaviors σ. For example, the *HourClock* module asserts that $HC \Rightarrow \Box HCini$ is a theorem, where HC and $HCini$ are the formulas defined in the module. This theorem expresses a property of the hour clock.

The formula $\Box HCini \Rightarrow HCini$ is also a theorem. However, it tells us nothing about the hour clock because it's true regardless of how $HCini$ is defined. For example, substituting $x > 7$ for $HCini$ yields the theorem $\Box(x > 7) \Rightarrow (x > 7)$. A formula like $\Box HCini \Rightarrow HCini$ that is true when any formulas are substituted for its identifiers is called a *tautology*. To distinguish them from the tautologies of ordinary logic, tautologies containing temporal operators are sometimes called *temporal* tautologies.

Let's prove that $\Box HCini \Rightarrow HCini$ is a temporal tautology. To avoid confusing the arbitrary identifier $HCini$ in this tautology with the formula $HCini$ defined in the *HourClock* module, let's replace it by F, so the tautology becomes $\Box F \Rightarrow F$. There are axioms and inference rules for temporal logic from which we can prove any temporal tautology that, like $\Box F \Rightarrow F$, contains no quantifiers. However, it's often easier and more instructive to prove them directly from the meanings of the operators. We prove that $\Box F \Rightarrow F$ is a tautology by proving that $\sigma \models (\Box F \Rightarrow F)$ equals TRUE, for any behavior σ and any formula F. The proof is simple:

$$
\begin{array}{lll}
\sigma \models (\Box F \Rightarrow F) & \equiv (\sigma \models \Box F) \Rightarrow (\sigma \models F) & \text{By the meaning of } \Rightarrow. \\
& \equiv (\forall\, n \in Nat : \sigma^{+n} \models F) \Rightarrow (\sigma \models F) & \text{By definition of } \Box. \\
& \equiv (\forall\, n \in Nat : \sigma^{+n} \models F) \Rightarrow (\sigma^{+0} \models F) & \text{By definition of } \sigma^{+0}. \\
& \equiv \text{TRUE} & \text{By predicate logic.}
\end{array}
$$

The temporal tautology $\Box F \Rightarrow F$ asserts the obvious fact that, if F is true at all times, then it's at time 0. Such a simple tautology should be obvious once you become accustomed to thinking in terms of temporal formulas. Here are three more simple tautologies, along with their English translations.

$\neg\Box F \equiv \Diamond\neg F$

F is not always true iff it is eventually false.

$\Box(F \wedge G) \equiv (\Box F) \wedge (\Box G)$

> F and G are both always true iff F is always true and G is always true. Another way of saying this is that $\Box$ distributes over $\wedge$.

$\Diamond(F \vee G) \equiv (\Diamond F) \vee (\Diamond G)$

> F or G is eventually true iff F is eventually true or G is eventually true. Another way of saying this is that $\Diamond$ distributes over $\vee$.

At the heart of the proof of each of these tautologies is a tautology of predicate logic. For example, the proof that $\Box$ distributes over $\wedge$ relies on the fact that $\forall$ distributes over $\wedge$:

$$\sigma \models (\Box(F \wedge G) \equiv (\Box F) \wedge (\Box G))$$

$\quad \equiv \quad (\sigma \models \Box(F \wedge G)) \equiv (\sigma \models (\Box F) \wedge (\Box G)) \qquad$ By the meaning of $\equiv$.

$\quad \equiv \quad (\sigma \models \Box(F \wedge G)) \equiv (\sigma \models \Box F) \wedge (\sigma \models \Box G) \qquad$ By the meaning of $\wedge$.

$\quad \equiv \quad (\forall\, n \in Nat : \sigma^{+n} \models (F \wedge G)) \equiv \qquad$ By definition of $\Box$.
$\qquad\qquad (\forall\, n \in Nat : \sigma^{+n} \models F) \wedge (\forall\, n \in Nat : \sigma^{+n} \models G)$

$\quad \equiv \quad \text{TRUE} \quad$ By the predicate-logic tautology $(\forall\, x \in S : P \wedge Q) \equiv (\forall\, x \in S : P) \wedge (\forall\, x \in S : Q)$.

The operator $\Box$ doesn't distribute over $\vee$, nor does $\Diamond$ distribute over $\wedge$. For example, $\Box((n \geq 0) \vee (n < 0))$ is not equivalent to $(\Box(n \geq 0) \vee \Box(n < 0))$; the first formula is true for any behavior in which n is always a number, but the second is false for a behavior in which n assumes both positive and negative values. However, the following two formulas are tautologies:

$$(\Box F) \vee (\Box G) \Rightarrow \Box(F \vee G) \qquad \Diamond(F \wedge G) \Rightarrow (\Diamond F) \wedge (\Diamond G)$$

Either of these tautologies can be derived from the other by substituting $\neg F$ for F and $\neg G$ for G. Making this substitution in the second tautology yields

$\text{TRUE} \quad \equiv \quad \Diamond((\neg F) \wedge (\neg G)) \Rightarrow (\Diamond\neg F) \wedge (\Diamond\neg G) \qquad$ By substitution in the second tautology.

$\qquad\quad \equiv \quad \Diamond\neg(F \vee G) \Rightarrow (\Diamond\neg F) \wedge (\Diamond\neg G) \qquad$ Because $(\neg P \wedge \neg Q) \equiv \neg(P \vee Q)$.

$\qquad\quad \equiv \quad \neg\Box(F \vee G) \Rightarrow (\neg\Box F) \wedge (\neg\Box G) \qquad$ Because $\Diamond\neg H \equiv \neg\Box H$.

$\qquad\quad \equiv \quad \neg\Box(F \vee G) \Rightarrow \neg((\Box F) \vee (\Box G)) \qquad$ Because $(\neg P \wedge \neg Q) \equiv \neg(P \vee Q)$.

$\qquad\quad \equiv \quad (\Box F) \vee (\Box G) \Rightarrow \Box(F \vee G) \qquad$ Because $(\neg P \Rightarrow \neg Q) \equiv (Q \Rightarrow P)$.

This pair of tautologies illustrates a general law: from any temporal tautology, we obtain a *dual* tautology by making the replacements

$$\Box \leftarrow \Diamond \qquad \Diamond \leftarrow \Box \qquad \wedge \leftarrow \vee \qquad \vee \leftarrow \wedge$$

and reversing the direction of all implications. (Any $\equiv$ or $\neg$ is left unchanged.) As in the example above, the dual tautology can be proved from the original by replacing each identifier with its negation and applying the (dual) tautologies $\Diamond\neg F \equiv \neg\Box F$ and $\neg\Diamond F \equiv \Box\neg F$ along with propositional-logic reasoning.

Another important pair of dual tautologies assert that $\Box\Diamond$ distributes over $\lor$ and $\Diamond\Box$ distributes over $\land$:

(8.1) $\Box\Diamond(F \lor G) \equiv (\Box\Diamond F) \lor (\Box\Diamond G)$ $\Diamond\Box(F \land G) \equiv (\Diamond\Box F) \land (\Diamond\Box G)$

The first asserts that F or G is true infinitely often iff F is true infinitely often or G is true infinitely often. Its truth should be fairly obvious, but let's prove it. To reason about $\Box\Diamond$, it helps to introduce the symbol $\exists_\infty$, which means *there exist infinitely many*. In particular, $\exists_\infty i \in Nat : P(i)$ means that $P(i)$ is true for infinitely many natural numbers i. On page 91, we showed that $\Box\Diamond F$ asserts that F is true infinitely often. Using $\exists_\infty$, we can express this as

(8.2) $(\sigma \models \Box\Diamond F) \equiv (\exists_\infty i \in Nat : \sigma^{+i} \models F)$

The same reasoning proves the following more general result, where P is any operator:

(8.3) $(\forall n \in Nat : \exists m \in Nat : P(n + m)) \equiv \exists_\infty i \in Nat : P(i)$

Here is another useful tautology involving $\exists_\infty$, where P and Q are arbitrary operators and S is an arbitrary set:

(8.4) $(\exists_\infty i \in S : P(i) \lor Q(i)) \equiv (\exists_\infty i \in S : P(i)) \lor (\exists_\infty i \in S : Q(i))$

Using these results, it's now easy to prove that $\Box\Diamond$ distributes over $\lor$:

$$
\begin{aligned}
\sigma &\models \Box\Diamond(F \lor G) \\
&\equiv \exists_\infty i \in Nat : \sigma^{+i} \models (F \lor G) &&\text{By (8.2).} \\
&\equiv (\exists_\infty i \in Nat : \sigma^{+i} \models F) \lor (\exists_\infty i \in Nat : \sigma^{+i} \models G) &&\text{By (8.4).} \\
&\equiv (\sigma \models \Box\Diamond F) \lor (\sigma \models \Box\Diamond G) &&\text{By (8.2).}
\end{aligned}
$$

From this, we deduce the dual tautology, that $\Diamond\Box$ distributes over $\land$.

In any TLA tautology, replacing a temporal formula by an action yields a tautology—a formula that is true for all behaviors—even if that formula isn't a legal TLA formula. (Remember that we have defined the meaning of nonTLA formulas like $\Box(x' = x + 1)$.) We can apply the rules of logic to transform those nonTLA tautologies into TLA tautologies. Among these rules are the following dual equivalences, which are easy to check:

$$[A \land B]_v \equiv [A]_v \land [B]_v \qquad \langle A \lor B \rangle_v \equiv \langle A \rangle_v \lor \langle B \rangle_v$$

(The second asserts that an $A \lor B$ step that changes v is either an A step that changes v or a B step that changes v.)

As an example of substituting actions for temporal formulas in TLA tautologies, let's substitute $\langle A \rangle_v$ and $\langle B \rangle_v$ for F and G in the first tautology of (8.1) to get

(8.5) $\Box\Diamond(\langle A \rangle_v \lor \langle B \rangle_v) \equiv (\Box\Diamond\langle A \rangle_v) \lor (\Box\Diamond\langle B \rangle_v)$

This isn't a TLA tautology, because $\Box\Diamond(\langle A\rangle_v \vee \langle B\rangle_v)$ isn't a TLA formula. However, a general rule of logic tells us that replacing a subformula by an equivalent one yields an equivalent formula. Substituting $\langle A \vee B\rangle_v$ for $\langle A\rangle_v \vee \langle B\rangle_v$ in (8.5) gives us the following TLA tautology:

$$\Box\Diamond\langle A \vee B\rangle_v \;\equiv\; (\Box\Diamond\langle A\rangle_v) \vee (\Box\Diamond\langle B\rangle_v)$$

8.3   Temporal Proof Rules

A proof rule is a rule for deducing true formulas from other true formulas. For example, the *Modus Ponens* Rule of propositional logic tells us that, for any formulas F and G, if we have proved F and $F \Rightarrow G$, then we can deduce G. Since the laws of propositional logic hold for temporal logic as well, we can apply the *Modus Ponens* Rule when reasoning about temporal formulas. Temporal logic also has some proof rules of its own. One is

> **Generalization Rule**   From F we can infer $\Box F$, for any temporal formula F.

This rule asserts that, if F is true for all behaviors, then so is $\Box F$. To prove it, we must show that, if $\sigma \models F$ is true for every behavior σ, then $\tau \models \Box F$ is true for every behavior τ. The proof is easy:

$$
\begin{aligned}
\tau \models \Box F \;&\equiv\; \forall\, n \in Nat \,:\, \tau^{+n} \models F && \text{By definition of } \Box. \\
&\equiv\; \forall\, n \in Nat \,:\, \text{TRUE} && \text{By the assumption that } \sigma \models F \text{ equals TRUE, for all } \sigma. \\
&\equiv\; \text{TRUE} && \text{By predicate logic.}
\end{aligned}
$$

Another temporal proof rule is

> **Implies Generalization Rule**   From $F \Rightarrow G$ we can infer $\Box F \Rightarrow \Box G$, for any temporal formulas F and G.

The Generalization Rule can be derived from the Implies Generalization Rule and the tautology TRUE $=$ $\Box$TRUE by substituting TRUE for F and F for G.

The difference between a temporal proof rule and a temporal tautology can be confusing. In propositional logic, every proof rule has a corresponding tautology. The *Modus Ponens* Rule, which asserts that we can deduce G by proving F and $F \Rightarrow G$, implies the tautology $F \wedge (F \Rightarrow G) \Rightarrow G$. But in temporal logic, a proof rule need not imply a tautology. The Generalization Rule, which states that we can deduce $\Box F$ by proving F, does not imply that $F \Rightarrow \Box F$ is a tautology. The rule means that, if $\sigma \models F$ is true for all σ, then $\sigma \models \Box F$ is true for all σ. That's different from the (false) assertion that $F \Rightarrow \Box F$ is a tautology, which would mean that $\sigma \models (F \Rightarrow \Box F)$ is true for all σ. For example, $\sigma \models (F \Rightarrow \Box F)$ equals FALSE if F is a state predicate that is true in the first state of σ and is false in some other state of σ. Forgetting the distinction between a proof rule and a tautology is a common source of mistakes when using temporal logic.

8.4   Weak Fairness

It's easy to specify liveness properties with the temporal operators $\Box$ and $\Diamond$. For example, consider the hour-clock specification of module *HourClock* in Figure 2.1 on page 20. We can require that the clock never stops by asserting that there must be infinitely many *HCnxt* steps. The obvious way to write this assertion is $\Box\Diamond HCnxt$, but that's not a legal TLA formula because *HCnxt* is an action, not a temporal formula. However, an *HCnxt* step advances the value *hr* of the clock, so it changes *hr*. Therefore, an *HCnxt* step is also an *HCnxt* step that changes *hr*—that is, it's an $\langle HCnxt \rangle_{hr}$ step. We can thus write the liveness property that the clock never stops as $\Box\Diamond\langle HCnxt \rangle_{hr}$. So, we can take $HC \wedge \Box\Diamond\langle HCnxt \rangle_{hr}$ to be the specification of a clock that never stops.

Before continuing, I must make a confession and then lead you on a brief digression about subscripts. Let me first confess that the argument I just gave, that we can write $\Box\Diamond\langle HCnxt \rangle_{hr}$ in place of $\Box\Diamond HCnxt$, was sloppy (a polite term for *wrong*). Not every *HCnxt* step changes *hr*. Consider a state in which *hr* has some value that is not a number—perhaps a value ∞. An *HCnxt* step that starts in such a state sets the new value of *hr* to $\infty + 1$. We don't know what $\infty + 1$ equals; it might or might not equal ∞. If it does, then the *HCnxt* step leaves *hr* unchanged, so it is not an $\langle HCnxt \rangle_{hr}$ step. Fortunately, states in which the value of *hr* is not a number are irrelevant. Because we are conjoining the liveness condition to the safety specification *HC*, we care only about behaviors that satisfy *HC*. In all such behaviors, *hr* is always a number, and every *HCnxt* step is an $\langle HCnxt \rangle_{hr}$ step. Therefore, $HC \wedge \Box\Diamond\langle HCnxt \rangle_{hr}$ is equivalent to the nonTLA formula $HC \wedge \Box\Diamond HCnxt$.[3]

When writing liveness properties, the syntax of TLA often forces us to write $\langle A \rangle_v$ instead of A, for some action A. As in the case of *HCnxt*, the safety specification usually implies that any A step changes some variable. To avoid having to think about which variables A actually changes, we generally take the subscript v to be the tuple of all variables, which is changed iff any variable changes. But what if A does allow stuttering steps? It's silly to assert that a stuttering step eventually occurs, since such an assertion is not invariant under stuttering. So, if A does allow stuttering steps, we want to require not that an A step eventually occurs, but that a nonstuttering A step occurs—that is, an $\langle A \rangle_v$ step, where v is the tuple of all the specification's variables. The syntax of TLA forces us to say what we should mean.

When discussing formulas, I will usually ignore the angle brackets and subscripts. For example, I might describe $\Box\Diamond\langle HCnxt \rangle_{hr}$ as the assertion that there are infinitely many *HCnxt* steps, rather than infinitely many $\langle Hnxt \rangle_{hr}$, which is what it really asserts. This finishes the digression; we now return to specifying liveness conditions.

---

[3]Even though $HC \wedge \Box\Diamond HCnxt$ is not a TLA formula, its meaning has been defined, so we can determine whether it is equivalent to a TLA formula.

Let's modify specification *Spec* of module *Channel* (Figure 3.2 on page 30) to require that every value sent is eventually received. We do this by conjoining a liveness condition to *Spec*. The analog of the liveness condition for the clock is $\Box\Diamond\langle Rcv\rangle_{chan}$, which asserts that there are infinitely many *Rcv* steps. However, only a value that has been sent can be received, so this condition would also require that infinitely many values be sent—a requirement we don't want to make. We want to permit behaviors in which no value is ever sent, so no value is ever received. We require only that any value that is sent is eventually received.

To assure that all values that should be received are eventually received, it suffices to require only that the next value to be received eventually is received. (When that value has been received, the one after it becomes the next value to be received, so it must eventually be received, and so on.) More precisely, we need only require it always to be the case that, if there is a value to be received, then the next value to be received eventually is received. The next value is received by a *Rcv* step, so the requirement is[4]

$$\Box(\text{There is an unreceived value} \;\Rightarrow\; \Diamond\langle Rcv\rangle_{chan})$$

There is an unreceived value iff action *Rcv* is enabled, meaning that it is possible to take a *Rcv* step. TLA$^+$ defines ENABLED *A* to be the predicate that is true iff action *A* is enabled. The liveness condition can then be written

$$(8.6) \quad \Box(\text{ENABLED}\,\langle Rcv\rangle_{chan} \;\Rightarrow\; \Diamond\langle Rcv\rangle_{chan})$$

In the ENABLED formula, it doesn't matter if we write *Rcv* or $\langle Rcv\rangle_{chan}$. We add the angle brackets so the two actions appearing in the formula are the same.

In any behavior satisfying the safety specification *HC*, it's always possible to take an *HCnxt* step that changes *hr*. Action $\langle HCnxt\rangle_{hr}$ is therefore always enabled, so ENABLED $\langle HCnxt\rangle_{hr}$ is true throughout such a behavior. Since TRUE $\Rightarrow \Diamond\langle HCnxt\rangle_{hr}$ is equivalent to $\Diamond\langle HCnxt\rangle_{hr}$, we can replace the liveness condition $\Box\Diamond\langle HCnxt\rangle_{hr}$ for the hour clock with

$$\Box(\text{ENABLED}\,\langle HCnxt\rangle_{hr} \Rightarrow \Diamond\langle HCnxt\rangle_{hr})$$

This suggests the following general liveness condition for an action *A*:

$$\Box(\text{ENABLED}\,\langle A\rangle_v \Rightarrow \Diamond\langle A\rangle_v)$$

This condition asserts that, if *A* ever becomes enabled, then an *A* step will eventually occur—even if *A* remains enabled for only a fraction of a nanosecond and is never again enabled. The obvious practical difficulty of implementing such a condition suggests that it's too strong. So, we replace it with the weaker formula $\text{WF}_v(A)$, defined to equal

$$(8.7) \quad \Box(\Box\text{ENABLED}\,\langle A\rangle_v \Rightarrow \Diamond\langle A\rangle_v)$$

---

[4]$\Box(F \Rightarrow \Diamond G)$ equals $F \rightsquigarrow G$, so we could write this formula more compactly with $\rightsquigarrow$. However, it's more convenient to keep it in the form $\Box(F \Rightarrow \Diamond G)$

This formula asserts that, if A ever becomes forever enabled, then an A step must eventually occur. WF stands for *W*eak *F*airness, and the condition $\mathrm{WF}_v(A)$ is called *weak fairness on* A. We'll soon see that our liveness conditions for the clock and the channel can be written as WF formulas. But first, let's examine (8.7) and the following two formulas, which turn out to be equivalent to it:

(8.8) $\Box\Diamond(\neg\mathrm{ENABLED}\,\langle A\rangle_v) \;\lor\; \Box\Diamond\langle A\rangle_v$

(8.9) $\Diamond\Box(\mathrm{ENABLED}\,\langle A\rangle_v) \;\Rightarrow\; \Box\Diamond\langle A\rangle_v$

These three formulas can be expressed in English as

(8.7) It's always the case that, if A is enabled forever, then an A step eventually occurs.

(8.8) A is infinitely often disabled, or infinitely many A steps occur.

(8.9) If A is eventually enabled forever, then infinitely many A steps occur.

The equivalence of these three formulas isn't obvious. Trying to deduce their equivalence from the English expressions often leads to confusion. The best way to avoid confusion is to use mathematics. We show that the three formulas are equivalent by proving that (8.7) is equivalent to (8.8) and that (8.8) is equivalent to (8.9). Instead of proving that they are equivalent for an individual behavior, we can use tautologies that we've already seen to prove their equivalence directly. Here's a proof that (8.7) is equivalent to (8.8). Studying it will help you learn to write liveness conditions.

$$\Box(\Box\mathrm{ENABLED}\,\langle A\rangle_v \Rightarrow \Diamond\langle A\rangle_v)$$
$$\equiv\; \Box(\neg\Box\mathrm{ENABLED}\,\langle A\rangle_v \lor \Diamond\langle A\rangle_v) \qquad \text{Because } (F\Rightarrow G)\equiv(\neg F\lor G).$$
$$\equiv\; \Box(\Diamond\neg\mathrm{ENABLED}\,\langle A\rangle_v \lor \Diamond\langle A\rangle_v) \qquad \text{Because } \neg\Box F\equiv\Diamond\neg F.$$
$$\equiv\; \Box\Diamond(\neg\mathrm{ENABLED}\,\langle A\rangle_v \lor \langle A\rangle_v) \qquad \text{Because } \Diamond F\lor\Diamond G\equiv\Diamond(F\lor G).$$
$$\equiv\; \Box\Diamond(\neg\mathrm{ENABLED}\,\langle A\rangle_v) \lor \Box\Diamond\langle A\rangle_v \quad \text{Because } \Box\Diamond(F\lor G)\equiv\Box\Diamond F\lor\Box\Diamond G.$$

The equivalence of (8.8) and (8.9) is proved as follows:

$$\Box\Diamond(\neg\mathrm{ENABLED}\,\langle A\rangle_v) \lor \Box\Diamond\langle A\rangle_v$$
$$\equiv\; \neg\Diamond\Box(\mathrm{ENABLED}\,\langle A\rangle_v) \lor \Box\Diamond\langle A\rangle_v \quad \text{Because } \Box\Diamond\neg F\equiv\Box\neg\Box F\equiv\neg\Diamond\Box F.$$
$$\equiv\; \Diamond\Box(\mathrm{ENABLED}\,\langle A\rangle_v) \Rightarrow \Box\Diamond\langle A\rangle_v \qquad \text{Because } (F\Rightarrow G)\equiv(\neg F\lor G).$$

We now show that the liveness conditions for the hour clock and the channel can be written as weak fairness conditions.

First, consider the hour clock. In any behavior satisfying HC, an $\langle HCnxt\rangle_{hr}$ step is always enabled, so $\Diamond\Box(\mathrm{ENABLED}\,\langle HCnxt\rangle_{hr})$ equals TRUE. Therefore, HC implies that $\mathrm{WF}_{hr}(HCnxt)$, which equals (8.9), is equivalent to formula $\Box\Diamond\langle HCnxt\rangle_{hr}$, our liveness condition for the hour clock.

Now, consider the channel. I claim that the liveness condition (8.6) can be replaced by WF$_{chan}(Rcv)$. More precisely, *Spec* implies that these two formulas are equivalent, so conjoining either of them to *Spec* yields equivalent specifications. The proof rests on the observation that, in any behavior satisfying *Spec*, once *Rcv* becomes enabled (because a value has been sent), it can be disabled only by a *Rcv* step (which receives the value). In other words, it's always the case that if *Rcv* is enabled, then it is enabled forever or a *Rcv* step eventually occurs. Stated formally, this observation asserts that *Spec* implies

(8.10) $\Box\,(\,\text{ENABLED}\,\langle Rcv\rangle_{chan}\;\Rightarrow\;\Box(\text{ENABLED}\,\langle Rcv\rangle_{chan})\,\lor\,\Diamond\langle Rcv\rangle_{chan}\,)$

We show that we can take WF$_{chan}(Rcv)$ as our liveness condition by showing that (8.10) implies the equivalence of (8.6) and WF$_{chan}(Rcv)$.

The proof is by purely temporal reasoning; we need no other facts about the channel specification. Both for compactness and to emphasize the generality of our reasoning, let's replace ENABLED $\langle Rcv\rangle_{chan}$ by E and $\langle Rcv\rangle_{chan}$ by A. Using version (8.7) of the definition of WF, we must prove

(8.11) $\Box(E\Rightarrow\Box E\lor\Diamond A)\;\Rightarrow\;(\Box(E\Rightarrow\Diamond A)\;\equiv\;\Box(\Box E\Rightarrow\Diamond A))$

So far, all our proofs have been by calculation. That is, we have proved that two formulas are equivalent, or that a formula is equivalent to TRUE, by proving a chain of equivalences. That's a good way to prove simple things, but it's usually better to tackle a complicated formula like (8.11) by splitting its proof into pieces. We have to prove that one formula implies the equivalence of two others. The equivalence of two formulas can be proved by showing that each implies the other. More generally, to prove that P implies $Q\equiv R$, we prove that $P\land Q$ implies R and that $P\land R$ implies Q. So, we prove (8.11) by proving the two formulas

(8.12) $\Box(E\Rightarrow\Box E\lor\Diamond A)\land\Box(E\Rightarrow\Diamond A)\;\Rightarrow\;\Box(\Box E\Rightarrow\Diamond A)$

(8.13) $\Box(E\Rightarrow\Box E\lor\Diamond A)\land\Box(\Box E\Rightarrow\Diamond A)\;\Rightarrow\;\Box(E\Rightarrow\Diamond A)$

Both (8.12) and (8.13) have the form $\Box F\land\Box G\Rightarrow\Box H$. We first show that, for any formulas F, G, and H, we can deduce $\Box F\land\Box G\Rightarrow\Box H$ by proving $F\land G\Rightarrow H$. We do this by assuming $F\land G\Rightarrow H$ and proving $\Box F\land\Box G\Rightarrow\Box H$ as follows:

1. $\Box(F\land G)\Rightarrow\Box H$

   PROOF: By the assumption $F\land G\Rightarrow H$ and the Implies Generalization Rule (page 95), substituting $F\land G$ for F and H for G in the rule.

2. $\Box F\land\Box G\Rightarrow\Box H$

   PROOF: By step 1 and the tautology $\Box(F\land G)\equiv\Box F\land\Box G$.

This shows that we can deduce $\Box F \wedge \Box G \Rightarrow \Box H$ by proving $F \wedge G \Rightarrow H$, for any F, G, and H. We can therefore prove (8.12) and (8.13) by proving

(8.14) $(E \Rightarrow \Box E \vee \Diamond A) \wedge (E \Rightarrow \Diamond A) \Rightarrow (\Box E \Rightarrow \Diamond A)$

(8.15) $(E \Rightarrow \Box E \vee \Diamond A) \wedge (\Box E \Rightarrow \Diamond A) \Rightarrow (E \Rightarrow \Diamond A)$

The proof of (8.14) is easy. In fact, we don't even need the first conjunct; we can prove $(E \Rightarrow \Diamond A) \Rightarrow (\Box E \Rightarrow \Diamond A)$ as follows:

$$(E \Rightarrow \Diamond A)$$
$$\equiv\ (\Box E \Rightarrow E) \wedge (E \Rightarrow \Diamond A) \quad \text{Because } \Box E \Rightarrow E \text{ is a temporal tautology.}$$
$$\Rightarrow\ (\Box E \Rightarrow \Diamond A) \quad\quad\quad\quad\ \text{By the tautology } (P \Rightarrow Q) \wedge (Q \Rightarrow R) \Rightarrow (P \Rightarrow R).$$

The proof of (8.15) uses only propositional logic. We deduce (8.15) by substituting E for P, $\Box E$ for Q, and $\Diamond A$ for R in the following propositional-logic tautology:

$$(P \Rightarrow Q \vee R) \wedge (Q \Rightarrow R) \Rightarrow (P \Rightarrow R)$$

A little thought should make the validity of this tautology seem obvious. If not, you can check it by constructing a truth table.

These proofs of (8.14) and (8.15) complete the proof that we can take $\text{WF}_{chan}(Rcv)$ instead of (8.7) as our liveness condition for the channel.

8.5 The Memory Specification

8.5.1 The Liveness Requirement

Let's now strengthen the specification of the linearizable memory of Section 5.3 with the liveness requirement that every request must receive a response. (We don't require that a request ever be issued.) The liveness requirement is conjoined to the internal memory specification, formula *ISpec* of the *InternalMemory* module (Figure 5.2 on pages 52–53).

We want to express the liveness requirement in terms of weak fairness. To do this, we must understand when actions are enabled. The action $Rsp(p)$ is enabled only if the action

(8.16) $Reply(p,\ buf[p],\ memInt,\ memInt')$

is enabled. Recall that the operator *Reply* is a constant parameter, declared in the *MemoryInterface* module (Figure 5.1 on page 48). Without knowing more about this operator, we can't say when action (8.16) is enabled.

Let's assume that *Reply* actions are always enabled. That is, for any processor p and reply r, and any old value *miOld* of *memInt*, there is a new value *miNew* of *memInt* such that $Reply(p, r, miOld, miNew)$ is true. For simplicity, we just assume that this is true for all p and r, and add the following assumption to the *MemoryInterface* module:

$$\text{ASSUME } \forall\, p,\, r,\, miOld\, :\, \exists\, miNew\, :\, Reply(p,\, r,\, miOld,\, miNew)$$

We should also make a similar assumption for *Send*, but we don't need it here.

We will subscript our weak fairness formulas with the tuple of all variables, so let's give that tuple a name:

$$vars \;\triangleq\; \langle\, memInt,\ mem,\ ctl,\ buf\,\rangle$$

When processor p issues a request, it enables the $Do(p)$ action, which remains enabled until a $Do(p)$ step occurs. The weak fairness condition $\text{WF}_{vars}(Do(p))$ implies that this $Do(p)$ step must eventually occur. A $Do(p)$ step enables the $Rsp(p)$ action, which remains enabled until a $Rsp(p)$ step occurs. The weak fairness condition $\text{WF}_{vars}(Rsp(p))$ implies that this $Rsp(p)$ step, which produces the desired response, must eventually occur. Hence, the requirement

$$(8.17) \quad \text{WF}_{vars}(Do(p)) \;\wedge\; \text{WF}_{vars}(Rsp(p))$$

assures that every request issued by processor p must eventually receive a reply. We want this condition to hold for every processor p, so we can take, as the liveness condition for the memory specification, the formula

$$(8.18) \quad Liveness \;\triangleq\; \forall\, p \in Proc\, :\, \text{WF}_{vars}(Do(p)) \;\wedge\; \text{WF}_{vars}(Rsp(p))$$

The internal memory specification is then *ISpec* ∧ *Liveness*.

8.5.2   Another Way to Write It

I find a single fairness condition simpler than the conjunction of fairness conditions. Seeing the conjunction of the two weak fairness formulas in the definition of *Liveness* leads me to ask if it can be replaced by a single weak fairness condition on $Do(p) \vee Rsp(p)$. Such a replacement isn't always possible; in general, the formulas $\text{WF}_v(A) \wedge \text{WF}_v(B)$ and $\text{WF}_v(A \vee B)$ are not equivalent. However, in this case, we can replace the two fairness conditions with one. If we define

$$(8.19) \quad Liveness2 \;\triangleq\; \forall\, p \in Proc\, :\, \text{WF}_{vars}(Do(p) \vee Rsp(p))$$

then *ISpec* ∧ *Liveness2* is equivalent to *ISpec* ∧ *Liveness*. As we will see, this equivalence holds because any behavior satisfying *ISpec* satisfies the following two properties:

DR1. Whenever $Do(p)$ is enabled, $Rsp(p)$ can never become enabled unless a $Do(p)$ step eventually occurs.

DR2. Whenever $Rsp(p)$ is enabled, $Do(p)$ can never become enabled unless a $Rsp(p)$ step eventually occurs.

These properties are satisfied because a request to p is issued by a $Req(p)$ step, executed by a $Do(p)$ step, and responded to by a $Rsp(p)$ step; and then, the next request to p can be issued by a $Req(p)$ step. Each of these steps becomes possible (the action enabled) only after the preceding one occurs.

Let's now show that DR1 and DR2 imply that the conjunction of weak fairness of $Do(p)$ and of $Rsp(p)$ is equivalent to weak fairness of $Do(p) \vee Rsp(p)$. For compactness, and to emphasize the generality of what we're doing, let's replace $Do(p)$, $Rsp(p)$, and *vars* by A, B, and v, respectively.

First, we must restate DR1 and DR2 as temporal formulas. The basic form of DR1 and DR2 is

Whenever F is true, G can never be true unless H is eventually true.

This is expressed in temporal logic as $\Box(F \Rightarrow \Box \neg G \vee \Diamond H)$. (The assertion "$P$ unless Q" just means $P \vee Q$.) Adding suitable subscripts, we can therefore write DR1 and DR2 in temporal logic as

$$DR1 \triangleq \Box(\text{ENABLED }\langle A \rangle_v \Rightarrow \Box \neg \text{ENABLED }\langle B \rangle_v \vee \Diamond \langle A \rangle_v)$$
$$DR2 \triangleq \Box(\text{ENABLED }\langle B \rangle_v \Rightarrow \Box \neg \text{ENABLED }\langle A \rangle_v \vee \Diamond \langle B \rangle_v)$$

Our goal is to prove

(8.20) $DR1 \wedge DR2 \Rightarrow (\text{WF}_v(A) \wedge \text{WF}_v(B) \equiv \text{WF}_v(A \vee B))$

This is complicated, so we split the proof into pieces. As in the proof of (8.11) in Section 8.4 above, we prove an equivalence by proving two implications. To prove (8.20), we prove the two theorems

$$DR1 \wedge DR2 \wedge \text{WF}_v(A) \wedge \text{WF}_v(B) \Rightarrow \text{WF}_v(A \vee B)$$
$$DR1 \wedge DR2 \wedge \text{WF}_v(A \vee B) \Rightarrow \text{WF}_v(A) \wedge \text{WF}_v(B)$$

We prove them by showing that they are true for an arbitrary behavior σ. In other words, we prove

(8.21) $(\sigma \models DR1 \wedge DR2 \wedge \text{WF}_v(A) \wedge \text{WF}_v(B)) \Rightarrow (\sigma \models \text{WF}_v(A \vee B))$

(8.22) $(\sigma \models DR1 \wedge DR2 \wedge \text{WF}_v(A \vee B)) \Rightarrow (\sigma \models \text{WF}_v(A) \wedge \text{WF}_v(B))$

These formulas seem daunting. Whenever you have trouble proving something, try a proof by contradiction; it gives you an extra hypothesis for free—namely, the negation of what you're trying to prove. Proofs by contradiction are especially useful in temporal logic. To prove (8.21) and (8.22) by contradiction, we need to compute $\neg(\sigma \models \text{WF}_v(C))$ for an action C. From the definition (8.7) of WF, we easily get

(8.23) $(\sigma \models \text{WF}_v(C)) \equiv$
$\qquad \forall n \in \textit{Nat} : (\sigma^{+n} \models \Box \text{ENABLED }\langle C \rangle_v) \Rightarrow (\sigma^{+n} \models \Diamond \langle C \rangle_v)$

This and the tautology

$$\neg(\forall\, x \in S\,:\, P \Rightarrow Q)\ \equiv\ (\exists\, x \in S\,:\, P \wedge \neg Q)$$

of predicate logic yields

(8.24) $\neg(\sigma \models \mathrm{WF}_v(C))\ \equiv$
$$\exists\, n \in Nat\,:\, (\sigma^{+n} \models \Box\ \mathrm{ENABLED}\ \langle C \rangle_v) \wedge \neg(\sigma^{+n} \models \Diamond\langle C \rangle_v)$$

We also need two further results, both of which are derived from the tautology $\langle A \vee B \rangle_v \equiv \langle A \rangle_v \vee \langle B \rangle_v$. Combining this tautology with the temporal tautology $\Diamond(F \vee G) \equiv \Diamond F \vee \Diamond G$ yields

(8.25) $\Diamond\langle A \vee B \rangle_v\ \equiv\ \Diamond\langle A \rangle_v \vee \Diamond\langle B \rangle_v$

Combining the tautology with the observation that an action $C \vee D$ is enabled iff action C or action D is enabled yields

(8.26) $\mathrm{ENABLED}\ \langle A \vee B \rangle_v\ \equiv\ \mathrm{ENABLED}\ \langle A \rangle_v \vee \mathrm{ENABLED}\ \langle B \rangle_v$

We can now prove (8.21) and (8.22). To prove (8.21), we assume that σ satisfies *DR1*, *DR2*, $\mathrm{WF}_v(A)$, and $\mathrm{WF}_v(B)$, but it does not satisfy $\mathrm{WF}_v(A \vee B)$, and we obtain a contradiction. By (8.24), the assumption that σ does not satisfy $\mathrm{WF}_v(A \vee B)$ means that there exists some number n such that

(8.27) $\sigma^{+n} \models \Box\ \mathrm{ENABLED}\ \langle A \vee B \rangle_v$

(8.28) $\neg(\sigma^{+n} \models \Diamond\langle A \vee B \rangle_v)$

We obtain a contradiction from (8.27) and (8.28) as follows:

1. $\neg(\sigma^{+n} \models \Diamond\langle A \rangle_v)\ \wedge\ \neg(\sigma^{+n} \models \Diamond\langle B \rangle_v)$
   PROOF: By (8.28) and (8.25), using the tautology $\neg(P \vee Q) \equiv (\neg P \wedge \neg Q)$.

2. (a) $(\sigma^{+n} \models \mathrm{ENABLED}\ \langle A \rangle_v)\ \Rightarrow\ (\sigma^{+n} \models \Box\,\neg\mathrm{ENABLED}\ \langle B \rangle_v)$
   (b) $(\sigma^{+n} \models \mathrm{ENABLED}\ \langle B \rangle_v)\ \Rightarrow\ (\sigma^{+n} \models \Box\,\neg\,\mathrm{ENABLED}\ \langle A \rangle_v)$
   PROOF: By definition of *DR1*, the assumption $\sigma \models DR1$ implies
   $$(\sigma^{+n} \models \mathrm{ENABLED}\ \langle A \rangle_v)\ \Rightarrow$$
   $$(\sigma^{+n} \models \Box\,\neg\,\mathrm{ENABLED}\ \langle B \rangle_v) \vee (\sigma^{+n} \models \Diamond\langle A \rangle_v)$$
   and part (a) then follows from 1. The proof of (b) is similar.

3. (a) $(\sigma^{+n} \models \mathrm{ENABLED}\ \langle A \rangle_v)\ \Rightarrow\ (\sigma^{+n} \models \Box\ \mathrm{ENABLED}\ \langle A \rangle_v)$
   (b) $(\sigma^{+n} \models \mathrm{ENABLED}\ \langle B \rangle_v)\ \Rightarrow\ (\sigma^{+n} \models \Box\ \mathrm{ENABLED}\ \langle B \rangle_v)$
   PROOF: Part (a) follows from 2(a), (8.27), (8.26), and the temporal tautology
   $$\Box(F \vee G) \wedge \Box\neg G\ \Rightarrow\ \Box F$$
   The proof of part (b) is similar.

4. (a) $(\sigma^{+n} \models \text{ENABLED}\,\langle A \rangle_v) \;\Rightarrow\; (\sigma^{+n} \models \Diamond \langle A \rangle_v)$
   (b) $(\sigma^{+n} \models \text{ENABLED}\,\langle B \rangle_v) \;\Rightarrow\; (\sigma^{+n} \models \Diamond \langle B \rangle_v)$

   PROOF: The assumption $\sigma \models \text{WF}_v(A)$ and (8.23) imply

   $$(\sigma^{+n} \models \Box \text{ENABLED}\,\langle A \rangle_v) \;\Rightarrow\; (\sigma^{+n} \models \Diamond \langle A \rangle_v)$$

   Part (a) follows from this and 3(a). The proof of part (b) is similar.

5. $(\sigma^{+n} \models \Diamond \langle A \rangle_v) \;\vee\; (\sigma^{+n} \models \Diamond \langle B \rangle_v)$

   PROOF: Since $\Box F$ implies F, for any F, (8.27) and (8.26) imply

   $$(\sigma^{+n} \models \text{ENABLED}\,\langle A \rangle_v) \;\vee\; (\sigma^{+n} \models \text{ENABLED}\,\langle B \rangle_v)$$

   Step 5 then follows by propositional logic from step 4.

Steps 1 and 5 provide the required contradiction.

   We can prove (8.22) by assuming that σ satisfies *DR1*, *DR2*, and $\text{WF}_v(A \vee B)$, and then proving that it satisfies $\text{WF}_v(A)$ and $\text{WF}_v(B)$. We prove only that it satisfies $\text{WF}_v(A)$; the proof for $\text{WF}_v(B)$ is similar. The proof is by contradiction; we assume that σ does not satisfy $\text{WF}_v(A)$ and obtain a contradiction. By (8.24), the assumption that σ does not satisfy $\text{WF}_v(A)$ means that there exists some number n such that

(8.29) $\sigma^{+n} \models \Box \text{ENABLED}\,\langle A \rangle_v$

(8.30) $\neg(\sigma^{+n} \models \Diamond \langle A \rangle_v)$

We obtain the contradiction as follows:

1. $\sigma^{+n} \models \Diamond \langle A \vee B \rangle_v$

   PROOF: From (8.29) and (8.26) we deduce $\sigma^{+n} \models \Box \text{ENABLED}\,\langle A \vee B \rangle_v$. By the assumption $\sigma \models \text{WF}_v(A \vee B)$ and (8.23), this implies $\sigma^{+n} \models \Diamond \langle A \vee B \rangle_v$.

2. $\sigma^{+n} \models \Box \neg \text{ENABLED}\,\langle B \rangle_v$

   PROOF: From (8.29) we deduce $\sigma^{+n} \models \text{ENABLED}\,\langle A \rangle_v$, which by the assumption $\sigma \models$ *DR1* and the definition of *DR1* implies

   $$(\sigma^{+n} \models \Box \neg \text{ENABLED}\,\langle B \rangle_v) \;\vee\; (\sigma^{+n} \models \Diamond \langle A \rangle_v)$$

   The assumption (8.30) then implies $\sigma^{+n} \models \Box \neg \text{ENABLED}\,\langle B \rangle_v$.

3. $\neg(\sigma^{+n} \models \Diamond \langle B \rangle_v)$

   PROOF: The definition of ENABLED implies $\neg \text{ENABLED}\,\langle B \rangle_v \Rightarrow \neg \langle B \rangle_v$. (A $\langle B \rangle_v$ step can occur only when it is enabled.) From this, simple temporal reasoning implies

   $$(\sigma^{+n} \models \Box \neg \text{ENABLED}\,\langle B \rangle_v) \;\Rightarrow\; \neg(\sigma^{+n} \models \Diamond \langle B \rangle_v)$$

   (A formal proof uses the Implies Generalization Rule and the tautology $\Box \neg F \equiv \neg \Diamond F$.) We then deduce $\neg(\sigma^{+n} \models \Diamond \langle B \rangle_v)$ from 2.

4. $\neg(\sigma^{+n} \models \Diamond \langle A \vee B \rangle_v)$

   PROOF: By (8.30), 3, and (8.25), using the tautology $\neg P \wedge \neg Q \equiv \neg(P \vee Q)$.

Steps 1 and 4 provide the necessary contradiction. This completes our proof of (8.22), which completes our proof of (8.20).

8.5.3 A Generalization

Formula (8.20) provides a rule for replacing the conjunction of weak fairness requirements on two actions with weak fairness of their disjunction. We now generalize it from two actions A and B to n actions $A_1, \ldots, A_n$. The generalization of *DR1* and *DR2* is

$$DR(i,j) \;\triangleq\; \Box\,(\,\text{ENABLED}\,\langle A_i\rangle_v \;\Rightarrow\; \Box\,\neg\,\text{ENABLED}\,\langle A_j\rangle_v \;\vee\; \Diamond\langle A_i\rangle_v\,)$$

If we substitute A_1 for A and A_2 for B, then *DR1* becomes $DR(1,2)$ and *DR2* becomes $DR(2,1)$. The generalization of (8.20) is

$$(8.31) \quad (\forall\, i,j \in 1 \mathrel{..} n : (i \neq j) \Rightarrow DR(i,j)) \;\Rightarrow$$
$$(\text{WF}_v(A_1) \wedge \ldots \wedge \text{WF}_v(A_n) \;\equiv\; \text{WF}_v(A_1 \vee \ldots \vee A_n))$$

To decide if you can replace the conjunction of weak fairness conditions by a single one in a specification, you will probably find it easier to use the following informal statement of (8.31):

> **WF Conjunction Rule** If $A_1, \ldots, A_n$ are actions such that, for any distinct i and j, whenever $\langle A_i\rangle_v$ is enabled, $\langle A_j\rangle_v$ cannot become enabled unless an $\langle A_i\rangle_v$ step occurs, then $\text{WF}_v(A_1) \wedge \ldots \wedge \text{WF}_v(A_n)$ is equivalent to $\text{WF}_v(A_1 \vee \ldots \vee A_n)$.

Perhaps the best way to think of this rule is as an assertion about an arbitrary individual behavior σ. Its hypothesis is then that $\sigma \models DR(i,j)$ holds for all distinct i and j; its conclusion is

$$\sigma \models (\text{WF}_v(A_1) \wedge \ldots \wedge \text{WF}_v(A_n) \;\equiv\; \text{WF}_v(A_1 \vee \ldots \vee A_n))$$

To replace $\text{WF}_v(A_1) \wedge \ldots \wedge \text{WF}_v(A_n)$ by $\text{WF}_v(A_1 \vee \ldots \vee A_n)$ in a specification, you have to check that any behavior satisfying the safety part of the specification also satisfies $DR(i,j)$, for all distinct i and j.

Conjunction and disjunction are special cases of quantification:

$$F_1 \vee \ldots \vee F_n \;\equiv\; \exists\, i \in 1 \mathrel{..} n : F_i$$
$$F_1 \wedge \ldots \wedge F_n \;\equiv\; \forall\, i \in 1 \mathrel{..} n : F_i$$

We can therefore easily restate the WF Conjunction Rule as a condition on when $\forall\, i \in S : \text{WF}_v(A_i)$ and $\text{WF}_v(\exists\, i \in S : A_i)$ are equivalent, for a finite set S. The resulting rule is actually valid for any set S:

> **WF Quantifier Rule** If, for all $i \in S$, the A_i are actions such that, for any distinct i and j in S, whenever $\langle A_i\rangle_v$ is enabled, $\langle A_j\rangle_v$ cannot become enabled unless an $\langle A_i\rangle_v$ step occurs, then $\forall\, i \in S : \text{WF}_v(A_i)$ is equivalent to $\text{WF}_v(\exists\, i \in S : A_i)$.

8.6   Strong Fairness

We define $\mathrm{SF}_v(A)$, *strong fairness* of action A, to be either of the following two equivalent formulas:

(8.32) $\Diamond\Box(\neg\,\mathrm{ENABLED}\,\langle A\rangle_v) \vee \Box\Diamond\langle A\rangle_v$

(8.33) $\Box\Diamond\mathrm{ENABLED}\,\langle A\rangle_v \Rightarrow \Box\Diamond\langle A\rangle_v$

Intuitively, these two formulas assert

(8.32) A is eventually disabled forever, or infinitely many A steps occur.

(8.33) If A is infinitely often enabled, then infinitely many A steps occur.

The proof that (8.32) and (8.33) are equivalent is similar to the proof on page 98 that the two formulations (8.8) and (8.9) of $\mathrm{WF}_v(A)$ are equivalent.

Definition (8.32) of $\mathrm{SF}_v(A)$ is obtained from definition (8.8) of $\mathrm{WF}_v(A)$ by replacing $\Box\Diamond(\neg\,\mathrm{ENABLED}\,\langle A\rangle_v)$ with $\Diamond\Box(\neg\,\mathrm{ENABLED}\,\langle A\rangle_v)$. Since $\Diamond\Box F$ (eventually always F) is stronger than (implies) $\Box\Diamond F$ (infinitely often F) for any formula F, strong fairness is stronger than weak fairness. We can express weak and strong fairness as follows:

- Weak fairness of A asserts that an A step must eventually occur if A is *continuously* enabled.

- Strong fairness of A asserts that an A step must eventually occur if A is *continually* enabled.

Continuously means without interruption. *Continually* means repeatedly, possibly with interruptions.

Strong fairness need not be strictly stronger than weak fairness. Weak and strong fairness of an action A are equivalent iff A infinitely often disabled implies that either A eventually becomes forever disabled, or else infinitely many A steps occur. This is expressed formally by the tautology

$$(\mathrm{WF}_v(A) \equiv \mathrm{SF}_v(A)) \equiv$$
$$(\,\Box\Diamond(\neg\,\mathrm{ENABLED}\,\langle A\rangle_v) \Rightarrow \Diamond\Box(\neg\,\mathrm{ENABLED}\,\langle A\rangle_v) \vee \Box\Diamond\langle A\rangle_v\,)$$

In the channel example, weak and strong fairness of Rcv are equivalent because $Spec$ implies that, once enabled, Rcv can be disabled only by a Rcv step. Hence, if Rcv is disabled infinitely often, then it either eventually remains disabled forever, or else it is disabled infinitely often by Rcv steps.

The analogs of the WF Conjunction and WF Quantifier Rules (page 105) hold for strong fairness—for example:

> **SF Conjunction Rule** If A_1, ..., A_n are actions such that, for any distinct i and j, whenever action A_i is enabled, action A_j cannot become enabled until an A_i step occurs, then $\mathrm{SF}_v(A_1) \wedge \ldots \wedge \mathrm{SF}_v(A_n)$ is equivalent to $\mathrm{SF}_v(A_1 \vee \ldots \vee A_n)$.

Strong fairness can be more difficult to implement than weak fairness, and it is a less common requirement. A strong fairness condition should be used in a specification only if it is needed. When strong and weak fairness are equivalent, the fairness property should be written as weak fairness.

Liveness properties can be subtle. Expressing them with *ad hoc* temporal formulas can lead to errors. We will specify liveness as the conjunction of weak and/or strong fairness properties whenever possible—and it almost always is possible. Having a uniform way of expressing liveness makes specifications easier to understand. Section 8.9.2 below discusses an even more compelling reason for using fairness to specify liveness.

8.7   The Write-Through Cache

Let's now add liveness to the write-through cache, specified in Figure 5.5 on pages 57–59. We want our specification to guarantee that every request eventually receives a response, without requiring that any requests are issued. This requires fairness on all the actions that make up the next-state action *Next* except for the following:

- A *Req(p)* action, which issues a request.

- An *Evict(p, a)* action, which evicts an address from the cache.

- A *MemQWr* action, if *memQ* contains only write requests and is not full (has fewer than *QLen* elements). Since a response to a write request can be issued before the value is written to memory, failing to execute a *MemQWr* action can prevent a response only if it prevents the dequeuing of a read operation in *memQ* or the enqueuing of an operation (because *memQ* is full).

For simplicity, let's require fairness for the *MemQWr* action too; we'll weaken this requirement later. Our liveness condition then has to assert fairness of the actions

$$MemQWr \quad MemQRd \quad Rsp(p) \quad RdMiss(p) \quad DoRd(p) \quad DoWr(p)$$

for all p in *Proc*. We now must decide whether to assert weak or strong fairness for these actions. Weak and strong fairness are equivalent for an action that, once enabled, remains enabled until it is executed. This is the case for all of these actions except $DoRd(p)$, $RdMiss(p)$, and $DoWr(p)$.

The $DoRd(p)$ action can be disabled by an *Evict* step that evicts the requested data from the cache. In this case, fairness of other actions should imply that the data will eventually be returned to the cache, re-enabling $DoRd(p)$.

The data cannot be evicted again until the $DoRd(p)$ action is executed, and weak fairness then suffices to ensure that the necessary $DoRd(p)$ step eventually occurs.

The $RdMiss(p)$ and $DoWr(p)$ actions append a request to the $memQ$ queue. They are disabled if that queue is full. A $RdMiss(p)$ or $DoWr(p)$ could be enabled and then become disabled because a $RdMiss(q)$ or $DoWr(q)$, for a different processor q, appends a request to $memQ$. We therefore need strong fairness for the $RdMiss(p)$ and $DoWr(p)$ actions. So, the fairness conditions we need are

    Weak Fairness   for $Rsp(p)$, $DoRd(p)$, $MemQWr$, and $MemQRd$

    Strong Fairness for $RdMiss(p)$ and $DoWr(p)$.

As before, let's define $vars$ to be the tuple of all variables.

$$vars \;\triangleq\; \langle memInt,\ wmem,\ buf,\ ctl,\ cache,\ memQ \rangle$$

We could just write the liveness condition as

$$(8.34)\ \land\ \forall\, p \in Proc\ :\ \land\ \mathrm{WF}_{vars}(Rsp(p))\ \land\ \mathrm{WF}_{vars}(DoRd(p))$$
$$\land\ \mathrm{SF}_{vars}(RdMiss(p))\ \land\ \mathrm{SF}_{vars}(DoWr(p))$$
$$\land\ \mathrm{WF}_{vars}(MemQWr)\ \land\ \mathrm{WF}_{vars}(MemQRd)$$

However, I prefer replacing the conjunction of fairness conditions by a single fairness condition on a disjunction, as we did in Section 8.5 for the memory specification. The WF and SF Conjunction Rules (pages 105 and 106) imply that the liveness condition (8.34) can be rewritten as

$$(8.35)\ \land\ \forall\, p \in Proc\ :\ \land\ \mathrm{WF}_{vars}(Rsp(p) \lor DoRd(p))$$
$$\land\ \mathrm{SF}_{vars}(RdMiss(p) \lor DoWr(p))$$
$$\land\ \mathrm{WF}_{vars}(MemQWr \lor MemQRd)$$

We can now try to simplify (8.35) by moving the quantifier inside the WF and SF formulas. First, because $\forall$ distributes over $\land$, we can rewrite the first conjunct of (8.35) as

$$(8.36)\ \land\ \forall\, p \in Proc\ :\ \mathrm{WF}_{vars}(Rsp(p) \lor DoRd(p))$$
$$\land\ \forall\, p \in Proc\ :\ \mathrm{SF}_{vars}(RdMiss(p) \lor DoWr(p))$$

We can now try to apply the WF Quantifier Rule (page 105) to the first conjunct of (8.36) and the corresponding SF Quantifier Rule to its second conjunct. However, the WF quantifier rule doesn't apply to the first conjunct. It's possible for both $Rsp(p) \lor DoRd(p)$ and $Rsp(q) \lor DoRd(q)$ to be enabled at the same time, for two different processors p and q. The formula

(8.37) $\mathrm{WF}_{vars}(\exists\, p \in Proc\ :\ Rsp(p) \vee DoRd(p))$

is satisfied by any behavior in which infinitely many $Rsp(p)$ and $DoRd(p)$ actions occur for some processor p. In such a behavior, $Rsp(q)$ could be enabled for some other processor q without an $Rsp(q)$ step ever occurring, making $\mathrm{WF}_{vars}(Rsp(q) \vee DoRd(q))$ false, which implies that the first conjunct of (8.36) is false. Hence, (8.37) is not equivalent to the first conjunct of (8.36). Similarly, the analogous rule for strong fairness cannot be applied to the second conjunct of (8.36). Formula (8.35) is as simple as we can make it.

Let's return to the observation that we don't have to execute $MemQWr$ if the $memQ$ queue contains only write requests and is not full. In other words, we have to execute $MemQWr$ only if $memQ$ is full or contains a read request. Let's define

$$QCond\ \stackrel{\Delta}{=}\ \begin{aligned}&\vee\ Len(memQ) = QLen \\ &\vee\ \exists\, i \in 1\,..\,Len(memQ)\ :\ memQ[i][2].op = \text{``Rd''}\end{aligned}$$

so we need eventually execute a $MemQWr$ action only when it's enabled and $QCond$ is true, which is the case iff the action $QCond \wedge MemQWr$ is enabled. In this case, a $MemQWr$ step is a $QCond \wedge MemQWr$ step. Hence, it suffices to require weak fairness of the action $QCond \wedge MemQWr$. We can therefore replace the second conjunct of (8.35) with

$$\mathrm{WF}_{vars}((QCond \wedge MemQWr) \vee MemQRd)$$

We would do this if we wanted the specification to describe the weakest liveness condition that implements the memory specification's liveness condition. However, if the specification were a description of an actual device, then that device would probably implement weak fairness on all $MemQWr$ actions, so we would take (8.35) as the liveness condition.

8.8   Quantification

Section 8.1 describes the meaning of ordinary quantification of temporal formulas. For example, the meaning of the formula $\forall\, r : F$, for any temporal formula F, is defined by

$$\sigma \models (\forall\, r\ :\ F)\ \stackrel{\Delta}{=}\ \forall\, r\ :\ (\sigma \models F)$$

where σ is any behavior.

The symbol r in $\exists\, r : F$ is usually called a bound variable. But we've been using the term *variable* to mean something else—something that's declared by a VARIABLE statement in a module. The bound "variable" r is actually a constant in these formulas—a value that is the same in every state of the behavior.[5] For example, the formula $\exists\, r : \Box(x = r)$ asserts that x has the same value in every state of a behavior.

Bounded quantification over a constant set S is defined by

$$\sigma \models (\forall\, r \in S \,:\, F) \;\triangleq\; (\forall\, r \in S \,:\, \sigma \models F)$$

$$\sigma \models (\exists\, r \in S \,:\, F) \;\triangleq\; (\exists\, r \in S \,:\, \sigma \models F)$$

The symbol r is declared to be a constant in formula F. The expression S lies outside the scope of the declaration of r, so the symbol r cannot occur in S. It's easy to define the meanings of these formulas even if S is not a constant—for example, by letting $\exists\, r \in S : F$ equal $\exists\, r : (r \in S) \wedge F$. However, for nonconstant S, it's better to write $\exists\, r : (r \in S) \wedge F$ explicitly.

It's also easy to define the meaning of CHOOSE as a temporal operator. We can just let $\sigma \models (\text{CHOOSE } r : F)$ be an arbitrary constant value r such that $\sigma \models F$ equals TRUE, if such an r exists. However, a temporal CHOOSE operator is not needed for writing specifications, so CHOOSE $r : F$ is not a legal TLA$^+$ formula if F is a temporal formula.

We now come to the temporal existential quantifier $\boldsymbol{\exists}$. In the formula $\boldsymbol{\exists}\, x : F$, the symbol x is declared to be a variable in F. Unlike $\exists\, r : F$, which asserts the existence of a single value r, the formula $\boldsymbol{\exists}\, x : F$ asserts the existence of a value for x in each state of a behavior. For example, if y is a variable, then the formula $\boldsymbol{\exists}\, x : \Box(x \in y)$ asserts that y always has some element x, so y is always a nonempty set. However, the element x could be different in different states, so the values of y in different states could be disjoint.

We have been using $\boldsymbol{\exists}$ as a hiding operator, thinking of $\boldsymbol{\exists}\, x : F$ as F with variable x hidden. The precise definition of $\boldsymbol{\exists}$ is a bit tricky because, as discussed in Section 8.1, the formula $\boldsymbol{\exists}\, x : F$ should be invariant under stuttering. Intuitively, $\boldsymbol{\exists}\, x : F$ is satisfied by a behavior σ iff F is satisfied by a behavior τ that is obtained from σ by adding and/or deleting stuttering steps and changing the value of x. A precise definition appears in Section 16.2.4 (page 314). However, for writing specifications, you can simply think of $\boldsymbol{\exists}\, x : F$ as F with x hidden.

TLA also has a temporal universal quantifier $\boldsymbol{\forall}$, defined by

$$\boldsymbol{\forall}\, x \,:\, F \;\triangleq\; \neg\,\boldsymbol{\exists}\, x \,:\, \neg F$$

This operator is hardly ever used. TLA$^+$ does not allow bounded versions of the operators $\boldsymbol{\exists}$ and $\boldsymbol{\forall}$.

---

[5]Logicians use the term *flexible variable* for a TLA variable, and the term *rigid variable* for a symbol like r that represents a constant.

8.9 Temporal Logic Examined

8.9.1 A Review

Let's look at the shapes of the specifications that we've written so far. We started with the simple form

(8.38) $Init \land \Box[Next]_{vars}$

where *Init* is the initial predicate, *Next* the next-state action, and *vars* the tuple of all variables. This kind of specification is, in principle, quite straightforward. We then introduced hiding, using $\exists$ to bind variables that should not appear in the specification. Those bound variables, also called *hidden* or *internal* variables, serve only to help describe how the values of the free variables (also called visible variables) change.

Hiding variables is easy enough, and it is mathematically elegant and philosophically satisfying. However, in practice, it doesn't make much difference to a specification. A comment can also tell a reader that a variable should be regarded as internal. Explicit hiding allows implementation to mean implication. A lower-level specification that describes an implementation can be expected to imply a higher-level specification only if the higher-level specification's internal variables, whose values don't really matter, are explicitly hidden. Otherwise, implementation means implementation under a refinement mapping. (See Section 5.8.) However, as explained in Section 10.8 below, implementation often involves a refinement of the visible variables as well.

To express liveness, the specification (8.38) is strengthened to the form

(8.39) $Init \land \Box[Next]_{vars} \land Liveness$

where *Liveness* is the conjunction of formulas of the form $\mathrm{WF}_{vars}(A)$ and/or $\mathrm{SF}_{vars}(A)$, for actions A. (I'm considering universal quantification to be a form of conjunction.)

8.9.2 Machine Closure

In the specifications of the form (8.39) we've written so far, the actions A whose fairness properties appear in formula *Liveness* have one thing in common: they are all *subactions* of the next-state action *Next*. An action A is a subaction of *Next* iff every A step is a *Next* step. Equivalently, A is a subaction of *Next* iff A implies *Next*.[6] In almost all specifications of the form (8.39), formula *Liveness*

---

[6] We can also use the following weaker definition of subaction: A is a subaction of formula (8.38) iff, for every state s of every behavior satisfying (8.38), if A is enabled in state s then *Next* $\land$ A is also enabled in s.

should be the conjunction of weak and/or strong fairness formulas for subactions of *Next*. I'll now explain why.

When we look at the specification (8.39), we expect *Init* to constrain the initial state, *Next* to constrain what steps may occur, and *Liveness* to describe only what must eventually happen. However, consider the following formula:

$$(8.40) \quad (x = 0) \,\wedge\, \Box[x' = x + 1]_x \,\wedge\, \mathrm{WF}_x((x > 99) \wedge (x' = x - 1))$$

The first two conjuncts of (8.40) assert that x is initially 0 and that any step either increments x by 1 or leaves it unchanged. Hence, they imply that if x ever exceeds 99, then it forever remains greater than 99. The weak fairness property asserts that, if this happens, then x must eventually be decremented by 1—contradicting the second conjunct. Hence, (8.40) implies that x can never exceed 99, so it is equivalent to

$$(x = 0) \,\wedge\, \Box[(x < 99) \wedge (x' = x + 1)]_x$$

Conjoining the weak fairness property to the first two conjuncts of (8.40) forbids an $x' = x + 1$ step when $x = 99$.

A specification of the form (8.39) is called *machine closed* iff the conjunct *Liveness* constrains neither the initial state nor what steps may occur. A more general way to express this is as follows. Let a finite behavior be a finite sequence of states.[7] We say that a finite behavior σ satisfies a safety property S iff the behavior obtained by adding infinitely many stuttering steps to the end of σ satisfies S. If S is a safety property, then we define the pair of formulas S, L to be machine closed iff every finite behavior that satisfies S can be extended to an infinite behavior that satisfies $S \wedge L$. We call (8.39) machine closed if the pair of formulas $Init \wedge \Box[Next]_{vars}$, *Liveness* is machine closed.

We seldom want to write a specification that isn't machine closed. If we do write one, it's usually by mistake. Specification (8.39) is guaranteed to be machine closed if *Liveness* is the conjunction of weak and/or strong fairness properties for subactions of *Next*.[8] This condition doesn't hold for specification (8.40), which is not machine closed, because $(x > 99) \wedge (x' = x - 1)$ is not a subaction of $x' = x + 1$.

Liveness requirements are philosophically satisfying. A specification of the form (8.38), which specifies only a safety property, allows behaviors in which the system does nothing. Therefore, the specification is satisfied by a system that does nothing. Expressing liveness requirements with fairness properties is less satisfying. These properties are subtle and it's easy to get them wrong.

---

[7]A finite behavior therefore isn't a behavior, which is an infinite sequence of states. Mathematicians often abuse language in this way.

[8]More precisely, this is the case for a finite or countably infinite conjunction of properties of the form $\mathrm{WF}_v(A)$ and/or $\mathrm{SF}_v(A)$, where each $\langle A \rangle_v$ is a subaction of *Next*. This result also holds for the weaker definition of subaction in the footnote on the preceding page.

It requires some thought to determine that the liveness condition for the write-through cache, formula (8.35) on page 108, does imply that every request receives a reply.

It's tempting to express liveness properties more directly, without using fairness properties. For example, it's easy to write a temporal formula asserting for the write-through cache that every request receives a response. When processor p issues a request, it sets $ctl[p]$ to "rdy". We just have to assert that, for every processor p, whenever a state in which $ctl[p] = $ "rdy" is true occurs, there will eventually be a $Rsp(p)$ step:

(8.41) $\forall p \in Proc : \Box((ctl[p] = \text{"rdy"}) \Rightarrow \Diamond \langle Rsp(p) \rangle_{vars})$

While such formulas are appealing, they are dangerous. It's very easy to make a mistake and write a specification that isn't machine closed.

Except in unusual circumstances, you should express liveness with fairness properties for subactions of the next-state action. These are the most straightforward specifications, and hence the easiest to write and to understand. Most system specifications, even if very detailed and complicated, can be written in this straightforward manner. The exceptions are usually in the realm of subtle, high-level specifications that attempt to be very general. An example of such a specification appears in Section 11.2.

8.9.3 Machine Closure and Possibility

Machine closure can be thought of as a possibility condition. For example, machine closure of the pair S, $\Box\Diamond\langle A \rangle_v$ asserts that for every finite behavior σ satisfying S, it is possible to extend σ to an infinite behavior satisfying S in which infinitely many $\langle A \rangle_v$ actions occur. If we regard S as a system specification, so a behavior that satisfies S represents a possible execution of the system, then we can restate machine closure of S, $\Box\Diamond\langle A \rangle_v$ as follows: in any system execution, it is always possible for infinitely many $\langle A \rangle_v$ actions to occur.

TLA specifications express safety and liveness properties, not possibility properties. A safety property asserts that something is impossible—for example, the system cannot take a step that doesn't satisfy the next-state action. A liveness property asserts that something must eventually happen. System requirements are sometimes stated informally in terms of what is possible. Most of the time, when examined rigorously, these requirements can be expressed with liveness and/or safety properties. (The most notable exceptions are statistical properties, such as assertions about the probability that something happens.) We are never interested in specifying that something *might* happen. It's never useful to know that the system *might* produce the right answer. We never have to specify that the user *might* type an "a"; we must specify what happens if he does.

Machine closure is a property of a pair of formulas, not of a system. Although a possibility property is never a useful assertion about a system, it can be a useful assertion about a specification. A specification S of a system with keyboard input should always allow the user to type an "a". So, every finite behavior satisfying S should be extendable to an infinite behavior satisfying S in which infinitely many "a"s are typed. If the action $\langle A \rangle_v$ represents the typing of an "a", then saying that the user should always be able to type infinitely many "a"s is equivalent to saying that the pair S, $\Box\Diamond\langle A \rangle_v$ should be machine closed. If S, $\Box\Diamond\langle A \rangle_v$ isn't machine closed, then it could become impossible for the user ever to type an "a". Unless the system is allowed to lock the keyboard, this would mean that there was something wrong with the specification.

This kind of possibility property can be proved. For example, to prove that it's always possible for the user to type infinitely many "a"s, we show that conjoining suitable fairness conditions on the input actions implies that the user *must* type infinitely many "a"s. However, proofs of this kind of simple property don't seem to be worth the effort. When writing a specification, you should make sure that possibilities allowed by the real system are allowed by the specification. Once you are aware of what should be possible, you will usually have little trouble ensuring that the specification makes it possible. You should also make sure that what the system *must* do is implied by the specification's fairness conditions. This can be more difficult.

8.9.4   Refinement Mappings and Fairness

Section 5.8 (page 62) describes how to prove that the write-through memory implements the memory specification. We have to prove $Spec \Rightarrow \overline{ISpec}$, where $Spec$ is the specification of the write-through memory, $ISpec$ is the internal specification of the memory (with the internal variables made visible), and, for any formula F, we let $\overline{F}$ mean F with expressions $omem$, $octl$, and $obuf$ substituted for the variables mem, ctl, and buf. We could rewrite this implication as (5.3) because substitution (overbarring) distributes over operators like $\wedge$ and $\Box$, so we had

$$
\overline{IInit \wedge \Box[INext]_{\langle memInt,\, mem,\, ctl,\, buf \rangle}}
$$

$$
\equiv \overline{IInit} \wedge \overline{\Box[INext]_{\langle memInt,\, mem,\, ctl,\, buf \rangle}} \qquad \text{Because } {}^{-} \text{ distributes over } \wedge.
$$

$$
\equiv \overline{IInit} \wedge \Box[INext]_{\overline{\langle memInt,\, mem,\, ctl,\, buf \rangle}} \qquad \text{Because } {}^{-} \text{ distributes over } \Box[\cdots]....
$$

$$
\equiv \overline{IInit} \wedge \Box[\overline{INext}]_{\langle \overline{memInt,\, mem,\, ctl,\, buf} \rangle} \qquad \text{Because } {}^{-} \text{ distributes over } \langle \ldots \rangle.
$$

$$
\equiv \overline{IInit} \wedge \Box[\overline{INext}]_{\langle memInt,\, \overline{mem},\, \overline{ctl},\, \overline{buf} \rangle} \qquad \text{Because } \overline{memInt} = memInt.
$$

Adding liveness to the specifications adds conjuncts to the formulas $Spec$ and $ISpec$. Suppose we take formula $Liveness2$, defined in (8.19) on page 101, as the liveness property of *ISpec*. Then $\overline{ISpec}$ has the additional term $\overline{Liveness2}$, which can be simplified as follows:

$$
\begin{aligned}
&\overline{Liveness2} \\
&\equiv \overline{\forall\, p \in Proc \,:\, \text{WF}_{vars}(Do(p) \vee Rsp(p))} \quad\quad \text{By definition of } Liveness2. \\
&\equiv \forall\, p \in Proc \,:\, \overline{\text{WF}_{vars}(Do(p) \vee Rsp(p))} \quad\quad \text{Because } ^{-} \text{ distributes over } \forall.
\end{aligned}
$$

But we cannot automatically move the $^{-}$ inside the WF because substitution does not, in general, distribute over ENABLED, and hence it does not distribute over WF or SF. For the specifications and refinement mappings that occur in practice, including this one, simply replacing each $\overline{\text{WF}_v(A)}$ by $\text{WF}_{\overline{v}}(\overline{A})$ and each $\overline{\text{SF}_v(A)}$ by $\text{SF}_{\overline{v}}(\overline{A})$ does give the right result. However, you don't have to depend on this. You can instead expand the definitions of WF and SF to get, for example:

$$
\begin{aligned}
\overline{\text{WF}_v(A)} &\equiv \overline{\Box\Diamond\neg\text{ENABLED}\,\langle A\rangle_v \vee \Box\Diamond\langle A\rangle_v} \quad\quad \text{By definition of WF.} \\
&\equiv \overline{\Box\Diamond\neg\text{ENABLED}\,\langle A\rangle_v} \vee \overline{\Box\Diamond\langle A\rangle_{\overline{v}}} \quad\quad \text{By distributivity of } ^{-}.
\end{aligned}
$$

You can compute the ENABLED predicates "by hand" and then perform the substitution. When computing ENABLED predicates, it suffices to consider only states satisfying the safety part of the specification, which usually means that ENABLED $\langle A\rangle_v$ equals ENABLED A. You can then compute ENABLED predicates using the following rules:

1. ENABLED $(A \vee B) \equiv (\text{ENABLED } A) \vee (\text{ENABLED } B)$, for any actions A and B.

2. ENABLED $(P \wedge A) \equiv P \wedge (\text{ENABLED } A)$, for any state predicate P and action A.

3. ENABLED $(A \wedge B) \equiv (\text{ENABLED } A) \wedge (\text{ENABLED } B)$, if A and B are actions such that the same variable does not appear primed in both A and B.

4. ENABLED $(x' = exp) \equiv \text{TRUE}$ and ENABLED $(x' \in exp) \equiv (exp \neq \{\})$, for any variable x and state function exp.

For example:

$$
\begin{aligned}
&\overline{\text{ENABLED}\,(Do(p) \vee Rsp(p))} \\
&\equiv \overline{(ctl[p] = \text{``rdy''}) \vee (ctl[p] = \text{``done''})} \quad\quad \text{By rules 1–4.} \\
&\equiv (octl[p] = \text{``rdy''}) \vee (octl[p] = \text{``done''}) \quad\quad \text{By the meaning of } ^{-}.
\end{aligned}
$$

8.9.5   The Unimportance of Liveness

While philosophically important, in practice the liveness property of (8.39) is not as important as the safety part, $Init \wedge \square[Next]_{vars}$. The ultimate purpose of writing a specification is to avoid errors. Experience shows that most of the benefit from writing and using a specification comes from the safety part. On the other hand, the liveness property is usually easy enough to write. It typically constitutes less than five percent of a specification. So, you might as well write the liveness part. However, when looking for errors, most of your effort should be devoted to examining the safety part.

8.9.6   Temporal Logic Considered Confusing

The most general type of specification I've discussed so far has the form

$$(8.42) \quad \exists\, v_1, \ldots, v_n \;:\; Init \wedge \square[Next]_{vars} \wedge Liveness$$

where *Liveness* is the conjunction of fairness properties of subactions of *Next*. This is a very restricted class of temporal-logic formulas. Temporal logic is quite expressive, and one can combine its operators in all sorts of ways to express a wide variety of properties. This suggests the following approach to writing a specification: express each property that the system must satisfy with a temporal formula, and then conjoin all these formulas. For example, formula (8.41) above expresses the property of the write-through cache that every request eventually receives a response.

This approach is philosophically appealing. It has just one problem: it's practical for only the very simplest of specifications—and even for them, it seldom works well. The unbridled use of temporal logic produces formulas that are hard to understand. Conjoining several of these formulas produces a specification that is impossible to understand.

The basic form of a TLA specification is (8.42). Most specifications should have this form. We can also use this kind of specification as a building block. Chapters 9 and 10 describe situations in which we write a specification as a conjunction of such formulas. Section 10.7 introduces an additional temporal operator $\overset{+}{\Rightarrow}$ and explains why we might want to write a specification $F \overset{+}{\Rightarrow} G$, where F and G have the form (8.42). But such specifications are of limited practical use. Most engineers need only know how to write specifications of the form (8.42). Indeed, they can get along quite well with specifications of the form (8.38) that express only safety properties and don't hide any variables.

Chapter 9

Real Time

With a liveness property, we can specify that a system must eventually respond to a request. We cannot specify that it must respond within the next 100 years. To specify timely response, we must use a real-time property.

A system that does not respond within our lifetime isn't very useful, so we might expect real-time specifications to be common. They aren't. Formal specifications are most often used to describe what a system does rather than how long it takes to do it. However, you may someday want to specify real-time properties of a system. This chapter tells you how.

9.1   The Hour Clock Revisited

Let's return to our specification of the simple hour clock in Chapter 2, which asserts that the variable *hr* cycles through the values 1 through 12. We now add the requirement that the clock keep correct time. For centuries, scientists have represented the real-time behavior of a system by introducing a variable, traditionally t, whose value is a real number that represents time. A state in which $t = -17.51$ represents a state of the system at time -17.51, perhaps measured in seconds elapsed since 00:00 UT on 1 January 2000. In TLA$^+$ specifications, I prefer to use the variable *now* rather than t. For linguistic convenience, I will usually assume that the unit of time is the second, though we could just as well choose any other unit.

Remember that a state is an assignment of values to all variables.

Unlike sciences such as physics and chemistry, computer science studies systems whose behavior can be described by a sequence of discrete states, rather than by states that vary continuously with time. We consider the hour clock's display to change directly from reading 12 to reading 1, and ignore the continuum of intermediate states that occur in the physical display. This means that we pretend that the change is instantaneous (happens in 0 seconds). So, a real-time specification of the clock might allow the step

$$\begin{bmatrix} hr & = & 12 \\ now & = & \sqrt{2.47} \end{bmatrix} \quad \rightarrow \quad \begin{bmatrix} hr & = & 1 \\ now & = & \sqrt{2.47} \end{bmatrix}$$

The value of *now* advances between changes to *hr*. If we wanted to specify how long it takes the display to change from 12 to 1, we would have to introduce an intermediate state that represents a changing display—perhaps by letting *hr* assume some intermediate value such as 12.5, or by adding a Boolean-valued variable *chg* whose value indicates whether the display is changing. We won't do this, but will be content to specify an hour clock in which we consider the display to change instantaneously.

The value of *now* changes between changes to *hr*. Just as we represent a continuously varying clock display by a variable whose value changes in discrete steps, we let the value of *now* change in discrete steps. A behavior in which *now* increases in femtosecond increments would be an accurate enough description of continuously changing time for our specification of the hour clock. In fact, there's no need to choose any particular granularity of time; we can let *now* advance by arbitrary amounts between clock ticks. (Since the value of *hr* is unchanged by steps that change *now*, the requirement that the clock keep correct time will rule out behaviors in which *now* changes by too much in a single step.)

What real-time condition should our hour clock satisfy? We might require that it always display the time correctly to within ρ seconds, for some real number ρ. However, this is not typical of the real-time requirements that arise in actual systems. Instead, we require that the clock tick approximately once per hour. More precisely, we require that the interval between ticks be one hour plus or minus ρ seconds, for some positive number ρ. Of course, this requirement allows the time displayed by the clock eventually to drift away from the actual time. But that's what real clocks do if they are not reset.

We could start our specification of the real-time clock from scratch. However, we still want the hour clock to satisfy the specification *HC* of module *HourClock* (Figure 2.1 on page 20). We just want to add an additional real-time requirement. So, we will write the specification as the conjunction of *HC* and a formula requiring that the clock tick every hour, plus or minus ρ seconds. This requirement is the conjunction of two separate conditions: that the clock tick at most once every $3600 - \rho$ seconds, and at least once every $3600 + \rho$ seconds.

To specify these requirements, we introduce a variable that records how much time has elapsed since the last clock tick. Let's call it *t* for *timer*. The value of *t* is set to 0 by a step that represents a clock tick—namely, by an *HCnxt* step. Any step that represents the passing of *s* seconds should advance *t* by *s*. A step represents the passing of time iff it changes *now*, and such a step represents the passage of $now' - now$ seconds. So, the change to the timer *t* is described by the action

$$TNext \quad \triangleq \quad t' = \text{IF } HCnxt \text{ THEN } 0 \text{ ELSE } t + (now' - now)$$

We let t initially equal 0, so we consider the initial state to be one in which the clock has just ticked. The specification of how t changes is then a formula asserting that t initially equals 0, and that every step is a *TNext* step or else leaves unchanged all relevant variables—namely, t, hr, and *now*. This formula is

$$Timer \ \triangleq \ (t = 0) \ \wedge \ \Box[TNext]_{\langle t, hr, now \rangle}$$

The requirement that the clock tick at least once every $3600 + \rho$ seconds means that it's always the case that at most $3600 + \rho$ seconds have elapsed since the last *HCnxt* step. Since t always equals the elapsed time since the last *HCnxt* step, this requirement is expressed by the formula

$$MaxTime \ \triangleq \ \Box(t \leq 3600 + \rho)$$

(Since we can't measure time with perfect accuracy, it doesn't matter whether we use $<$ or $\leq$ in this formula. When we generalize from this example, it is a bit more convenient to use $\leq$.)

The requirement that the clock tick at most once every $3600 - \rho$ seconds means that, whenever an *HCnxt* step occurs, at least $3600 - \rho$ seconds have elapsed since the previous *HCnxt* step. This suggests the condition

In the generalization, $\geq$ will be more convenient than $>$.

$$(9.1) \quad \Box(HCnxt \Rightarrow (t \geq 3600 - \rho))$$

However, (9.1) isn't a legal TLA formula because $HCnxt \Rightarrow \ldots$ is an action (a formula containing primes), and a TLA formula asserting that an action is always true must have the form $\Box[A]_v$. We don't care about steps that leave hr unchanged, so we can replace (9.1) by the TLA formula

$$MinTime \ \triangleq \ \Box[HCnxt \Rightarrow (t \geq 3600 - \rho)]_{hr}$$

The desired real-time constraint on the clock is expressed by the conjunction of these three formulas:

$$HCTime \ \triangleq \ Timer \wedge MaxTime \wedge MinTime$$

Formula *HCTime* contains the variable t, and the specification of the real-time clock should describe only the changes to hr (the clock display) and *now* (the time). So, we have to hide t. Hiding is expressed in TLA$^+$ by the temporal existential quantifier $\exists$, introduced in Section 4.3 (page 41). However, as explained in that section, we can't simply write $\exists t : HCTime$. We must define *HCTime* in a module that declares t, and then use a parametrized instantiation of that module. This is done in Figure 9.1 on page 121. Instead of defining *HCTime* in a completely separate module, I have defined it in a submodule named *Inner* of the module *RealTimeHourClock* containing the specification of the real-time hour clock. Note that all the symbols declared and defined in the main module up to that point can be used in the submodule. Submodule *Inner* is instantiated in the main module with the statement

$$I(t) \triangleq \text{INSTANCE } Inner$$

The *t* in *HCTime* can then be hidden by writing $\exists\, t : I(t)!HCTime$.

The formula $HC \wedge (\exists\, t : I(t)!HCTime)$ describes the possible changes to the value of *hr*, and relates those changes to the value of *now*. But it says very little about how the value of *now* can change. For example, it allows the following behavior:

$$\begin{bmatrix} hr & = 11 \\ now & = 23.5 \end{bmatrix} \rightarrow \begin{bmatrix} hr & = 11 \\ now & = 23.4 \end{bmatrix} \rightarrow \begin{bmatrix} hr & = 11 \\ now & = 23.5 \end{bmatrix} \rightarrow \begin{bmatrix} hr & = 11 \\ now & = 23.4 \end{bmatrix} \rightarrow \cdots$$

Because time can't go backwards, such a behavior doesn't represent a physical possibility. Everyone knows that time only increases, so there's no need to forbid this behavior if the only purpose of our specification is to describe the hour clock. However, a specification should also allow us to reason about a system. If the clock ticks approximately once per hour, then it can't stop. However, as the behavior above shows, the formula $HC \wedge (\exists\, t : I(t)!HCTime)$ by itself allows the clock to stop. To infer that it can't, we also need to state how *now* changes.

We define a formula *RTnow* that specifies the possible changes to *now*. This formula does not specify the granularity of the changes to *now*; it allows a step to advance *now* by a microsecond or by a century. However, we have decided that a step that changes *hr* should leave *now* unchanged, which implies that a step that changes *now* should leave *hr* unchanged. Therefore, steps that change *now* are described by the following action, where *Real* is the set of all real numbers:

$$NowNext \triangleq \wedge\ now' \in \{r \in Real\ :\ r > now\} \quad \textit{now' can equal any real number} > now. $$
$$\wedge \text{ UNCHANGED } hr$$

Formula *RTnow* should also allow steps that leave *now* unchanged. The initial value of *now* is an arbitrary real number (we can start the system at any time), so the safety part of *RTnow* is

$$(now \in Real) \wedge \Box[NowNext]_{now}$$

The liveness condition we want is that *now* should increase without bound. Simple weak fairness of the *NowNext* action isn't good enough, because it allows "Zeno" behaviors such as

$$[now = .9] \rightarrow [now = .99] \rightarrow [now = .999] \rightarrow [now = .9999] \rightarrow \cdots$$

in which the value of *now* remains bounded. Weak fairness of the action $NowNext \wedge (now' > r)$ implies that eventually a *NowNext* step will occur in which the new value of *now* is greater than *r*. (This action is always enabled, so weak fairness implies that infinitely many such actions must occur.) Asserting

Weak fairness is discussed in Chapter 8.

$\qquad\qquad\qquad$ MODULE *RealTimeHourClock* $\qquad\qquad\qquad$

EXTENDS *Reals, HourClock*

VARIABLE *now* $\quad$ The current time, measured in seconds.

CONSTANT *Rho* $\quad$ A positive real number.

ASSUME $(Rho \in Real) \wedge (Rho > 0)$

$\qquad\qquad\qquad$ MODULE *Inner* $\qquad\qquad\qquad$

VARIABLE *t*

$TNext \triangleq t' = \text{IF } HCnxt \text{ THEN } 0 \text{ ELSE } t + (now' - now)$

$Timer \triangleq (t = 0) \wedge \Box[TNext]_{\langle t,\,hr,\,now\rangle}$ $\qquad$ *t* is the elapsed time since the last *HCnxt* step.

$MaxTime \triangleq \Box(t \leq 3600 + Rho)$ $\qquad\qquad$ *t* is always at most $3600 + Rho$.

$MinTime \triangleq \Box[HCnxt \Rightarrow t \geq 3600 - Rho]_{hr}$ $\quad$ An *HCnxt* step can occur only if $t \geq 3600 - Rho$.

$HCTime \triangleq Timer \wedge MaxTime \wedge MinTime$

$I(t) \triangleq \text{INSTANCE } Inner$

$NowNext \triangleq \wedge now' \in \{r \in Real : r > now\}$ $\qquad$ A *NowNext* step can advance *now* by any amount
$\qquad\qquad\qquad \wedge \text{UNCHANGED } hr$ $\qquad\qquad\qquad\qquad$ while leaving *hr* unchanged.

$RTnow \triangleq \wedge now \in Real$ $\qquad\qquad\qquad\qquad\qquad\qquad$ *RTnow* specifies how time may change.
$\qquad\qquad\quad \wedge \Box[NowNext]_{now}$
$\qquad\qquad\quad \wedge \forall r \in Real : \text{WF}_{now}(NowNext \wedge (now' > r))$

$RTHC \triangleq HC \wedge RTnow \wedge (\exists t : I(t)!HCTime)$ $\quad$ The complete specification.

Figure 9.1: The real-time specification of an hour clock that ticks every hour, plus or minus *Rho* seconds.

this for all real numbers *r* implies that *now* grows without bound, so we take as the fairness condition[1]

$$\forall r \in Real : \text{WF}_{now}(NowNext \wedge (now' > r))$$

The complete specification *RTHC* of the real-time hour clock, with the definition of formula *RTnow*, is in the *RealTimeHourClock* module of Figure 9.1 on this page. That module extends the standard *Reals* module, which defines the set *Real* of real numbers.

---

[1]An equivalent condition is $\forall r \in Real : \Diamond(now > r)$, but I like to express fairness with WF and SF formulas.

9.2   Real-Time Specifications in General

In Section 8.4 (page 96), we saw that the appropriate generalization of the liveness requirement that the hour clock tick infinitely often is weak fairness of the clock-tick action. There is a similar generalization for real-time specifications. Weak fairness of an action A asserts that if A is continuously enabled, then an A step must eventually occur. The real-time analog is that if A is continuously enabled for ϵ seconds, then an A step must occur. Since an *HCnxt* action is always enabled, the requirement that the clock tick at least once every $3600 + \rho$ seconds can be expressed in this way by letting A be *HCnxt* and ϵ be $3600 + \rho$.

The requirement that an *HCnxt* action occur at most once every $3600 - \rho$ seconds can be similarly generalized to the condition that an action A must be continuously enabled for at least δ seconds before an A step can occur.

The first condition, the upper bound ϵ on how long A can be enabled without an A step occurring, is vacuously satisfied if ϵ equals *Infinity*—a value defined in the *Reals* module to be greater than any real number. The second condition, the lower bound δ on how long A must be enabled before an A step can occur, is vacuously satisfied if δ equals 0. So, nothing is lost by combining both of these conditions into a single formula containing δ and ϵ as parameters. I now define such a formula, which I call a *real-time bound condition*.

The weak fairness formula $\mathrm{WF}_v(A)$ actually asserts weak fairness of the action $\langle A \rangle_v$, which equals $A \wedge (v' \neq v)$. The subscript v is needed to rule out stuttering steps. Since the truth of a meaningful formula can't depend on whether or not there are stuttering steps, it makes no sense to say that an A step did or did not occur if that step could be a stuttering step. For this reason, the corresponding real-time condition must also be a condition on an action $\langle A \rangle_v$, not on an arbitrary action A. In most cases of interest, v is the tuple of all variables that occur in A. I therefore define the real-time bound formula $RTBound(A,\, v,\, \delta,\, \epsilon)$ to assert that

- An $\langle A \rangle_v$ step cannot occur until $\langle A \rangle_v$ has been continuously enabled for at least δ time units since the last $\langle A \rangle_v$ step—or since the beginning of the behavior.

- $\langle A \rangle_v$ can be continuously enabled for at most ϵ time units before an $\langle A \rangle_v$ step occurs.

$RTBound(A,\, v,\, \delta,\, \epsilon)$ generalizes the formula $\exists\, t : I(t)\,!\,HCTime$ of the real-time hour-clock specification, and it can be defined in the same way, using a submodule. However, the definition can be structured a little more compactly as

$$RTBound(A,\, v,\, D,\, E) \;\triangleq\; \text{LET } Timer(t) \;\triangleq\; \dots$$
$$\dots$$
$$\text{IN } \quad \exists\, t : Timer(t) \wedge \dots$$

For the TLA$^+$ specification, I have replaced δ and ϵ by D and E.

We first define $Timer(t)$ to be a temporal formula asserting that t always equals the length of time that $\langle A \rangle_v$ has been continuously enabled since the last $\langle A \rangle_v$ step. The value of t should be set to 0 by an $\langle A \rangle_v$ step or a step that disables $\langle A \rangle_v$. A step that advances *now* should increment t by $now' - now$ iff $\langle A \rangle_v$ is enabled. Changes to t are therefore described by the action

$$TNext(t) \;\triangleq\; t' = \text{IF } \langle A \rangle_v \vee \neg(\text{ENABLED } \langle A \rangle_v)'$$
$$\text{THEN } 0$$
$$\text{ELSE } \; t + (now' - now)$$

We are interested in the meaning of $Timer(t)$ only when v is a tuple whose components include all the variables that may appear in A. In this case, a step that leaves v unchanged cannot enable or disable $\langle A \rangle_v$. So, the formula $Timer(t)$ should allow steps that leave t, v, and *now* unchanged. Letting the initial value of t be 0, we define

$$Timer(t) \;\triangleq\; (t = 0) \,\wedge\, \Box[TNext(t)]_{\langle t, v, now \rangle}$$

Formulas *MaxTime* and *MinTime* of the real-time hour clock's specification have the obvious generalizations:

- $MaxTime(t)$ asserts that t is always less than or equal to E:

$$MaxTime(t) \;\triangleq\; \Box(t \leq E)$$

- $MinTime(t)$ asserts that an $\langle A \rangle_v$ step can occur only if $t \geq D$:

$$MinTime(t) \;\triangleq\; \Box[A \Rightarrow (t \geq D)]_v$$

  (An equally plausible definition of $MinTime(t)$ is $\Box[\langle A \rangle_v \Rightarrow (t \geq D)]_v$, but the two are, in fact, equivalent.)

We then define $RTBound(A, v, D, E)$ to equal

$$\exists\, t \,:\, Timer(t) \,\wedge\, MaxTime(t) \,\wedge\, MinTime(t)$$

We must also generalize formula $RTnow$ of the real-time hour clock's specification. That formula describes how *now* changes, and it asserts that hr remains unchanged when *now* changes. The generalization is the formula $RTnow(v)$, which replaces hr with an arbitrary state function v that will usually be the tuple of all variables, other than *now*, appearing in the specification. Using these definitions, the specification $RTHC$ of the real-time hour clock can be written

$$HC \,\wedge\, RTnow(hr) \,\wedge\, RTBound(HCnxt, hr, 3600 - Rho, 3600 + Rho)$$

The *RealTime* module, with its definitions of $RTBound$ and $RTnow$, appears in Figure 9.2 on page 125.

Strong fairness strengthens weak fairness by requiring an A step to occur not just if action A is continuously enabled, but if it is repeatedly enabled. Being repeatedly enabled includes the possibility that it is also repeatedly disabled. We can similarly strengthen our real-time bound conditions by defining a stronger formula $SRTBound(A, v, \delta, \epsilon)$ to assert that

- An $\langle A \rangle_v$ step cannot occur until $\langle A \rangle_v$ has been enabled for a total of at least δ time units since the last $\langle A \rangle_v$ step—or since the beginning of the behavior.

- $\langle A \rangle_v$ can be enabled for a total of at most ϵ time units before an $\langle A \rangle_v$ step occurs.

If $\epsilon < Infinity$, then $RTBound(A, v, \delta, \epsilon)$ implies that an $\langle A \rangle_v$ step must occur if $\langle A \rangle_v$ is continuously enabled for ϵ seconds. Hence, if $\langle A \rangle_v$ is ever enabled forever, infinitely many $\langle A \rangle_v$ steps must occur. Thus, $RTBound(A, v, \delta, \epsilon)$ implies weak fairness of A. More precisely, $RTBound(A, v, \delta, \epsilon)$ and $RTnow(v)$ together imply $\mathrm{WF}_v(A)$. However, $SRTBound(A, v, \delta, \epsilon)$ does not similarly imply strong fairness of A. It allows behaviors in which $\langle A \rangle_v$ is enabled infinitely often but never executed—for example, A can be enabled for $\epsilon/2$ seconds, then for $\epsilon/4$ seconds, then for $\epsilon/8$ seconds, and so on. For this reason, $SRTBound$ does not seem to be of much practical use, so I won't bother defining it formally.

9.3   A Real-Time Caching Memory

Let's now use the *RealTime* module to write a real-time versions of the linearizable memory specification of Section 5.3 (page 51) and the write-through cache specification of Section 5.6 (page 54). We obtain the real-time memory specification by strengthening the specification in module *Memory* (Figure 5.3 on page 53) to require that the memory responds to a processor's requests within *Rho* seconds. The complete memory specification *Spec* of module *Memory* was obtained by hiding the variables *mem*, *ctl*, and *buf* in the internal specification *ISpec* of module *InternalMemory*. It's generally easier to add a real-time constraint to an internal specification, where the constraints can mention the internal (hidden) variables. So, we first add the timing constraint to *ISpec* and then hide the internal variables.

To specify that the system must respond to a processor request within *Rho* seconds, we add an upper-bound timing constraint for an action that becomes enabled when a request is issued, and that becomes disabled (possibly by being executed) only when the processor responds to the request. In specification *ISpec*, responding to a request requires two actions—$Do(p)$ to perform the operation internally, and $Rsp(p)$ to issue the response. Neither of these actions is the one we want; we have to define a new action for the purpose. There is a pending request for processor p iff $ctl[p]$ equals "rdy". So, we assert that the

---

——————————— MODULE *RealTime* ———————————

This module declares the variable *now*, which represents real time, and defines operators for writing real-time specifications. Real-time constraints are added to a specification by conjoining it with $RTnow(v)$ and formulas of the form $RTBound(A, v, \delta, \epsilon)$ for actions A, where v is the tuple of all specification variables and $0 \leq \delta \leq \epsilon \leq Infinity$.

EXTENDS *Reals*

VARIABLE *now*   The value of *now* is a real number that represents the current time, in unspecified units.

$RTBound(A, v, \delta, \epsilon)$ asserts that an $\langle A \rangle_v$ step can occur only after $\langle A \rangle_v$ has been continuously enabled for δ time units since the last $\langle A \rangle_v$ step (or the beginning of the behavior), and it must occur before $\langle A \rangle_v$ has been continuously enabled for more than ϵ time units since the last $\langle A \rangle_v$ step (or the beginning of the behavior).

$RTBound(A, v, D, E) \triangleq$

$\quad$ LET $TNext(t) \triangleq t' =$ IF $\langle A \rangle_v \vee \neg(\text{ENABLED } \langle A \rangle_v)'$ $\qquad$ $Timer(t)$ asserts that t is the length
$\qquad\qquad\qquad\qquad\qquad$ THEN 0 $\qquad\qquad\qquad\qquad\qquad\qquad\qquad$ of time $\langle A \rangle_v$ has been continuously
$\qquad\qquad\qquad\qquad\qquad$ ELSE $\quad t + (now' - now)$ $\qquad\qquad\qquad\qquad$ enabled without an $\langle A \rangle_v$ step occur-
$\qquad\qquad\qquad\qquad\qquad\qquad\qquad\qquad\qquad\qquad\qquad\qquad\qquad\qquad\qquad\qquad$ ring.

$\qquad Timer(t) \triangleq (t = 0) \wedge \Box[TNext(t)]_{\langle t, v, now \rangle}$

$\qquad MaxTime(t) \triangleq \Box(t \leq E)$ Asserts that t is always $\leq E$.

$\qquad MinTime(t) \triangleq \Box[A \Rightarrow (t \geq D)]_v$ Asserts that an $\langle A \rangle_v$ step can occur only if $t \geq D$.

$\quad$ IN $\quad \exists\, t : Timer(t) \wedge MaxTime(t) \wedge MinTime(t)$

$RTnow(v)$ asserts that *now* is a real number that is increased without bound, in arbitrary increments, by steps that leave v unchanged.

$RTnow(v) \triangleq$ LET $NowNext \triangleq \wedge now' \in \{r \in Real : r > now\}$
$\qquad\qquad\qquad\qquad\qquad\qquad\quad \wedge$ UNCHANGED v

$\qquad\qquad$ IN $\quad \wedge now \in Real$
$\qquad\qquad\qquad\quad \wedge \Box[NowNext]_{now}$
$\qquad\qquad\qquad\quad \wedge \forall\, r \in Real : \text{WF}_{now}(NowNext \wedge (now' > r))$

Figure 9.2: The *RealTime* module for writing real-time specifications.

following action cannot be enabled for more than *Rho* seconds without being executed:

$\qquad Respond(p) \triangleq (ctl[p] \neq \text{"rdy"}) \wedge (ctl'[p] = \text{"rdy"})$

The complete specification is formula *RTSpec* of module *RTMemory* in Figure 9.3 on the next page. To permit variables *mem*, *ctl*, and *buf* to be hidden, the *RTMemory* module contains a submodule *Inner* that extends module *InternalMemory*.

Having added a real-time constraint to the specification of a linearizable memory, let's strengthen the specification of the write-through cache so it sat-

─────────────────────────── MODULE *RTMemory* ───────────────────────────

A specification that strengthens the linearizable memory specification of Section 5.3 by requiring that a response be sent to every processor request within *Rho* seconds.

EXTENDS *MemoryInterface*, *RealTime*

CONSTANT *Rho*

ASSUME $(Rho \in Real) \wedge (Rho > 0)$

─────────────────────────── MODULE *Inner* ───────────────────────────

We introduce a submodule so we can hide the variables *mem*, *ctl*, and *buf*.

EXTENDS *InternalMemory*

$Respond(p) \triangleq$
$\quad (ctl[p] \neq \text{"rdy"}) \wedge (ctl'[p] = \text{"rdy"})$

Respond(p) is enabled when a request is received from *p*; it is disabled when a *Respond(p)* step issues the response.

$RTISpec \triangleq \wedge ISpec$
$\quad\quad\quad\quad\;\; \wedge \forall\, p \in Proc : RTBound(Respond(p), ctl, 0, Rho)$
$\quad\quad\quad\quad\;\; \wedge RTnow(\langle memInt, mem, ctl, buf\rangle)$

We assert an upper-bound delay of *Rho* on *Respond(p)*, for all processors *p*.

───

$Inner(mem, ctl, buf) \triangleq$ INSTANCE *Inner*

$RTSpec \triangleq \exists\, mem, ctl, buf : Inner(mem, ctl, buf)!RTISpec$

───

Figure 9.3: A real-time version of the linearizable memory specification.

isfies that constraint. The object is not just to add any real-time constraint that does the job—that's easy to do by using the same constraint that we added to the memory specification. We want to write a specification of a real-time algorithm—a specification that tells an implementer how to meet the real-time constraints. This is generally done by placing real-time bounds on the original actions of the untimed specification, not by adding time bounds on a new action, as we did for the memory specification. An upper-bound constraint on the response time should be achieved by enforcing upper-bound constraints on the system's actions.

If we try to achieve a bound on response time by adding real-time bounds to the write-through cache specification's actions, we encounter the following problem. Operations by different processors "compete" with one another to enqueue operations on the finite queue *memQ*. For example, when servicing a write request for processor *p*, the system must execute a *DoWr(p)* action to enqueue the operation to the tail of *memQ*. That action is not enabled if *memQ* is full. The *DoWr(p)* action can be continually disabled by the system performing *DoWr* or *RdMiss* actions for other processors. That's why, to guarantee liveness—that each request eventually receives a response—in Section 8.7 (page 107) we had to assert strong fairness of *DoWr* and *RdMiss* actions. The only way to ensure that a $DoWr(p)$ action is executed within some length of time is to use lower-bound constraints on the actions of other processors to ensure that they cannot perform $DoWr$ or $RdMiss$ actions too frequently. Although such a specification is possible, it is not the kind of approach anyone is likely to take in practice.

The usual method of enforcing real-time bounds on accesses to a shared resource is to schedule the use of the resource by different processors. So, let's modify the write-through cache to add a scheduling discipline to actions that enqueue operations on $memQ$. We use round-robin scheduling, which is probably the easiest one to implement. Suppose processors are numbered from 0 through $N - 1$. Round-robin scheduling means that an operation for processor p is the next one to be enqueued after an operation for processor q iff there is not an operation for any of the processors $(q + 1) \% N$, $(q + 2) \% N$, ..., $(p - 1) \% N$ waiting to be put on $memQ$.

To express this formally, we first let the set *Proc* of processors equal the set $0 .. (N - 1)$ of integers. We normally do this by defining *Proc* to equal $0 .. (N-1)$. However, we want to reuse the parameters and definitions from the write-through cache specification, and that's easiest to do by extending module *WriteThroughCache*. Since *Proc* is a parameter of that module, we can't define it. We therefore let N be a new constant parameter and let $Proc = 0 .. (N-1)$ be an assumption.[2]

To implement round-robin scheduling, we use a variable *lastP* that equals the last processor whose operation was enqueued to $memQ$. We define the operator *position* so that p is the $position(p)^{\text{th}}$ processor after *lastP* in the round-robin order:

$$position(p) \ \triangleq \ \text{CHOOSE } i \in 1 .. N : p = (lastP + i) \% N$$

(Thus, $position(lastP)$ equals N.) An operation for processor p can be the next to access $memQ$ iff there is no operation for a processor q with $position(q) < position(p)$ ready to access it—that is, iff $canGoNext(p)$ is true, where

$$canGoNext(p) \ \triangleq \ \forall\, q \in Proc : (position(q) < position(p)) \ \Rightarrow$$
$$\neg\, \text{ENABLED}\, (RdMiss(q) \vee DoWr(q))$$

We then define $RTRdMiss(p)$ and $RTDoWr(p)$ to be the same as $RdMiss(p)$ and $DoWr(p)$, respectively, except that they have the additional enabling condition $canGoNext(p)$, and they set *lastP* to p. The other subactions of the next-state action are the same as before, except that they must also leave *lastP* unchanged.

For simplicity, we assume a single upper bound of *Epsilon* on the length of time any of the actions of processor p can remain enabled without being executed—except for the $Evict(p, a)$ action, which we never require to happen. In general, suppose A_1, ..., A_k are actions such that (i) no two of them are

---

[2]We could also instantiate module *WriteThroughCache* with $0 .. (N - 1)$ substituted for *Proc*; but that would require declaring the other parameters of *WriteThroughCache*, including the ones from the *MemoryInterface* module.

ever simultaneously enabled, and (ii) once any A_i becomes enabled, it must be executed before another A_j can be enabled. In this case, a single *RTBound* constraint on $A_1 \vee \ldots \vee A_k$ is equivalent to separate constraints on all the A_i. We can therefore place a single constraint on the disjunction of all the actions of processor p, except that we can't use the same constraint for both *DoRd(p)* and *RTRdMiss(p)* because an *Evict(p, a)* step could disable *DoRd(p)* and enable *RTRdMiss(p)*. We therefore use a separate constraint for *RTRdMiss(p)*.

We assume an upper bound of *Delta* on the time *MemQWr* or *MemQRd* can be enabled without dequeuing an operation from *memQ*. The variable *memQ* represents a physical queue between the bus and the main memory, and *Delta* must be large enough so an operation inserted into an empty queue will reach the memory and be dequeued within *Delta* seconds.

We want the real-time write-through cache to implement the real-time memory specification. This requires an assumption relating *Delta*, *Epsilon*, and *Rho* to assure that the memory specification's timing constraint is satisfied—namely, that the delay between when the memory receives a request from processor p and when it responds is at most *Rho*. Determining this assumption requires computing an upper bound on that delay. Finding the smallest upper bound is hard; it's easier to show that

$$2 * (N + 1) * Epsilon + (N + QLen) * Delta$$

is an upper bound. So we assume that this value is less than or equal to *Rho*.

The complete specification appears in Figure 9.4 on the following two pages. The module also asserts as a theorem that the specification *RTSpec* of the real-time write-through cache implements (implies) the real-time memory specification, formula *RTSpec* of module *RTMemory*.

9.4   Zeno Specifications

I have described the formula $RTBound(HCnxt, hr, \delta, \epsilon)$ as asserting that an *HCnxt* step must occur within ϵ seconds of the previous *HCnxt* step. However, implicit in this description is a notion of causality that is not present in the formula. It would be just as accurate to describe the formula as asserting that *now* cannot advance by more than ϵ seconds before the next *HCnxt* step occurs. The formula doesn't tell us whether this condition is met by causing the clock to tick or by preventing time from advancing. Indeed, the formula is satisfied by a "Zeno" behavior:[3]

$$\begin{bmatrix} hr & = & 11 \\ now & = & 0 \end{bmatrix} \rightarrow \begin{bmatrix} hr & = & 11 \\ now & = & \epsilon/2 \end{bmatrix} \rightarrow \begin{bmatrix} hr & = & 11 \\ now & = & 3\epsilon/4 \end{bmatrix} \rightarrow \begin{bmatrix} hr & = & 11 \\ now & = & 7\epsilon/8 \end{bmatrix} \rightarrow \cdots$$

---

[3]The Greek philosopher Zeno posed the paradox that an arrow first had to travel half the distance to its target, then the next quarter of the distance, then the next eighth, and so on; thus it should not be able to land within a finite length of time.

───────────────── MODULE *RTWriteThroughCache* ─────────────────

EXTENDS *WriteThroughCache*, *RealTime*

CONSTANT N We assume that the set *Proc* of processors

ASSUME $(N \in Nat) \land (Proc = 0 .. N - 1)$ equals $0 .. N - 1$.

CONSTANTS *Delta*, *Epsilon*, *Rho*   Some real-time bounds on actions.

ASSUME $\land (Delta \in Real) \land (Delta > 0)$
$\quad\quad\;\; \land (Epsilon \in Real) \land (Epsilon > 0)$
$\quad\quad\;\; \land (Rho \in Real) \land (Rho > 0)$
$\quad\quad\;\; \land 2 * (N + 1) * Epsilon + (N + QLen) * Delta \leq Rho$

───

We modify the write-through cache specification to require that operations for different processors are enqueued on *memQ* in round-robin order.

VARIABLE *lastP*   The last processor to enqueue an operation on *memQ*.

$RTInit \triangleq Init \land (lastP \in Proc)$ Initially, *lastP* can equal any processor.

$position(p) \triangleq$ p is the $position(p)$th processor after *lastP* in the round-robin order.

    CHOOSE $i \in 1 .. N : p = (lastP + i) \% N$

$canGoNext(p) \triangleq$ True if processor p can be the next to enqueue an operation on *memQ*.

   $\forall q \in Proc : (position(q) < position(p)) \Rightarrow \neg \text{ENABLED} (RdMiss(q) \lor DoWr(q))$

$RTRdMiss(p) \triangleq \land canGoNext(p)$ Actions $RTRdMiss(p)$ and $RTDoWr(p)$ are the same as $RdMiss(p)$
$\quad\quad\quad\quad\quad\quad\;\; \land RdMiss(p)$ and $DoWr(p)$ except that they are not enabled unless p is the next
$\quad\quad\quad\quad\quad\quad\;\; \land lastP' = p$ processor in the round-robin order ready to enqueue an operation
                           on *memQ*, and they set *lastP* to p.

$RTDoWr(p) \triangleq \land canGoNext(p)$
$\quad\quad\quad\quad\quad\quad\; \land DoWr(p)$
$\quad\quad\quad\quad\quad\quad\; \land lastP' = p$

$RTNext \triangleq \lor \exists p \in Proc : RTRdMiss(p) \lor RTDoWr(p)$ The next-state action *RTNext*
$\quad\quad\quad\;\; \lor \land \lor \exists p \in Proc : \lor Req(p) \lor Rsp(p) \lor DoRd(p)$ is the same as *Next* except with
$\quad\quad\quad\quad\quad\quad\quad\quad\quad\quad\quad\;\; \lor \exists a \in Adr : Evict(p, a)$ $RTRdMiss(p)$ and $RTDoWr(p)$
$\quad\quad\quad\quad\quad\quad\; \lor MemQWr \lor MemQRd$ replaced by $RdMiss(p)$ and
$\quad\quad\quad\quad\quad\quad\; \land \text{UNCHANGED } lastP$ $DoWr(p)$, and with other
                                           actions modified to leave *lastP*
                                           unchanged.

$vars \triangleq \langle memInt, wmem, buf, ctl, cache, memQ, lastP \rangle$

Figure 9.4a: A real-time version of the write-through cache (beginning).

$RTSpec \triangleq$
  $\wedge\ RTInit \wedge \Box[RTNext]_{vars}$
  $\wedge\ RTBound(MemQWr \vee MemQRd,\ vars,\ 0,\ Delta)$
  $\wedge\ \forall\, p \in Proc\ :\ \wedge\ RTBound(RTDoWr(p) \vee DoRd(p) \vee Rsp(p),$
               $vars,\ 0,\ Epsilon)$
          $\wedge\ RTBound(RTRdMiss(p),\ vars,\ 0,\ Epsilon)$
  $\wedge\ RTnow(vars)$

> We put an upper-bound delay of *Delta* on *MemQWr* and *MemQRd* actions (which dequeue operations from *memQ*), and an upper-bound delay of *Epsilon* on other actions.

$RTM \triangleq$ INSTANCE $RTMemory$
THEOREM $RTSpec \Rightarrow RTM\,!\,RTSpec$

Figure 9.4b: A real-time version of the write-through cache (end).

in which ϵ seconds never pass. We rule out such Zeno behaviors by conjoining to our specification the formula $RTnow(hr)$—more precisely by conjoining its liveness conjunct

$$\forall\, r \in Real\ :\ \mathrm{WF}_{now}(Next \wedge (now' > r))$$

which implies that time advances without bound. Let's call this formula NZ (for Non-Zeno).

Zeno behaviors pose no problem; they are trivially forbidden by conjoining NZ. A problem does exist if a specification allows *only* Zeno behaviors. For example, suppose we conjoined to the untimed hour-clock's specification the condition $RTBound(HCnxt, hr, \delta, \epsilon)$ for some δ and ϵ with $\delta > \epsilon$. This would assert that the clock must wait at least δ seconds before ticking, but must tick within a shorter length of time. In other words, the clock could never tick. Only a Zeno behavior, in which ϵ seconds never elapsed, can satisfy this specification. Conjoining NZ to this specification yields a formula that allows no behaviors—that is, a formula equivalent to FALSE.

This example is an extreme case of what is called a *Zeno specification*. A Zeno specification is one for which there exists a finite behavior σ that satisfies the safety part but cannot be extended to an infinite behavior that satisfies both the safety part and NZ.[4] In other words, the only complete behaviors satisfying the safety part that extend σ are Zeno behaviors. A specification that is not Zeno is, naturally enough, said to be *non-Zeno*. By the definition of machine closure (in Section 8.9.2 on page 111), a specification is non-Zeno iff it is machine closed. More precisely, it is non-Zeno iff the pair of properties consisting of the safety part of the specification (the conjunction of the untimed specification, the real-time bound conditions, and the safety part of the $RTnow$ formula) and NZ is machine closed.

---

[4]Recall that, on page 112, a finite behavior σ was defined to satisfy a safety property P iff adding infinitely many stuttering steps to the end of σ produces a behavior that satisfies P.

A Zeno specification is one in which the requirement that time increases without bound rules out some finite behaviors that would otherwise be allowed. Such a specification is likely to be incorrect because the real-time bound conditions are probably constraining the system in unintended ways. In this respect, Zeno specifications are much like other non-machine-closed specifications.

Section 8.9.2 mentions that the conjunction of fairness conditions on subactions of the next-state relation produces a machine closed specification. There is an analogous result for *RTBound* conditions and non-Zeno specifications. A specification is non-Zeno if it is the conjunction of (i) a formula of the form $Init \land \Box[Next]_{vars}$, (ii) the formula *RTnow(vars)*, and (iii) a finite number of formulas of the form $RTBound(A_i, vars, \delta_i, \epsilon_i)$, where for each i

- $0 \le \delta_i \le \epsilon_i \le Infinity$

- A_i is a subaction of the next-state action *Next*.

- No step is both an A_i and an A_j step, for any A_j with $j \neq i$.

The definition of a subaction appears on page 111.

In particular, this implies that the specification *RTSpec* of the real-time write-through cache in module *RTWriteThroughCache* is non-Zeno.

This result does not apply to the specification of the real-time memory in module *RTMemory* (Figure 9.3 on page 126) because the action *Respond(p)* is not a subaction of the next-state action *INext* of formula *ISpec*. The specification is nonetheless non-Zeno, because any finite behavior σ that satisfies the specification can be extended to one in which time advances without bound. For example, we can first extend σ to respond to all pending requests immediately (in 0 time), and then extend it to an infinite behavior by adding steps that just increase *now*.

INext is defined on page 53

It's easy to construct an example in which conjoining an *RTBound* formula for an action that is not a subaction of the next-state action produces a Zeno specification. For example, consider the formula

(9.2) $HC \land RTBound(hr' = hr - 1, hr, 0, 3600) \land RTnow(hr)$

where *HC* is the specification of the hour clock. The next-state action *HCnxt* of *HC* asserts that *hr* is either incremented by 1 or changes from 12 to 1. The *RTBound* formula asserts that *now* cannot advance for 3600 or more seconds without an $hr' = hr - 1$ step occurring. Since *HC* asserts that every step that changes *hr* is an *HCnxt* step, the safety part of (9.2) is satisfied only by behaviors in which *now* increases by less than 3600 seconds. Since the complete specification (9.2) contains the conjunct *NZ*, which asserts that *now* increases without bound, it is equivalent to FALSE, and is thus a Zeno specification.

When a specification describes how a system is implemented, the real-time constraints are likely to be expressed as *RTBound* formulas for subactions of the next-state action. These are the kinds of formulas that correspond fairly directly to an implementation. For example, module *RTWriteThroughCache* describes an algorithm for implementing a memory, and it has real-time bounds on subactions of the next-state action. On the other hand, more abstract, higher-level specifications—ones describing what a system is supposed to do rather than how to do it—are less likely to have real-time constraints expressed in this way. Thus, the high-level specification of the real-time memory in module *RTMemory* contains an *RTBound* formula for an action that is not a subaction of the next-state action.

9.5   Hybrid System Specifications

A system described by a TLA$^+$ specification is a physical entity. The specification's variables represent some part of the physical state—the display of a clock, or the distribution of charge in a piece of silicon that implements a memory cell. In a real-time specification, the variable *now* is different from the others because we are not abstracting away the continuous nature of time. The specification allows *now* to assume any of a continuum of values. The discrete states in a behavior mean that we are observing the state of the system, and hence the value of *now*, at a sequence of discrete instants.

There may be physical quantities other than time whose continuous nature we want to represent in a specification. For an air traffic control system, we might want to represent the positions and velocities of the aircraft. For a system controlling a nuclear reactor, we might want to represent the physical parameters of the reactor itself. A specification that represents such continuously varying quantities is called a *hybrid system specification*.

As an example, consider a system that, among other things, controls a switch that influences the one-dimensional motion of some object. Suppose the object's position p obeys one of the following laws, depending on whether the switch is off or on:

$$(9.3)\quad d^2p/dt^2 \,+\, c * dp/dt \,+\, f[t] \;=\; 0$$
$$d^2p/dt^2 \,+\, c * dp/dt \,+\, f[t] \,+\, k * p \;=\; 0$$

where c and k are constants, f is some function, and t represents time. At any instant, the future position of the object is determined by the object's current position and velocity. So, the state of the object is described by two variables—namely, its position p and its velocity w. These variables are related by $w = dp/dt$.

We describe this system with a TLA$^+$ specification in which the variables p and w are changed only by steps that change *now*—that is, steps representing the passage of time. We specify the changes to the discrete system state and any real-time constraints as before. However, we replace $RTnow(v)$ with a formula having the following next-state action, where *Integrate* and D are explained below, and v is the tuple of all discrete variables:

$$\wedge \; now' \in \{r \in Real \; : \; r > now\}$$
$$\wedge \; \langle p', w' \rangle \; = \; Integrate(D, now, now', \langle p, w \rangle)$$
$$\wedge \; \text{UNCHANGED} \; v \quad \text{The discrete variables change instantaneously.}$$

The second conjunct asserts that p' and w' equal the expressions obtained by solving the appropriate differential equation for the object's position and velocity at time now', assuming that their values at time now are p and w. The differential equation is specified by D, while *Integrate* is a general operator for solving (integrating) an arbitrary differential equation.

To specify the differential equation satisfied by the object, let's suppose that *switchOn* is a Boolean-valued state variable that describes the position of the switch. We can then rewrite the pair of equations (9.3) as

$$d^2 p / dt^2 \; + \; c * dp / dt \; + \; f[t] \; + \; (\text{IF} \; switchOn \; \text{THEN} \; k * p \; \text{ELSE} \; 0) \; = \; 0$$

We then define the function D so this equation can be written as

$$D[t, \; p, \; dp/dt, \; d^2 p / dt^2] \; = \; 0$$

Using the TLA$^+$ notation for defining functions of multiple arguments, which is explained in Section 16.1.7 on page 301, the definition is

$$D[t, \; p0, \; p1, \; p2 \in Real] \;\; \stackrel{\Delta}{=}$$
$$p2 \; + \; c * p1 \; + \; f[t] \; + \; (\text{IF} \; switchOn \; \text{THEN} \; k * p0 \; \text{ELSE} \; 0)$$

We obtain the desired specification if the operator *Integrate* is defined so that $Integrate(D, t_0, t_1, \langle x_0, \ldots, x_{n-1} \rangle)$ is the value at time t_1 of the n-tuple

$$\langle x, \; dx/dt, \; \ldots, \; d^{n-1}/dt^{n-1} \rangle$$

where x is a solution to the differential equation

$$D[t, \; x, \; dx/dt, \; \ldots, \; d^n x / ct^n] \; = \; 0$$

whose 0th through $(n-1)^{\text{st}}$ derivatives at time t_0 are $x_0, \ldots, x_{n-1}$. The definition of *Integrate* appears in the *DifferentialEquations* module of Section 11.1.3 (page 174).

In general, a hybrid-system specification is similar to a real-time specification, except that the formula $RTnow(v)$ is replaced by one that describes the changes to all variables that represent continuously changing physical quantities. The *Integrate* operator will allow you to specify those changes for many hybrid systems. Some systems will require different operators. For example, describing the evolution of some physical quantities might require an operator for describing the solution to a partial differential equation. However, if you can describe the evolution mathematically, then it can be specified in TLA$^+$.

Hybrid system specifications still seem to be of only academic interest, so I won't say any more about them. If you do have occasion to write one, this brief discussion should indicate how you can do it.

9.6 Remarks on Real Time

Real-time constraints are used most often to place an upper bound on how long it can take the system to do something. In this capacity, they can be considered a strong form of liveness, specifying not just that something must eventually happen, but when it must happen. In very simple specifications, such as the hour clock and the write-through cache, real-time constraints usually replace liveness conditions. More complicated specifications can assert both real-time constraints and liveness properties.

The real-time specifications I have seen have not required very complicated timing constraints. They have been specifications either of fairly simple algorithms in which timing constraints are crucial to correctness, or of more complicated systems in which real time appears only through the use of simple timeouts to ensure liveness. I suspect that people don't build systems with complicated real-time constraints because it's too hard to get them right.

I've described how to write a real-time specification by conjoining $RTnow$ and $RTBound$ formulas to an untimed specification. One can prove that all real-time specifications can be written in this form. In fact, it suffices to use $RTBound$ formulas only for subactions of the next-state action. However, this result is of theoretical interest only because the resulting specification can be incredibly complicated. The operators $RTnow$ and $RTBound$ solve all the real-time specification problems that I have encountered; but I haven't encountered enough to say with confidence that they're all you will ever need. Still, I am quite confident that, whatever real-time properties you have to specify, it will not be hard to express them in TLA$^+$.

Chapter 10

Composing Specifications

Systems are usually described in terms of their components. In the specifications we've written so far, the components have been represented as separate disjuncts of the next-state action. For example, the FIFO system pictured on page 35 is specified in module *InnerFIFO* on page 38 by representing the three components with the following disjuncts of the next-state action:

Sender: $\exists\, msg \in Message\, :\, SSend(msg)$

Buffer: $BufRcv \lor BufSend$

Receiver: $RRcv$

In this chapter, we learn how to specify the components separately and compose their specifications to form a single system specification. Most of the time, there's no point doing this. The two ways of writing the specification differ by only a few lines—a trivial difference in a specification of hundreds or thousands of lines. Still, you may encounter a situation in which it's better to specify a system as a composition.

First, we must understand what it means to compose specifications. We usually say that a TLA formula specifies the correct behavior of a system. However, as explained in Section 2.3 (page 18), a behavior actually represents a possible history of the entire universe, not just of the system. So, it would be more accurate to say that a TLA formula specifies a universe in which the system behaves correctly. Building a system that implements a specification F means constructing the universe so it satisfies F. (Fortunately, correctness of the system depends on the behavior of only a tiny part of the universe, and that's the only part we must build.) Composing two systems whose specifications are F and G means making the universe satisfy both F and G, which is the same as making it satisfy $F \land G$. Thus, the specification of the composition of two systems is the conjunction of their specifications.

Writing a specification as the composition of its components therefore means writing the specification as a conjunction, each conjunct of which can be viewed as the specification of a component. While the basic idea is simple, the details are not always obvious. To simplify the exposition, I begin by considering only safety properties, ignoring liveness and largely ignoring hiding. Liveness and hiding are discussed in Section 10.6.

10.1 Composing Two Specifications

Let's return once again to the simple hour clock, with no liveness or real-time requirement. In Chapter 2, we specified such a clock whose display is represented by the variable hr. We can write that specification as

$$(hr \in 1 \,.\,.\, 12) \,\wedge\, \Box[HCN(hr)]_{hr}$$

where HCN is defined by

$$HCN(h) \;\triangleq\; h' = (h \,\%\, 12) + 1$$

Now let's write a specification $TwoClocks$ of a system composed of two separate hour clocks, whose displays are represented by the variables x and y. (The two clocks are not synchronized and are completely independent of one another.) We can just define $TwoClocks$ to be the conjunction of the two clock specifications

$$TwoClocks \;\triangleq\; \begin{array}{l} \wedge\, (x \in 1 \,.\,.\, 12) \,\wedge\, \Box[HCN(x)]_x \\ \wedge\, (y \in 1 \,.\,.\, 12) \,\wedge\, \Box[HCN(y)]_y \end{array}$$

The following calculation shows how we can rewrite $TwoClocks$ in the usual form as a "monolithic" specification with a single next-state action:[1]

$TwoClocks$
$\quad \equiv\; \wedge\, (x \in 1 \,.\,.\, 12) \,\wedge\, (y \in 1 \,.\,.\, 12)$
$\qquad\quad \wedge\, \Box[HCN(x)]_x \,\wedge\, \Box[HCN(y)]_y$

$\quad \equiv\; \wedge\, (x \in 1 \,.\,.\, 12) \,\wedge\, (y \in 1 \,.\,.\, 12) \qquad$ Because $\Box(F \wedge G) \equiv (\Box F) \wedge (\Box G)$.
$\qquad\quad \wedge\, \Box(\,[HCN(x)]_x \,\wedge\, [HCN(y)]_y\,)$

$\quad \equiv\; \wedge\, (x \in 1 \,.\,.\, 12) \,\wedge\, (y \in 1 \,.\,.\, 12) \qquad$ By definition of $[\ldots]_x$ and $[\ldots]_y$.
$\qquad\quad \wedge\, \Box(\wedge\, HCN(x) \,\vee\, x' = x$
$\qquad\qquad\qquad \wedge\, HCN(y) \,\vee\, y' = y\,)$

---

[1]This calculation is informal because it contains formulas that are not legal TLA—namely, ones of the form $\Box A$ where A is an action that doesn't have the syntactic form $[B]_v$. However, it can be done rigorously.

$$
\begin{aligned}
\equiv \ & \wedge \ (x \in 1 \ .. \ 12) \wedge (y \in 1 \ .. \ 12) \\
& \wedge \ \Box \ (\vee \ HCN(x) \wedge HCN(y) \\
& \qquad\quad \vee \ HCN(x) \wedge (y' = y) \\
& \qquad\quad \vee \ HCN(y) \wedge (x' = x) \\
& \qquad\quad \vee \ (x' = x) \wedge (y' = y) \ )
\end{aligned}
$$

Because:

$$
\begin{pmatrix} \wedge \vee A_1 \\ \vee A_2 \\ \wedge \vee B_1 \\ \vee B_2 \end{pmatrix} \equiv \begin{pmatrix} \vee A_1 \wedge B_1 \\ \vee A_1 \wedge B_2 \\ \vee A_2 \wedge B_1 \\ \vee A_2 \wedge B_2 \end{pmatrix}
$$

$$
\begin{aligned}
\equiv \ & \wedge \ (x \in 1 \ .. \ 12) \wedge (y \in 1 \ .. \ 12) \\
& \wedge \ \Box \ [\vee \ HCN(x) \wedge HCN(y) \\
& \qquad\quad \vee \ HCN(x) \wedge (y' = y) \\
& \qquad\quad \vee \ HCN(y) \wedge (x' = x) \ ]_{\langle x, y \rangle}
\end{aligned}
$$

By definition of $[\ldots]_{\langle x, y \rangle}$.

Thus, *TwoClocks* is equivalent to $Init \wedge \Box[TCNxt]_{\langle x, y \rangle}$ where the next-state action *TCNxt* is

$$
\begin{aligned}
TCnxt \ \triangleq \ & \vee \ HCN(x) \wedge HCN(y) \\
& \vee \ HCN(x) \wedge (y' = y) \\
& \vee \ HCN(y) \wedge (x' = x)
\end{aligned}
$$

This next-state action differs from the ones we are used to writing because of the disjunct $HCN(x) \wedge HCN(y)$, which represents the simultaneous advance of the two displays. In the specifications we have written so far, different components never act simultaneously.

Up until now, we have been writing what are called *interleaving* specifications. In an interleaving specification, each step represents an operation of only one component. For example, in our FIFO specification, a (nonstuttering) step represents an action of either the sender, the buffer, or the receiver. For want of a better term, we describe as *noninterleaving* a specification that, like *TwoClocks*, does permit simultaneous actions by two components.

Suppose we want to write an interleaving specification of the two-clock system as the conjunction of two component specifications. One way is to replace the next-state actions $HCN(x)$ and $HCN(y)$ of the two components by two actions $HCNx$ and $HCNy$ so that, when we perform the analogous calculation to the one above, we get

$$
\begin{pmatrix} \wedge \ (x \in 1 \ .. \ 12) \wedge \Box[HCNx]_x \\ \wedge \ (y \in 1 \ .. \ 12) \wedge \Box[HCNy]_y \end{pmatrix} \ \equiv \ \begin{pmatrix} \wedge \ (x \in 1 \ .. \ 12) \wedge (y \in 1 \ .. \ 12) \\ \wedge \ \Box \ [\vee \ HCNx \wedge (y' = y) \\ \qquad\quad \vee \ HCNy \wedge (x' = x) \ ]_{\langle x, y \rangle} \end{pmatrix}
$$

From the calculation above, we see that this equivalence holds if the following three conditions are satisfied: (i) $HCNx$ implies $HCN(x)$, (ii) $HCNy$ implies $HCN(y)$, and (iii) $HCNx \wedge HCNy$ implies $x' = x$ or $y' = y$. (Condition (iii) implies that the disjunct $HCNx \wedge HCNy$ of the next-state action is subsumed by one of the disjuncts $HCNx \wedge (y' = y)$ and $HCNy \wedge (x' = x)$.) The common way

of satisfying these conditions is to let the next-state action of each clock assert
that the other clock's display is unchanged. We do this by defining

$$HCNx \;\triangleq\; HCN(x) \wedge (y' = y) \qquad HCNy \;\triangleq\; HCN(y) \wedge (x' = x)$$

Another way to write an interleaving specification is simply to disallow simultaneous changes to both clock displays. We can do this by taking as our
specification the formula

$$TwoClocks \;\wedge\; \square[(x' = x) \vee (y' = y)]_{\langle x, y \rangle}$$

The second conjunct asserts that any step must leave x or y (or both) unchanged.

Everything we have done for the two-clock system generalizes to any system
comprising two components. The same calculation as above shows that if

$$(v_1{}' = v_1) \wedge (v_2{}' = v_2) \;\equiv\; (v' = v) \quad \text{This asserts that } v \text{ is unchanged iff both } v_1 \text{ and } v_2 \text{ are.}$$

then

$$(10.1) \quad \begin{pmatrix} \wedge\, I_1 \,\wedge\, \square[N_1]_{v_1} \\ \wedge\, I_2 \,\wedge\, \square[N_2]_{v_2} \end{pmatrix} \;\equiv\; \begin{pmatrix} \wedge\, I_1 \,\wedge\, I_2 \\ \wedge\, \square[\;\vee\, N_1 \wedge N_2 \\ \qquad \vee\, N_1 \wedge (v_2{}' = v_2) \\ \qquad \vee\, N_2 \wedge (v_1{}' = v_1)\,]_v \end{pmatrix}$$

for any state predicates I_1 and I_2 and any actions N_1 and N_2. The left-hand side
of this equivalence represents the composition of two component specifications
if v_k is a tuple containing the variables that describe the kth component, for
$k = 1, 2$, and v is the tuple of all the variables.

The equivalent formulas in (10.1) represent an interleaving specification if
the first disjunct in the next-state action of the right-hand side is redundant, so
it can be removed. This is the case if $N_1 \wedge N_2$ implies that v_1 or v_2 is unchanged.
The usual way to ensure that this condition is satisfied is by defining each N_k so
it implies that the other component's tuple is left unchanged. Another way to
obtain an interleaving specification is by conjoining the formula $\square[(v_1{}' = v_1) \vee$
$(v_2{}' = v_2)]_v$.

10.2 Composing Many Specifications

We can generalize (10.1) to the composition of any set C of components. Because universal quantification generalizes conjunction, the following rule is a
generalization of (10.1):

Composition Rule For any set C, if

$$(\forall\, k \in C \;:\; v_k{}' = v_k) \;\equiv\; (v' = v) \quad \text{This asserts that } v \text{ is unchanged iff all the } v_k \text{ are.}$$

then

$$(\forall\, k \in C \,:\, I_k \,\wedge\, \Box[N_k]_{v_k}) \;\equiv$$
$$\wedge\; \forall\, k \in C \,:\, I_k$$
$$\wedge\; \Box \left[\begin{array}{l} \vee\; \exists\, k \in C \,:\, N_k \,\wedge\, (\forall\, i \in C \setminus \{k\} \,:\, v_i{'} = v_i) \\ \vee\; \exists\, i,j \in C \,:\, (i \neq j) \wedge N_i \wedge N_j \wedge F_{ij} \end{array} \right]_v$$

for some actions F_{ij}.

The second disjunct of the next-state action is redundant, and we have an interleaving specification, if each N_i implies that v_j is unchanged, for all $j \neq i$. However, for this to hold, N_i must mention v_j for components j other than i. You might object to this approach—either on philosophical grounds, because you feel that the specification of one component should not mention the state of another component, or because mentioning other component's variables complicates the component's specification. An alternative approach is simply to assert interleaving. You can do this by conjoining the following formula, which states that no step changes both v_i and v_j, for any i and j with $i \neq j$:

$$\Box[\, \exists\, k \in C \,:\, \forall\, i \in C \setminus \{k\} \,:\, v_i{'} = v_i\,]_v$$

This conjunct can be viewed as a global condition, not attached to any component's specification.

For the left-hand side of the conclusion of the Composition Rule to represent the composition of separate components, the v_k need not be composed of separate variables. They could contain different "parts" of the same variable that describe different components. For example, our system might consist of a set *Clock* of separate, independent clocks, where clock k's display is described by the value of $hr[k]$. Then v_k would equal $hr[k]$. It's easy to specify such an array of clocks as a composition. Using the definition of *HCN* on page 136 above, we can write the specification as

$$(10.2) \quad ClockArray \;\stackrel{\Delta}{=}\; \forall\, k \in Clock \,:\, (hr[k] \in 1 \,..\, 12) \,\wedge\, \Box[HCN(hr[k])]_{hr[k]}$$

This is a noninterleaving specification, since it allows simultaneous steps by different clocks.

Suppose we wanted to use the Composition Rule to express *ClockArray* as a monolithic specification. What would we substitute for v? Our first thought is to substitute hr for v. However, the hypothesis of the rule requires that v must be left unchanged iff $hr[k]$ is left unchanged, for all $k \in Clock$. However, as explained in Section 6.5 on page 72, specifying the values of $hr[k]{'}$ for all $k \in Clock$ does not specify the value of hr. It doesn't even imply that hr is a function. We must substitute for v the function *hrfcn* defined by

$$(10.3) \quad hrfcn \;\stackrel{\Delta}{=}\; [k \in Clock \mapsto hr[k]]$$

The function *hrfcn* equals *hr* iff *hr* is a function with domain *Clock*. Formula *ClockArray* does not imply that *hr* is always a function. It specifies the possible values of $hr[k]$, for all $k \in Clock$, but it doesn't specify the value of *hr*. Even if we changed the initial condition to imply that *hr* is initially a function with domain *Clock*, formula *ClockArray* would not imply that *hr* is always a function. For example, it would still allow "stuttering" steps that leave each $hr[k]$ unchanged, but change *hr* in unknown ways.

We might prefer to write a specification in which *hr* is a function with domain *Clock*. One way of doing this is to conjoin to the specification the formula $\Box IsFcnOn(hr,\ Clock)$, where $IsFcnOn(hr,\ Clock)$ asserts that *hr* is an arbitrary function with domain *Clock*. The operator *IsFcnOn* is defined by

$$IsFcnOn(f,\ S) \;\triangleq\; f = [x \in S \mapsto f[x]]$$

We can view the formula $\Box IsFcnOn(hr,\ Clock)$ as a global constraint on *hr*, while the value of $hr[k]$ for each component *k* is described by that component's specification.

Now, suppose we want to write an interleaving specification of the array of clocks as the composition of specifications of the individual clocks. In general, the conjunction in the Composition Rule is an interleaving specification if each N_k implies that v_i is unchanged, for all $i \neq k$. So, we want the next-state action N_k of clock *k* to imply that $hr[i]$ is unchanged for every clock *i* other than *k*. The most obvious way to do this is to define N_k to equal

$$\begin{aligned} &\land\ hr'[k] = (hr[k]\ \%\ 12) + 1 \\ &\land\ \forall\, i \in Clock \setminus \{k\}\ :\ hr'[i] = hr[i] \end{aligned}$$

We can express this formula more compactly using the EXCEPT construct. This construct applies only to functions, so we must choose whether or not to require *hr* to be a function. If *hr* is a function, then we can let N_k equal

The EXCEPT construct is explained in Section 5.2 on page 48.

$$(10.4)\quad hr' = [hr \text{ EXCEPT } ![k] = (hr[k]\ \%\ 12) + 1]$$

As noted above, we can ensure that *hr* is a function by conjoining the formula $\Box IsFcnOn(hr,\ Clock)$ to the specification. Another way is to define the state function *hrfcn* by (10.3) on the preceding page and let $N(k)$ equal

$$hrfcn' = [hrfcn \text{ EXCEPT } ![k] = (hr[k]\ \%\ 12) + 1]$$

A specification is just a mathematical formula; as we've seen before, there are often many equivalent ways of writing a formula. Which one you choose is usually a matter of taste.

10.3 The FIFO

Let's now specify the FIFO, described in Chapter 4, as the composition of its three components—the Sender, the Buffer, and the Receiver. We start with the internal specification, in which the variable q occurs—that is, q is not hidden. First, we decide what part of the state describes each component. The variables *in* and *out* are channels. Recall that the *Channel* module (page 30) specifies a channel *chan* to be a record with *val*, *rdy*, and *ack* components. The *Send* action, which sends a value, modifies the *val* and *rdy* components; the *Rcv* action, which receives a value, modifies the *ack* component. So, the components' states are described by the following state functions:

Sender: $\langle in.val,\ in.rdy \rangle$

Buffer: $\langle in.ack,\ q,\ out.val,\ out.rdy \rangle$

Receiver: *out.ack*

Unfortunately, we can't reuse the definitions from the *InnerFIFO* module on page 38 for the following reason. The variable q, which is hidden in the final specification, is part of the Buffer component's internal state. Therefore, it should not appear in the specifications of the Sender or Receiver component. The Sender and Receiver actions defined in the *InnerFIFO* module all mention q, so we can't use them. We therefore won't bother reusing that module. However, instead of starting completely from scratch, we can make use of the *Send* and *Rcv* actions from the *Channel* module on page 30 to describe the changes to *in* and *out*.

Let's write a noninterleaving specification. The next-state actions of the components are then the same as the corresponding disjuncts of the *Next* action in module *InnerFIFO*, except that they do not mention the parts of the states belonging to the other components. These contain *Send* and *Rcv* actions, instantiated from the *Channel* module, which use the EXCEPT construct. As noted above, we can apply EXCEPT only to functions—and to records, which are functions. We therefore add to our specification the conjunct

Section 5.2 on page 48 explains why records are functions.

$$\Box(IsChannel(in)\ \wedge\ IsChannel(out))$$

where *IsChannel(c)* asserts that c is a channel—that is a record with *val*, *ack*, and *rdy* fields. Since a record with *val*, *ack*, and *rdy* fields is a function whose domain is { "val", "ack", "rdy" }, we can define *IsChannel(c)* to equal *IsFcnOn(c,* { "val", "ack", "rdy" }*)*. However, it's just as easy to define formula *IsChannel(c)* directly by

$$IsChannel(c)\ \triangleq\ c = [ack \mapsto c.ack,\ val \mapsto c.val,\ rdy \mapsto c.rdy]$$

In writing this specification, we face the same problem as in our original FIFO specification of introducing the variable q and then hiding it. In Chapter 4, we solved this problem by introducing q in a separate *InnerFIFO* module, which is instantiated by the *FIFO* module that defines the final specification. We do essentially the same thing here, except that we introduce q in a submodule instead of in a completely separate module. All the symbols declared and defined at the point where the submodule appears can be used within it. The submodule itself can be instantiated in the containing module anywhere after it appears. (Submodules are used in the *RealTimeHourClock* and *RTMemory* specifications on pages 121 and 126 of Chapter 9.)

There is one small problem to be solved before we can write a composite specification of the FIFO—how to specify the initial predicates. It makes sense for the initial predicate of each component's specification to specify the initial values of its part of the state. However the initial condition includes the requirements $in.ack = in.rdy$ and $out.ack = out.rdy$, each of which relates the initial states of two different components. (These requirements are stated in module *InnerFIFO* by the conjuncts *InChan!Init* and *OutChan!Init* of the initial predicate *Init*.) There are three ways of expressing a requirement that relates the initial states of multiple components:

- Assert it in the initial conditions of all the components. Although symmetric, this seems needlessly redundant.

- Arbitrarily assign the requirement to one of the components. This intuitively suggests that we are assigning to that component the responsibility of ensuring that the requirement is met.

- Assert the requirement as a conjunct separate from either of the component specifications. This intuitively suggests that it is an assumption about how the components are put together, rather than a requirement of either component.

When we write an open-system specification, as described in Section 10.7 below, the intuitive suggestions of the last two approaches can be turned into formal requirements. I've taken the last approach and added

$$(in.ack = in.rdy) \land (out.ack = out.rdy)$$

as a separate condition. The complete specification is in module *CompositeFIFO* of Figure 10.1 on the next page. Formula *Spec* of this module is a noninterleaving specification; for example, it allows a single step that is both an *InChan!Send* step (the sender sends a value) and an *OutChan!Rcv* step (the receiver acknowledges a value). Hence, it is not equivalent to the interleaving specification *Spec* of the *FIFO* module on page 41, which does not allow such a step.

10.4   Composition with Shared State

Thus far, we have been considering *disjoint-state compositions*—ones in which the components are represented by disjoint parts of the state, and a compo-

---

———— MODULE *CompositeFIFO* ————

EXTENDS *Naturals, Sequences*
CONSTANT *Message*
VARIABLES *in, out*

$InChan \;\triangleq\;$ INSTANCE *Channel* WITH $Data \leftarrow Message,\ chan \leftarrow in$
$OutChan \;\triangleq\;$ INSTANCE *Channel* WITH $Data \leftarrow Message,\ chan \leftarrow out$

$SenderInit \;\triangleq\; (in.rdy \in \{0, 1\}) \wedge (in.val \in Message)$ The Sender's
$Sender \;\triangleq\; SenderInit \wedge \Box [\exists\, msg \in Message\, :\, InChan!Send(msg)]_{\langle in.val,\, in.rdy \rangle}$ specification.

———— MODULE *InnerBuf* ————

VARIABLE q

$BufferInit \;\triangleq\; \wedge\ in.ack \in \{0, 1\}$ The Buffer's internal
$\qquad\qquad\quad \wedge\ q = \langle\,\rangle$ specification, with q
$\qquad\qquad\quad \wedge\ (out.rdy \in \{0, 1\}) \wedge (out.val \in Message)$ visible.

$BufRcv \;\triangleq\; \wedge\ InChan!Rcv$
$\qquad\qquad \wedge\ q' = Append(q,\, in.val)$
$\qquad\qquad \wedge\ $UNCHANGED $\langle out.val,\, out.rdy \rangle$

$BufSend \;\triangleq\; \wedge\ q \neq \langle\,\rangle$
$\qquad\qquad\ \ \wedge\ OutChan!Send(Head(q))$
$\qquad\qquad\ \ \wedge\ q' = Tail(q)$
$\qquad\qquad\ \ \wedge\ $UNCHANGED $in.ack$

$InnerBuffer \;\triangleq\; BufferInit \wedge \Box[BufRcv \vee BufSend]_{\langle in.ack,\, q,\, out.val,\, out.rdy \rangle}$

$Buf(q) \;\triangleq\;$ INSTANCE *InnerBuf*   The Buffer's external specification
$Buffer \;\triangleq\; \exists\, q\ :\ Buf(q)!InnerBuffer$ with q hidden.

$ReceiverInit \;\triangleq\; out.ack \in \{0, 1\}$ The Receiver's
$Receiver \;\triangleq\; ReceiverInit \wedge \Box[OutChan!Rcv]_{out.ack}$ specification.

$IsChannel(c) \;\triangleq\; c = [ack \mapsto c.ack,\ val \mapsto c.val,\ rdy \mapsto c.rdy]$

$Spec \;\triangleq\; \wedge\ \Box(IsChannel(in) \wedge IsChannel(out))$ Asserts that *in* and *out* are always records.
$\qquad\quad \wedge\ (in.ack = in.rdy) \wedge (out.ack = out.rdy)$ Relates different components' initial states.
$\qquad\quad \wedge\ Sender \wedge Buffer \wedge Receiver$ Conjoins the three specifications.

Figure 10.1: A noninterleaving composite specification of the FIFO.

nent's next-state action describes changes only to its part of the state.[2] We now consider the case when this may not be possible.

10.4.1 Explicit State Changes

We first examine the situation in which some part of the state cannot be partitioned among the different components, but the state change that each component performs is completely described by the specification. As an example, let's again consider a Sender and a Receiver that communicate with a FIFO buffer. In the system we studied in Chapter 4, sending or receiving a value required two steps. For example, the Sender executes a *Send* step to send a value, and it must then wait until the buffer executes a *Rcv* step before it can send another value. We simplify the system by replacing the Buffer component with a variable *buf* whose value is the sequence of values sent by the Sender but not yet received by the Receiver. This replaces the three-component system pictured on page 35 with this two-component one:

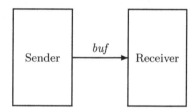

The Sender sends a value by appending it to the end of *buf*; the Receiver receives a value by removing it from the head of *buf*.

In general, the Sender performs some computation to produce the values that it sends, and the Receiver does some computation on the values that it receives. The system state consists of *buf* and two tuples *s* and *r* of variables that describe the Sender and Receiver states. In a monolithic specification, the system's next-state action is a disjunction $Sndr \lor Rcvr$, where $Sndr$ and $Rcvr$ describe steps taken by the Sender and Receiver, respectively. These actions are defined by

$$
\begin{aligned}
Sndr \;\triangleq\;\; &\lor\;\land\; buf' = Append(buf, \ldots) \\
&\;\land\; SComm \\
&\;\land\; \text{UNCHANGED } r \\
&\lor\;\land\; SCompute \\
&\;\land\; \text{UNCHANGED } \langle buf, r\rangle
\end{aligned}
\qquad
\begin{aligned}
Rcvr \;\triangleq\;\; &\lor\;\land\; buf \neq \langle\rangle \\
&\;\land\; buf' = Tail(buf) \\
&\;\land\; RComm \\
&\;\land\; \text{UNCHANGED } s \\
&\lor\;\land\; RCompute \\
&\;\land\; \text{UNCHANGED } \langle buf, s\rangle
\end{aligned}
$$

---

[2]In an interleaving composition, a component specification may assert that the state of other components is *not* changed.

for some actions *SComm*, *SCompute*, *RComm*, and *RCompute*. For simplicity, we assume that neither *Sndr* nor *Rcvr* allows stuttering actions, so *SCompute* changes s and *RCompute* changes r. We now write the specification as the composition of separate specifications of the Sender and Receiver.

Splitting the initial predicate is straightforward. The initial conditions on s belong to the Sender's initial predicate; those on r belong to the Receiver's initial predicate; and the initial condition $buf = \langle\rangle$ can be assigned arbitrarily to either of them.

Now let's consider the next-state actions *NS* and *NR* of the Sender and Receiver components. The trick is to define them by

$$NS \stackrel{\Delta}{=} Sndr \vee (\sigma \wedge (s' = s)) \qquad NR \stackrel{\Delta}{=} Rcvr \vee (\rho \wedge (r' = r))$$

where σ and ρ are actions containing only the variable buf. Think of σ as describing possible changes to buf that are not caused by the Sender, and ρ as describing possible changes to buf that are not caused by the Receiver. Thus, *NS* permits any step that is either a *Sndr* step or one that leaves s unchanged and is a change to buf that can't be "blamed" on the Sender.

Suppose σ and ρ satisfy the following three conditions:

- $\forall\, d : (buf' = Append(buf, d)) \Rightarrow \rho$
  A step that appends a value to buf is not caused by the Receiver.

- $(buf \neq \langle\rangle) \wedge (buf' = Tail(buf)) \Rightarrow \sigma$
  A step that removes a value from the head of buf is not caused by the Sender.

- $(\sigma \wedge \rho) \Rightarrow (buf' = buf)$
  A step that is caused by neither the Sender nor the Receiver cannot change buf.

Using obvious relations such as[3]

$$(buf' = buf) \wedge (buf \neq \langle\rangle) \wedge (buf' = Tail(buf)) \equiv \text{FALSE}$$

a computation like the one by which we derived (10.1) shows

$$\Box[NS]_{\langle buf,\, s\rangle} \wedge \Box[NR]_{\langle buf,\, r\rangle} \equiv \Box[Sndr \vee Rcvr]_{\langle buf,\, s,\, r\rangle}$$

Thus, *NS* and *NR* are suitable next-state actions for the components, if we choose σ and ρ to satisfy the three conditions above. There is considerable freedom in that choice. The strongest possible choices of σ and ρ are ones that describe exactly the changes permitted by the other component:

$$\sigma \stackrel{\Delta}{=} (buf \neq \langle\rangle) \wedge (buf' = Tail(buf))$$
$$\rho \stackrel{\Delta}{=} \exists\, d : buf' = Append(buf, d)$$

---

[3]These relations are true only if buf is a sequence. A rigorous calculation requires the use of an invariant to assert that buf actually is a sequence.

We can weaken these definitions any way we want, so long as we maintain the condition that $\sigma \wedge \rho$ implies that buf is unchanged. For example, we can define σ as above and let ρ equal $\neg\sigma$. The choice is a matter of taste.

I've been describing an interleaving specification of the Sender/Receiver system. Now let's consider a noninterleaving specification—one that allows steps in which both the Sender and the Receiver are computing. In other words, we want the specification to allow $SCompute \wedge RCompute$ steps that leave buf unchanged. Let $SSndr$ be the action that is the same as $Sndr$ except it doesn't mention r, and let $RRcvr$ be defined analogously. We then have

$$Sndr \;\equiv\; SSndr \wedge (r' = r) \qquad Rcvr \;\equiv\; RRcvr \wedge (s' = s)$$

A monolithic noninterleaving specification has the next-state action

$$Sndr \;\vee\; Rcvr \;\vee\; (SSndr \wedge RRcvr \wedge (buf' = buf))$$

It is the conjunction of component specifications having the next-state actions NS and NR defined by

$$NS \;\stackrel{\Delta}{=}\; SSndr \vee (\sigma \wedge (s' = s)) \qquad NR \;\stackrel{\Delta}{=}\; RRcvr \vee (\rho \wedge (r' = r))$$

where σ and ρ are as above.

This two-process situation generalizes to the composition of any set C of components that share a variable or tuple of variables w. The interleaving case generalizes to the following rule, in which N_k is the next-state action of component k, the action μ_k describes all changes to w that are attributed to some component other than k, the tuple v_k describes the private state of k, and v is the tuple formed by all the v_k:

Shared-State Composition Rule The four conditions

1. $(\forall k \in C : v_k' = v_k) \;\equiv\; (v' = v)$

   v is unchanged iff the private state v_k of every component is unchanged.

2. $\forall i, k \in C : N_k \wedge (i \neq k) \Rightarrow (v_i' = v_i)$

   The next-state action of any component k leaves the private state v_i of all other components i unchanged.

3. $\forall i, k \in C : N_k \wedge (w' \neq w) \wedge (i \neq k) \Rightarrow \mu_i$

   A step of any component k that changes w is a μ_i step, for any other component i.

4. $(\forall k \in C : \mu_k) \;\equiv\; (w' = w)$

   A step is caused by no component iff it does not change w.

imply

$$(\forall k \in C : I_k \wedge \Box[N_k \vee (\mu_k \wedge (v_k' = v_k))]_{\langle w, v_k \rangle})$$

$$\equiv\; (\forall k \in C : I_k) \wedge \Box[\exists k \in C : N_k]_{\langle w, v \rangle}$$

Assumption 2 asserts that we have an interleaving specification. If we drop that assumption, then the right-hand side of the conclusion may not be a sensible specification, since a disjunct N_k may allow steps in which a variable of some other component assumes arbitrary values. However, if each N_k correctly determines the new values of component k's private state v_k, then the left-hand side will be a reasonable specification, though possibly a noninterleaving one (and not necessarily equivalent to the right-hand side).

10.4.2 Composition with Joint Actions

Consider the linearizable memory of Chapter 5. As shown in the picture on page 45, it is a system consisting of a collection of processors, a memory, and an interface represented by the variable *memInt*. We now take it to be a two-component system, where the set of processors forms one component, called the *environment*, and the memory is the other component. Let's neglect hiding for now and consider only the internal specification, with all variables visible. We want to write the specification in the form

(10.5) $(IE \land \Box[NE]_{vE}) \land (IM \land \Box[NM]_{vM})$

where E refers to the environment component (the processors) and M to the memory component. The tuple vE of variables includes *memInt* and the variables of the environment component; the tuple vM includes *memInt* and the variables of the memory component. We must choose the formulas IE, NE, etc. so that (10.5), with internal variables hidden, is equivalent to the memory specification *Spec* of module *Memory* on page 53.

In the memory specification, communication between the environment and the memory is described by an action of the form

$Send(p, d, memInt, memInt')$ or $Reply(p, d, memInt, memInt')$

where *Send* and *Reply* are unspecified operators declared in the *MemoryInterface* module (page 48). The specification says nothing about the actual value of *memInt*. So, not only do we not know how to split *memInt* into two parts that are each changed by only one of the components, we don't even know exactly how *memInt* changes.

The trick to writing the specification as a composition is to put the *Send* and *Reply* actions in the next-state actions of both components. We represent the sending of a value over *memInt* as a *joint action* performed by both the memory and the environment. The next-state actions have the following form:

$NM \triangleq \exists p \in Proc : MRqst(p) \lor MRsp(p) \lor MInternal(p)$

$NE \triangleq \exists p \in Proc : ERqst(p) \lor ERsp(p)$

where an $MRqst(p) \wedge ERqst(p)$ step represents the sending of a request by processor p (part of the environment) to the memory, an $MRsp(p) \wedge ERsp(p)$ step represents the sending of a reply by the memory to processor p, and an $MInternal(p)$ step is an internal step of the memory component that performs the request. (There are no internal steps of the environment.)

The sending of a reply is controlled by the memory, which chooses what value is sent and when it is sent. The enabling condition and the value sent are therefore specified by the $MRsp(p)$ action. Let's take the internal variables of the memory component to be the same variables *mem*, *ctl*, and *buf* as in the internal monolithic memory specification of module *InternalMemory* on pages 52 and 53. We can then let $MRsp(p)$ be the same as the action $Rsp(p)$ defined in that module. The $ERsp(p)$ action should always be enabled, and it should allow any legal response to be sent. A legal response is an element of *Val* or the special value *NoVal*, so we can define $ERsp(p)$ to equal[4]

$$\wedge \; \exists \, rsp \in Val \cup \{NoVal\} \; : \; Reply(p, \, rsp, \, memInt, \, memInt')$$
$$\wedge \ldots$$

where the "..." describes the new values of the environment's internal variables.

The sending of a request is controlled by the environment, which chooses what value is sent and when it is sent. Hence, the enabling condition should be part of the $ERqst(p)$ action. In the monolithic specification of the *Internal-Memory* module, that enabling condition was $ctl[p] = $ "rdy". However, if *ctl* is an internal variable of the memory, it can't also appear in the environment specification. We therefore have to add a new variable whose value indicates whether a processor is allowed to send a new request. Let's use a Boolean variable *rdy*, where $rdy[p]$ is true iff processor p can send a request. The value of $rdy[p]$ is set false when p sends a request and is set true again when the corresponding response to p is sent. We can therefore define $ERqst(p)$, and complete the definition of $ERsp(p)$, as follows:

$$ERqst(p) \; \stackrel{\Delta}{=} \; \wedge \; rdy[p]$$
$$\wedge \; \exists \, req \in MReq \; : \; Send(p, \, req, \, memInt, \, memInt')$$
$$\wedge \; rdy' = [rdy \; \text{EXCEPT} \; ![p] = \text{FALSE}]$$

$$ERsp(p) \; \stackrel{\Delta}{=} \; \wedge \; \exists \, rsp \in Val \cup \{NoVal\} \; :$$
$$Reply(p, \, rsp, \, memInt, \, memInt')$$
$$\wedge \; rdy' = [rdy \; \text{EXCEPT} \; ![p] = \text{TRUE}]$$

The memory's $MRqst(p)$ action is the same as the $Req(p)$ action of the *Internal-Memory* module, except without the enabling condition $ctl[p] = $ "rdy".

---

[4]The bound on the $\exists$ isn't necessary. We can let the processor accept any value, not just a legal one, by taking $\exists \, rsp : Reply(p, \, rsp, \, memInt, \, memInt')$ as the first conjunct. However, it's generally better to use bounded quantifiers when possible.

Finally, the memory's internal action $MInternal(p)$ is the same as the $Do(p)$ action of the *InternalMemory* module.

The rest of the specification is easy. The tuples vE and vM are $\langle memInt, rdy \rangle$ and $\langle memInt, mem, ctl, buf \rangle$, respectively. Defining the initial predicates IE and IM is straightforward, except for the decision of where to put the initial condition $memInt \in InitMemInt$. We can put it in either IE or IM, in both, or else in a separate conjunct that belongs to neither component's specification. Let's put it in IM, which then equals the initial predicate $IInit$ from the *InternalMemory* module. The final environment specification is obtained by hiding rdy in its internal specification; the final memory component specification is obtained by hiding mem, ctl, and buf in its internal specification. The complete specification appears in Figure 10.2 on the next page. I have not bothered to define IM, $MRsp(p)$, or $MInternal(p)$, since they equal $IInit$, $Rsp(p)$, and $Do(p)$ from the *InternalMemory* module, respectively.

What we've just done for the environment-memory system generalizes naturally to joint-action specifications of any two-component system in which part of the state cannot be considered to belong to either component. It also generalizes to systems in which any number of components share some part of the state. For example, suppose we want to write a composite specification of the linearizable memory system in which each processor is a separate component. The specification of the memory component would be the same as before. The next-state action of processor p would now be

$$ERqst(p) \lor ERsp(p) \lor OtherProc(p)$$

where $ERqst(p)$ and $ERsp(p)$ are the same as above, and an $OtherProc(p)$ step represents the sending of a request by, or a response to, some processor other than p. Action $OtherProc(p)$ represents p's participation in the joint action by which another processor q communicates with the memory component. It is defined to equal

$$\exists q \in Proc \setminus \{p\} : \lor \exists req \in MReq : Send(q, req, memInt, memInt')$$
$$\lor \exists rsp \in Val \cup \{NoVal\} :$$
$$Reply(q, rsp, memInt, memInt')$$

This example is rather silly because each processor must participate in communication actions that concern only other components. It would be better to change the interface to make $memInt$ an array, with communication between processor p and the memory represented by a change to $memInt[p]$. A sensible example would require that a joint action represent a true interaction between all the components—for example, a barrier synchronization operation in which the components wait until they are all ready and then perform a synchronization step together.

—————————— MODULE *JointActionMemory* ——————————

EXTENDS *MemoryInterface*

————————— MODULE *InnerEnvironmentComponent* —————————

VARIABLE *rdy*

$IE \triangleq rdy = [p \in Proc \mapsto \text{TRUE}]$

$ERqst(p) \triangleq \land rdy[p]$
$\qquad\qquad \land \exists req \in MReq : Send(p, req, memInt, memInt')$
$\qquad\qquad \land rdy' = [rdy \text{ EXCEPT } ![p] = \text{FALSE}]$

$ERsp(p) \triangleq \land \exists rsp \in Val \cup \{NoVal\} : Reply(p, rsp, memInt, memInt')$
$\qquad\qquad \land rdy' = [rdy \text{ EXCEPT } ![p] = \text{TRUE}]$

$NE \triangleq \exists p \in Proc : ERqst(p) \lor ERsp(p)$

$IESpec \triangleq IE \land \Box[NE]_{\langle memInt, rdy \rangle}$

————————— MODULE *InnerMemoryComponent* —————————

EXTENDS *InternalMemory*

$MRqst(p) \triangleq \land \exists req \in MReq : \land Send(p, req, memInt, memInt')$
$\qquad\qquad\qquad\qquad\qquad\quad \land buf' = [buf \text{ EXCEPT } ![p] = req]$
$\qquad\qquad\qquad\qquad\qquad\quad \land ctl' = [ctl \text{ EXCEPT } ![p] = \text{"busy"}]$
$\qquad\qquad \land \text{UNCHANGED } mem$

$NM \triangleq \exists p \in Proc : MRqst(p) \lor Do(p) \lor Rsp(p)$

$IMSpec \triangleq IInit \land \Box[NM]_{\langle memInt, mem, ctl, buf \rangle}$

$IEnv(rdy) \triangleq \text{INSTANCE } InnerEnvironmentComponent$

$IMem(mem, ctl, buf) \triangleq \text{INSTANCE } InnerMemoryComponent$

$Spec \triangleq \land \exists rdy : IEnv(rdy)!IESpec$
$\qquad\quad \land \exists mem, ctl, buf : IMem(mem, ctl, buf)!IMSpec$

Figure 10.2: A joint-action specification of a linearizable memory.

10.5 A Brief Review

The basic idea of composing specifications is simple: a composite specification is the conjunction of formulas, each of which can be considered to be the specification of a separate component. This chapter has presented several techniques for writing a specification as a composition. Before going further, let's put these techniques in perspective.

10.5.1   A Taxonomy of Composition

We have seen three different ways of categorizing composite specifications:

Interleaving versus noninterleaving. An interleaving specification is one in which each (nonstuttering) step can be attributed to exactly one component. A noninterleaving specification allows steps that represent simultaneous operations of two or more different components.

Disjoint-state versus shared-state. A disjoint-state specification is one in which the state can be partitioned, with each part belonging to a separate component. A part of the state can be a variable v, or a "piece" of that variable such as $v.c$ or $v[c]$ for some fixed c. Any change to a component's part of the state is attributed to that component. In a shared-state specification, some part of the state can be changed by steps attributed to more than one component.

Joint-action versus separate-action. A joint-action specification is a noninterleaving one in which some step attributed to one component must occur simultaneously with a step attributed to another component. A separate-action specification is simply one that is not a joint-action specification.

These are independent ways of classifying specifications, except that a joint-action specification must be noninterleaving.

The word *interleaving* is standard; there is no common terminology for the other concepts.

10.5.2   Interleaving Reconsidered

Should we write interleaving or noninterleaving specifications? We might try to answer this question by asking: can different components really take simultaneous steps? However, this question makes no sense. A step is a mathematical abstraction; real components perform operations that take a finite amount of time. Operations performed by two different components could overlap in time. We are free to represent this physical situation either with a single simultaneous step of the two components, or with two separate steps. In the latter case, the specification usually allows the two steps to occur in either order. (If the two operations must occur simultaneously, then we have written a joint-action specification.) It's up to you whether to write an interleaving or a noninterleaving specification. You should choose whichever is more convenient.

The choice is not completely arbitrary if you want one specification to implement another. A noninterleaving specification will not, in general, implement an interleaving one because the noninterleaving specification will allow simultaneous actions that the interleaving specification prohibits. So, if you want to write a lower-level specification that implements a higher-level interleaving specification, then you'll have to use an interleaving specification. As we've seen, it's easy

to turn a noninterleaving specification into an interleaving one by conjoining an
interleaving assumption.

10.5.3   Joint Actions Reconsidered

The reason for writing a composite specification is to separate the specifications
of the different components. The mixing of actions from different components
in a joint-action specification destroys this separation. So, why should we write
such a specification?

Joint-action specifications arise most often in highly abstract descriptions of
inter-component communication. In writing a composite specification of the linearizable memory, we were led to use joint actions because of the abstract nature
of the interface. In real systems, communication occurs when one component
changes the state and another component later observes that change. The interface described by the *MemoryInterface* module abstracts away those two steps,
replacing them with a single one that represents instantaneous communication—
a fiction that does not exist in the real world. Since each component must remember that the communication has occurred, the single communication step
has to change the private state of both components. That's why we couldn't
use the approach of Section 10.4.1 (page 144), which requires that any change
to the shared interface change the nonshared state of just one component.

The abstract memory interface simplifies the specification, allowing communication to be represented as one step instead of two. But this simplification
comes at the cost of blurring the distinction between the two components. If we
blur this distinction, it may not make sense to write the specification as the conjunction of separate component specifications. As the memory system example
illustrates, decomposing the system into separate components communicating
with joint actions may require the introduction of extra variables. There may
occasionally be a good reason for adding this kind of complexity to a specification, but it should not be done as a matter of course.

10.6   Liveness and Hiding

10.6.1   Liveness and Machine Closure

Thus far, the discussion of composition has neglected liveness. In composite
specifications, it is usually easy to specifying liveness by placing fairness conditions on the actions of individual components. For example, to specify an
array of clocks that all keep ticking forever, we would modify the specification

ClockArray of (10.2) on page 139 to equal

$$\forall\, k \in Clock\,:$$
$$(hr[k] \in 1\,..\,12)\,\wedge\,\Box[HCN(hr[k])]_{hr[k]}\,\wedge\,\mathrm{WF}_{hr[k]}(HCN(hr[k]))$$

When writing a weak or strong fairness formula for an action A of a component c, there arises the question of what the subscript should be. The obvious choices are (i) the tuple v describing the entire specification state, and (ii) the tuple v_c describing that component's state. The choice can matter only if the safety part of the specification allows the system to reach some state in which an A step could leave v_c unchanged while changing v. Although unlikely, this could conceivably be the case in a joint-action specification. If it is, we probably don't want the fairness condition to be satisfied by a step that leaves the component's state unchanged, so we would use the subscript v_c.

Fairness conditions for composite specifications do raise one important question: if each component specification is machine closed, is the composite specification necessarily machine closed? Suppose we write the specification as $\forall\, k \in C : S_k \wedge L_k$, where each pair S_k, L_k is machine closed. Let S be the conjunction of the S_k and L the conjunction of the L_k, so the specification equals $S \wedge L$. The conjunction of safety properties is a safety property,[5] so S is a safety property. Hence, we can ask if the pair S, L is machine closed.

(margin note: Machine closure is defined in Section 8.9.2 on page 111.)

In general, S, L need not be machine closed. But, for an interleaving composition, it usually is. Liveness properties are usually expressed as the conjunction of weak and strong fairness properties of actions. As stated on page 112, a specification is machine closed if its liveness property is the conjunction of fairness properties for subactions of the next-state action. In an interleaving composition, each S_k usually has the form $I_k \wedge \Box[N_k]_{v_k}$ where the v_k satisfy the hypothesis of the Composition Rule (page 138), and each N_k implies $v_i{}' = v_i$, for all i in $C \setminus \{k\}$. In this case, the Composition Rule implies that a subaction of N_k is also a subaction of the next-state action of S. Hence, if we write an interleaving composition in the usual way, and we write machine-closed component specifications in the usual way, then the composite specification is machine closed.

It is not so easy to obtain a machine-closed noninterleaving composition—especially with a joint-action composition. We have actually seen an example of a joint-action specification in which each component is machine closed but the composition is not. In Chapter 9, we wrote a real-time specification by conjoining one or more *RTBound* formulas and an *RTnow* formula to an untimed specification. A pathological example was the following, which is formula (9.2) on page 131:

(margin note: HC is the hour-clock specification from Chapter 2.)

$$HC\,\wedge\,RTBound(hr' = hr - 1,\,hr,\,0,\,3600)\,\wedge\,RTnow(hr)$$

---

[5] Recall that a safety property is one that is violated by a behavior iff it is violated at some particular point in the behavior. A behavior violates a conjunction of safety properties S_k iff it violates some particular S_k, and that S_k is violated at some specific point.

We can view this formula as the conjunction of three component specifications:

1. HC specifies a clock, represented by the variable hr.

2. $RTBound(hr' = hr - 1, hr, 0, 3600)$ specifies a timer, represented by the hidden (existentially quantified) timer variable.

3. $RTnow(hr)$ specifies real time, represented by the variable now.

The formula is a joint-action composition, with two kinds of joint actions:

- Joint actions of the first and second components that change both hr and the timer.

- Joint actions of the second and third components that change both the timer and now.

The first two specifications are trivially machine closed because they assert no liveness condition, so their liveness property is TRUE. The third specification's safety property asserts that now is a real number that is changed only by steps that increment it and leave hr unchanged; its liveness property NZ asserts that now increases without bound. Any finite behavior satisfying the safety property can easily be extended to an infinite behavior satisfying the entire specification, so the third specification is also machine closed. However, as we observed in Section 9.4, the composite specification is Zeno, meaning that it's not machine closed.

10.6.2   Hiding

Suppose we can write a specification S as the composition of two component specifications S_1 and S_2. Can we write $\exists\, h : S$, the specification S with variable h hidden, as a composition—that is, as the conjunction of two separate component specifications? If h represents state that is accessed by both components, then the answer is no. If the two components communicate through some part of the state, then that part of the state cannot be made internal to the separate components.

   The simplest situation in which h doesn't represent shared state is when it occurs in only one of the component specifications—say, S_2. If h doesn't occur in S_1, then the equivalence

$$(\exists\, h : S_1 \land S_2) \;\equiv\; S_1 \land (\exists\, h : S_2)$$

provides the desired decomposition.

   Now suppose that h occurs in both component specifications, but does not represent state accessed by both components. This can be the case only if different "parts" of h occur in the two component specifications. For example,

h might be a record with components $h.c1$ and $h.c2$, where S_1 mentions only $h.c1$ and S_2 mentions only $h.c2$. In this case, we have

$$(\exists\, h\,:\, S_1 \wedge S_2) \;\equiv\; (\exists\, h1\,:\, T_1) \wedge (\exists\, h2\,:\, T_2)$$

where T_1 is obtained from S_1 by substituting the variable $h1$ for the expression $h.c1$, and T_2 is defined similarly. Of course we can use any variables in place of $h1$ and $h2$; in particular, we can replace them both by the same variable.

We can generalize this result as follows to the composition of any finite number[6] of components:

> **Compositional Hiding Rule** If the variable h does not occur in the formula T_i, and S_i is obtained from T_i by substituting $h[i]$ for q, then
>
> $$(\exists\, h\,:\, \forall\, i \in C\,:\, S_i) \;\equiv\; (\forall\, i \in C\,:\, \exists\, q\,:\, T_i)$$
>
> for any finite set C.

The assumption that h does not occur in T_i means that the variable h occurs in formula S_i only in the expression $h[i]$. This in turn implies that the composition $\forall\, i \in C : S_i$ does not determine the value of h, just of its components $h[i]$ for $i \in C$. As noted in Section 10.2 on page 138, we can make the composite specification determine the value of h by conjoining the formula $\Box\mathit{IsFcnOn}(h, C)$ to it, where $\mathit{IsFcnOn}$ is defined on page 140. The hypotheses of the Compositional Hiding Rule imply

$$(\exists\, h\,:\, \Box\mathit{IsFcnOn}(h, C) \wedge \forall\, i \in C\,:\, S_i) \;\equiv\; (\forall\, i \in C\,:\, \exists\, q\,:\, T_i)$$

Now consider the common case in which $\forall\, i \in C : S_i$ is an interleaving composition, where each specification S_i describes changes to $h[i]$ and asserts that steps of component i leave $h[j]$ unchanged for $j \neq i$. We cannot apply the Compositional Hiding Rule because S_i must mention other components of h besides $h[i]$. For example, it probably contains an expression of the form

(10.6) $\quad h' = [h \text{ EXCEPT } ![i] = exp]$

which mentions all of h. However, we can transform S_i into a specification $\widehat{S_i}$ that describes only the changes to $h[i]$ and makes no assertions about other components. For example, we can replace (10.6) with $h'[i] = exp$, and we can replace an assertion that h is unchanged by the assertion that $h[i]$ is unchanged. The composition $\forall\, i \in C : \widehat{S_i}$ may allow steps that change two different components $h[i]$ and $h[j]$, while leaving all other variables unchanged, making it a noninterleaving specification. It will then not be equivalent to $\forall\, i \in C : S_i$, which requires that the changes to $h[i]$ and $h[j]$ be performed by different steps. However, it can be shown that hiding h hides this difference, making the two specifications equivalent. We can then apply the Compositional Hiding Rule with S_i replaced by $\widehat{S_i}$.

---

[6] The Compositional Hiding Rule is not true in general if C is an infinite set; but the examples in which it doesn't hold are pathological and don't arise in practice.

10.7   Open-System Specifications

A specification describes the interaction between a system and its environment. For example, the FIFO buffer specification of Chapter 4 specifies the interaction between the buffer (the system) and an environment consisting of the sender and receiver. So far, all the specifications we have written have been complete-system specifications, meaning that they are satisfied by a behavior that represents the correct operation of both the system and its environment. When we write such a specification as the composition of an environment specification E and a system specification M, it has the form $E \wedge M$.

An open-system specification is one that can serve as a contract between a user of the system and its implementer. An obvious choice of such a specification is the formula M that describes the correct behavior of the system component by itself. However, such a specification is unimplementable. It asserts that the system acts correctly no matter what the environment does. A system cannot behave as expected in the face of arbitrary behavior of its environment. It would be impossible to build a buffer that satisfies the buffer component's specification regardless of what the sender and receiver did. For example, if the sender sends a value before the previous value has been acknowledged, then the buffer could read the value while it is changing, causing unpredictable results.

> Open-system specifications are sometimes called *rely-guarantee* or *assume-guarantee* specifications.

A contract between a user and an implementer should require the system to act correctly only if the environment does. If M describes correct behavior of the system and E describes correct behavior of the environment, such a specification should require that M be true if E is. This suggests that we take as our open-system specification the formula $E \Rightarrow M$, which is true if the system behaves correctly or the environment behaves incorrectly. However, $E \Rightarrow M$ is too weak a specification for the following reason. Consider again the example of a FIFO buffer, where M describes the buffer and E the sender and receiver. Suppose now that the buffer sends a new value before the receiver has acknowledged the previous one. This could cause the receiver to act incorrectly, possibly modifying the output channel in some way not allowed by the receiver's specification. This situation is described by a behavior in which both E and M are false—a behavior that satisfies the specification $E \Rightarrow M$. However, the buffer should not be considered to act correctly in this case, since it was the buffer's error that caused the receiver to act incorrectly. Hence, this behavior should not satisfy the buffer's specification.

An open-system specification should assert that the system behaves correctly at least as long as the environment does. To express this, we introduce a new temporal operator $\stackrel{+}{\rhd}$, where $E \stackrel{+}{\rhd} M$ asserts that M remains true at least one step longer than E does, remaining true forever if E does. Somewhat more precisely, $E \stackrel{+}{\rhd} M$ asserts that

- E implies M.

- If the safety property of E is not violated by the first n states of a behavior, then the safety property of M is not violated by the first $n+1$ states, for any natural number n. (Recall that a safety property is one that, if violated, is violated at some definite point in the behavior.)

A more precise definition of $\xrightarrow{+}$ appears in Section 16.2.4 (page 314). If E describes the desired behavior of the environment and M describes the desired behavior of the system, then we take as our open-system specification the formula $E \xrightarrow{+} M$.

Once we write separate specifications of the components, we can usually transform a complete-system specification into an open-system one by simply replacing conjunction with $\xrightarrow{+}$. This requires first deciding whether each conjunct of the complete-system specification belongs to the specification of the environment, of the system, or of neither. As an example, consider the composite specification of the FIFO buffer in module *CompositeFIFO* on page 143. We take the system to consist of just the buffer, with the sender and receiver forming the environment. The closed-system specification *Spec* has three main conjuncts:

Sender $\wedge$ *Buffer* $\wedge$ *Receiver*
> The conjuncts *Sender* and *Receiver* are clearly part of the environment specification, and *Buffer* is part of the system specification.

$(in.ack = in.rdy) \wedge (out.ack = out.rdy)$
> These two initial conjuncts can be assigned to either, depending on which component we want to blame if they are violated. Let's assign to the component sending on a channel c the responsibility for establishing that $c.ack = c.rdy$ holds initially. We then assign $in.ack = in.rdy$ to the environment and $out.ack = out.rdy$ to the system.

$\square(IsChannel(in) \wedge IsChannel(out))$
> This formula is not naturally attributed to either the system or the environment. We regard it as a property inherent in our way of modeling the system, which assumes that in and out are records with ack, val, and rdy components. We therefore take the formula to be a separate conjunct of the complete specification, not belonging to either the system or the environment.

We then have the following open-system specification for the FIFO buffer:

$$\wedge \ \square(IsChannel(in) \ \wedge \ IsChannel(out))$$
$$\wedge \ (in.ack = in.rdy) \ \wedge \ Sender \ \wedge \ Receiver \ \xrightarrow{+}$$
$$(out.ack = out.rdy) \ \wedge \ Buffer$$

As this example suggests, there is little difference between writing a composite complete-system specification and an open-system specification. Most of the specification doesn't depend on which we choose. The two differ only at the very end, when we put the pieces together.

10.8   Interface Refinement

An *interface refinement* is a method of obtaining a lower-level specification by refining the variables of a higher-level specification. Let's start with two examples and then discuss interface refinement in general.

10.8.1   A Binary Hour Clock

In specifying an hour clock, we described its display with a variable hr whose value (in a behavior satisfying the specification) is an integer from 1 to 12. Suppose we want to specify a *binary hour clock*. This is an hour clock for use in a computer, where the display consists of a four-bit register that displays the hour as one of the twelve values 0001, 0010, ..., 1100. We can easily specify such a clock from scratch. But suppose we want to describe it informally to someone who already knows what an hour clock is. We would simply say that a binary hour clock is the same as an ordinary hour clock, except that the value of the display is represented in binary. We now formalize that description.

We begin by describing what it means for a four-bit value to represent a number. There are several reasonable ways to represent a four-bit value mathematically. We could use a four-element sequence, which in TLA$^+$ is a function whose domain is $1 \mathrel{..} 4$. However, a mathematician would find it more natural to represent an $(n+1)$-bit number as a function from $0 \mathrel{..} n$ to $\{0,1\}$, the function b representing the number $b[0] * 2^0 + b[1] * 2^1 + \ldots + b[n] * 2^n$. In TLA$^+$, we can define $BitArrayVal(b)$ to be the numerical value of such a function b by

> We can also write $\{0,1\}$ as $0 \mathrel{..} 1$.

$$
\begin{aligned}
BitArrayVal(b) \;\triangleq\; \\
\text{LET}\quad n \;\triangleq\; \text{CHOOSE } i \in Nat : \text{DOMAIN } b = 0 \mathrel{..} i \\
val[i \in 0 \mathrel{..} n] \;\triangleq\; \quad \text{Defines } val[i] \text{ to equal } b[0]*2^0 + \ldots + b[i]*2^i. \\
\text{IF } i = 0 \text{ THEN } b[0]*2^0 \text{ ELSE } b[i]*2^i + val[i-1] \\
\text{IN}\quad val[n]
\end{aligned}
$$

To specify a binary hour clock whose display is described by the variable $bits$, we would simply say that $BitArrayVal(bits)$ changes the same way that the specification HC of the hour clock allows hr to change. Mathematically, this means that we obtain the specification of the binary hour clock by substituting $BitArrayVal(bits)$ for the variable hr in HC. In TLA$^+$, substitution is expressed with the INSTANCE statement. Writing

$$ B \;\triangleq\; \text{INSTANCE } HourClock \text{ WITH } hr \leftarrow BitArrayVal(bits) $$

defines (among other things) $B!HC$ to be the formula obtained from HC by substituting $BitArrayVal(bits)$ for hr.

Unfortunately, this specification is wrong. The value of $BitArrayVal(b)$ is specified only if b is a function with domain $0 .. n$ for some natural number n. We don't know what $BitArrayVal(\{\text{"abc"}\})$ equals. It might equal 7. If it did, then $B!HC$ would allow a behavior in which the initial value of $bits$ is $\{\text{"abc"}\}$. We must rule out this possibility by substituting for hr not $BitArrayVal(bits)$, but some expression $HourVal(bits)$ whose value is an element of $1 .. 12$ only if b is a function in $[(0 .. 3) \to \{0, 1\}]$. For example, we can write

$$HourVal(b) \;\triangleq\; \text{IF}\;\; b \in [(0 .. 3) \to \{0, 1\}] \;\;\text{THEN}\;\; BitArrayVal(b)$$
$$\text{ELSE}\;\; 99$$

$$B \;\triangleq\; \text{INSTANCE}\;\; HourClock \;\;\text{WITH}\;\; hr \leftarrow HourVal(bits)$$

This defines $B!HC$ to be the desired specification of the binary hour clock. Because HC is not satisfied by a behavior in which hr ever assumes the value 99, $B!HC$ is not satisfied by any behavior in which $bits$ ever assumes a value not in the set $[(0 .. 3) \to \{0, 1\}]$.

There is another way to use the specification HC of the hour clock to specify the binary hour clock. Instead of substituting for hr in the hour-clock specification, we first specify a system consisting of both an hour clock and a binary hour clock that keep the same time, and we then hide the hour clock. This specification has the form

(10.7) $\exists\, hr : IR \land HC$

where IR is a temporal formula that is true iff $bits$ is always the four-bit value representing the value of hr. This formula asserts that $bits$ is the representation of hr as a four-bit array, for some choice of values for hr that satisfies HC. Using the definition of $HourVal$ given above, we can define IR simply to equal $\Box(h = HourVal(b))$.

If HC is defined as in module $HourClock$, then (10.7) can't appear in a TLA$^+$ specification. For HC to be defined in the context of the formula, the variable hr must be declared in that context. If hr is already declared, then it can't be used as the bound variable of the quantifier $\exists$. As usual, this problem is solved with parametrized instantiation. The complete TLA$^+$ specification BHC of the binary hour clock appears in module $BinaryHourClock$ of Figure 10.3 on the next page.

10.8.2   Refining a Channel

As our second example of interface refinement, consider a system that interacts with its environment by sending numbers from 1 through 12 over a channel. We refine it to a lower-level system that is the same, except it sends a number

```
┌──────────────────────── MODULE BinaryHourClock ────────────────────────┐
│                                                                         │
│  EXTENDS Naturals                                                       │
│  VARIABLE bits                                                          │
│                                                                         │
│  H(hr)  ≜  INSTANCE HourClock                                           │
│                                                                         │
│  BitArrayVal(b)  ≜  LET  n  ≜  CHOOSE i ∈ Nat : DOMAIN b = 0 .. i       │
│                          val[i ∈ 0 .. n]  ≜   Defines val[i] to equal b[0] * 2^0 + ... + b[i] * 2^i. │
│                            IF  i = 0  THEN  b[0] * 2^0  ELSE  (b[i] * 2^i) + val[i − 1]  │
│                     IN   val[n]                                         │
│  HourVal(b)  ≜  IF  b ∈ [(0 .. 3) → {0,1}]  THEN  BitArrayVal(b)        │
│                                             ELSE  99                    │
│                                                                         │
│  IR(b, h)  ≜  □(h = HourVal(b))                                         │
│  BHC  ≜  ∃ hr : IR(bits, hr) ∧ H(hr)!HC                                 │
└─────────────────────────────────────────────────────────────────────┘
```

Figure 10.3: A specification of a binary hour clock.

as a sequence of four bits. Each bit is sent separately, starting with the left-most (most significant) one. For example, to send the number 5, the lower-level system sends the sequence of bits 0, 1, 0, 1. We specify both channels with the *Channel* module of Figure 3.2 on page 30, so each value that is sent must be acknowledged before the next one can be sent.

Suppose *HSpec* is the higher-level system's specification, and its channel is represented by the variable h. Let l be the variable representing the lower-level channel. We write the lower-level system's specification as

(10.8) $\exists\, h \,:\, IR \wedge HSpec$

where *IR* specifies the sequence of values sent over h as a function of the values sent over l. The sending of the fourth bit on l is interpreted as the sending of the complete number on h; the next acknowledgment on l is interpreted as the sending of the acknowledgment on h; and any other step is interpreted as a step that doesn't change h.

To define *IR*, we instantiate module *Channel* for each of the channels:

> $H \;\triangleq\;$ INSTANCE *Channel* WITH $chan \leftarrow h,\ Data \leftarrow 1\,..\,12$
>
> $L \;\triangleq\;$ INSTANCE *Channel* WITH $chan \leftarrow l,\ Data \leftarrow \{0,1\}$

Data is the set of values that can be sent over the channel.

Sending a value d over channel l is thus represented by an $L!Send(d)$ step, and acknowledging receipt of a value on channel h is represented by an $H!Rcv$ step. The following behavior represents sending and acknowledging a 5, where I have omitted all steps that don't change l:

$$s_0 \xrightarrow{L!Send(0)} s_1 \xrightarrow{L!Rcv} s_2 \xrightarrow{L!Send(1)} s_3 \xrightarrow{L!Rcv} s_4 \xrightarrow{L!Send(0)}$$

$$s_5 \xrightarrow{L!Rcv} s_6 \xrightarrow{L!Send(1)} s_7 \xrightarrow{L!Rcv} s_8 \longrightarrow \cdots$$

This behavior will satisfy *IR* iff $s_6 \to s_7$ is an $H!Send(5)$ step, $s_7 \to s_8$ is an $H!Rcv$ step, and all the other steps leave h unchanged.

We want to make sure that (10.8) is not satisfied unless l represents a correct lower-level channel—for example, (10.8) should be false if l is set to some bizarre value. We will therefore define *IR* so that, if the sequence of values assumed by l does not represent a channel over which bits are sent and acknowledged, then the sequence of values of h does not represent a correct behavior of a channel over which numbers from 1 to 12 are sent. Formula *HSpec*, and hence (10.8), will then be false for such a behavior.

Formula *IR* will have the standard form for a TLA specification, with an initial condition and a next-state action. However, it specifies h as a function of l; it does not constrain l. Therefore, the initial condition does not specify the initial value of l, and the next-state action does not specify the value of l'. (The value of l is constrained implicitly by *IR*, which asserts a relation between the values of h and l, together with the conjunct *HSpec* in (10.8), which constrains the value of h.) For the next-state action to specify the value sent on h, we need an internal variable that remembers what has been sent on l since the last complete number. We let the variable *bitsSent* contain the sequence of bits sent so far for the current number. For convenience, *bitsSent* contains the sequence of bits in reverse order, with the most recently-sent bit at the head. This means that the high-order bit of the number, which is sent first, is at the tail of *bitsSent*.

The definition of *IR* appears in module *ChannelRefinement* of Figure 10.4 on the next page. The module first defines *ErrorVal* to be an arbitrary value that is not a legal value of h. Next comes the definition of the function *BitSeqToNat*. If s is a sequence of bits, then $BitSeqToNat[s]$ is its numeric value interpreted as a binary number whose low-order bit is at the head of s. For example $BitSeqToNat[\langle 0, 1, 1 \rangle]$ equals 6. Then come the two instantiations of module *Channel*.

There follows a submodule that defines the internal specification—the one with the internal variable *bitsSent* visible. The internal specification's initial predicate *Init* asserts that if l has a legal initial value, then h can have any legal initial value; otherwise h has an illegal value. Initially *bitsSent* is the empty sequence $\langle \rangle$. The internal specification's next-state action is the disjunction of three actions:

The use of a submodule to define an internal specification was introduced in the real-time hour-clock specification of Section 9.1.

SendBit A *SendBit* step is one in which a bit is sent on l. If *bitsSent* has fewer than three elements, so fewer than three bits have already been sent, then the bit is prepended to the head of *bitsSent* and h

──────── MODULE *ChannelRefinement* ────────

This module defines an interface refinement from a higher-level channel h, over which numbers in $1 .. 12$ are sent, to a lower-level channel l in which a number is sent as a sequence of four bits, each separately acknowledged. (See the *Channel* module in Figure 3.2 on page 30.) Formula *IR* is true iff the sequence of values assumed by h represents the higher-level view of the sequence of values sent on l. If the sequence of values assumed by l doesn't represent the sending and acknowledging of bits, then h assumes an illegal value.

EXTENDS *Naturals*, *Sequences*

VARIABLES h, l

$ErrorVal \triangleq$ CHOOSE $v : v \notin [val : 1 .. 12, \ rdy : \{0, 1\}, \ ack : \{0, 1\}]$

$BitSeqToNat[s \in Seq(\{0, 1\})] \triangleq$ $BitSeqToNat[\langle b_0, \ b_1, \ b_2, \ b_3 \rangle] = b_0 + 2 * (b_1 + 2 * (b_2 + 2 * b_3))$

   IF $s = \langle \rangle$ THEN 0 ELSE $Head(s) + 2 * BitSeqToNat[Tail(s)]$

$H \triangleq$ INSTANCE *Channel* WITH $chan \leftarrow h, \ Data \leftarrow 1 .. 12$ *H is a channel for sending numbers*

$L \triangleq$ INSTANCE *Channel* WITH $chan \leftarrow l, \ Data \leftarrow \{0, 1\}$ *in 1 .. 12; L is a channel for sending bits.*

──────── MODULE *Inner* ────────

VARIABLE *bitsSent*   *The sequence of the bits sent so far for the current number.*

$Init \triangleq \ \wedge \ bitsSent = \langle \rangle$

       $\wedge$ IF $L!Init$ THEN $H!Init$ *Defines the initial value of h as a function of l.*

            ELSE $h = ErrorVal$

$SendBit \triangleq \ \exists \, b \in \{0, 1\} :$

       $\wedge \ L!Send(b)$ *Sending one of the first three bits*

       $\wedge$ IF $Len(bitsSent) < 3$ *on l prepends it to the front of*

            THEN $\wedge \ bitsSent' = \langle b \rangle \circ bitsSent$ *bitsSent and leaves h unchanged;*

                   $\wedge$ UNCHANGED h *sending the fourth bit resets*

            ELSE $\wedge \ bitsSent' = \langle \rangle$ *bitsSent and sends the complete*

                   $\wedge \ H!Send(BitSeqToNat[\langle b \rangle \circ bitsSent])$ *number on h.*

$RcvBit \triangleq \ \wedge \ L!Rcv$ *A Rcv action on l causes a Rcv*

       $\wedge$ IF $bitsSent = \langle \rangle$ THEN $H!Rcv$ *action on h iff it follows the*

                  ELSE UNCHANGED h *sending of the fourth bit.*

       $\wedge$ UNCHANGED $bitsSent$

$Error \triangleq \ \wedge \ l' \neq l$ *An illegal action on l sets h to ErrorVal.*

       $\wedge \ \neg((\exists \, b \in \{0, 1\} : L!Send(b)) \vee L!Rcv)$

       $\wedge \ h' = ErrorVal$

$Next \triangleq \ SendBit \vee RcvBit \vee Error$

$InnerIR \triangleq \ Init \wedge \Box[Next]_{\langle l, \, h, \, bitsSent \rangle}$

$I(bitsSent) \triangleq$ INSTANCE *Inner*

$IR \triangleq \ \exists \, bitsSent : I(bitsSent)!InnerIR$

Figure 10.4: Refining a channel.

is left unchanged. Otherwise, the value represented by the four bits sent so far, including the current bit, is sent on h and *bitsSent* is reset to $\langle\,\rangle$.

RcvBit   A *RcvBit* step is one in which an acknowledgment is sent on l. It represents the sending of an acknowledgment on h iff this is an acknowledgment of the fourth bit, which is true iff *bitsSent* is the empty sequence.

Error   An *Error* step is one in which an illegal change to l occurs. It sets h to an illegal value.

The inner specification *InnerIR* has the usual form. (There is no liveness requirement.) The outer module then instantiates the inner submodule with *bitsSent* as a parameter, and it defines *IR* to equal *InnerIR* with *bitsSent* hidden.

Now suppose we have a module *HigherSpec* that defines a specification *HSpec* of a system that communicates by sending numbers from 1 through 12 over a channel *hchan*. We obtain, as follows, a lower-level specification *LSpec* in which the numbers are sent as sequences of bits on a channel *lchan*. We first declare *lchan* and all the variables and constants of the *HigherSpec* module except *hchan*. We then write

$$HS(hchan) \;\triangleq\; \text{INSTANCE } HigherSpec$$

$$CR(h) \;\triangleq\; \text{INSTANCE } ChannelRefinement \text{ WITH } l \leftarrow lchan$$

$$LSpec \;\triangleq\; \exists\, h : CR(h)!IR \,\wedge\, HS(h)!HSpec$$

10.8.3   Interface Refinement in General

In the examples of the binary clock and of channel refinement, we defined a lower-level specification *LSpec* in terms of a higher-level one *HSpec* as

$$(10.9)\quad LSpec \;\triangleq\; \exists\, h : IR \wedge HSpec$$

where h is a free variable of *HSpec* and *IR* is a relation between h and the lower-level variable l of *LSpec*. We can view the internal specification $IR \wedge HSpec$ as the composition of two components, as shown here:

We can regard *IR* as the specification of a component that transforms the lower-level behavior of l into the higher-level behavior of h. Formula *IR* is called an *interface refinement*.

In both examples, the interface refinement is independent of the system specification. It depends only on the representation of the interface—that is, on how the interaction between the system and its environment is represented. In general, for an interface refinement IR to be independent of the system using the interface, it should ascribe a behavior of the higher-level interface variable h to any behavior of the lower-level variable l. In other words, for any sequence of values for l, there should be some sequence of values for h that satisfy IR. This is expressed mathematically by the requirement that the formula $\exists h : IR$ should be valid—that is, true for all behaviors.

So far, I have discussed refinement of a single interface variable h by a single variable l. This generalizes in the obvious way to the refinement of a collection of higher-level variables $h_1, \ldots, h_n$ by the variables $l_1, \ldots, l_m$. The interface refinement IR specifies the values of the h_i in terms of the values of the l_j and perhaps of other variables as well. Formula (10.9) is replaced by

$$LSpec \;\triangleq\; \exists\, h_1, \ldots, h_n \;:\; IR \wedge HSpec$$

A particularly simple type of interface refinement is a *data refinement*, in which IR has the form $\Box P$, where P is a state predicate that expresses the values of the higher-level variables $h_1, \ldots, h_n$ as functions of the values of the lower-level variables $l_1, \ldots, l_m$. The interface refinement in our binary clock specification is a data refinement, where P is the predicate $hr = HourVal(bits)$. As another example, the two specifications of an asynchronous channel interface in Chapter 3 can each be obtained from the other by an interface refinement. The specification $Spec$ of the $Channel$ module (page 30) is equivalent to the specification obtained as a data refinement of the specification $Spec$ of the $AsynchInterface$ module (page 27) by letting P equal

(10.10) $\quad chan = [val \mapsto val,\ rdy \mapsto rdy,\ ack \mapsto ack]$

This formula asserts that $chan$ is a record whose val field is the value of the variable val, whose rdy field is the value of the variable rdy, and whose ack field is the value of the variable ack. Conversely, specification $Spec$ of the $AsynchInterface$ module is equivalent to a data refinement of the specification $Spec$ of the $Channel$ module. In this case, defining the state predicate P is a little tricky. The obvious choice is to let P be the formula $GoodVals$ defined by

$$GoodVals \;\triangleq\; \begin{aligned} &\wedge\ val = chan.val \\ &\wedge\ rdy = chan.rdy \\ &\wedge\ ack = chan.ack \end{aligned}$$

However, this can assert that val, rdy, and ack have good values even if $chan$ has an illegal value—for example, if it is a record with more than three fields. Instead, we let P equal

$$\text{IF}\ \ chan \in [val : Data,\ rdy : \{0,1\},\ ack : \{0,1\}]\ \ \text{THEN}\quad GoodVals$$
$$\text{ELSE}\quad BadVals$$

where *BadVals* asserts that *val*, *rdy*, and *ack* have some illegal values—that is, values that are impossible in a behavior satisfying formula *Spec* of module *AsynchInterface*. (We don't need such a trick when defining *chan* as a function of *val*, *rdy*, and *ack* because (10.10) implies that the value of *chan* is legal iff the values of all three variables *val*, *rdy*, and *ack* are legal.)

Data refinement is the simplest form of interface refinement. In a more complicated interface refinement, the value of the higher-level variables cannot be expressed as a function of the current values of the lower-level variables. In the channel refinement example of Section 10.8.2, the number being sent on the higher-level channel depends on the values of bits that were previously sent on the lower-level channel, not just on the lower-level channel's current state.

We often refine both a system and its interface at the same time. For example, we may implement a specification *H* of a system that communicates by sending numbers over a channel with a lower-level specification *LImpl* of a system that sends individual bits. In this case, *LImpl* is not itself obtained from *HSpec* by an interface refinement. Rather, *LImpl* implements some specification *LSpec* that is obtained from *HSpec* by an interface refinement *IR*. In that case, we say that *LImpl* implements *HSpec under the interface refinement IR*.

10.8.4 Open-System Specifications

So far, we have considered interface refinement for complete-system specifications. Let's now consider what happens if the higher-level specification *HSpec* is the kind of open-system specification discussed in Section 10.7 above. For simplicity, we consider the refinement of a single higher-level interface variable *h* by a single lower-level variable *l*. The generalization to more variables will be obvious.

Let's suppose first that *HSpec* is a safety property, with no liveness condition. As explained in Section 10.7, the specification attributes each change to *h* either to the system or to the environment. Any change to a lower-level interface variable *l* that produces a change to *h* is therefore attributed to the system or the environment. A bad change to *h* that is attributed to the environment makes *HSpec* true; a bad change that is attributed to the system makes *HSpec* false. Thus, (10.9) defines *LSpec* to be an open-system specification. For this to be a sensible specification, the interface refinement *IR* must ensure that the way changes to *l* are attributed to the system or environment is sensible.

If *HSpec* contains liveness conditions, then interface refinement can be more subtle. Suppose *IR* is the interface refinement defined in the *ChannelRefinement* module of Figure 10.4 on page 162, and suppose that *HSpec* requires that the system eventually send some number on *h*. Consider a behavior in which the system sends the first bit of a number on *l*, but the environment never acknowledges it. Under the interface refinement *IR*, this behavior is interpreted as one in which *h* never changes. Such a behavior fails to satisfy the liveness condition of *HSpec*. Thus, if *LSpec* is defined by (10.9), then failure of the environment to do something can cause *LSpec* to be violated, through no fault of the system.

In this example, we want the environment to be at fault if it causes the system to halt by failing to acknowledge any of the first three bits of a number sent by the system. (The acknowledgment of the fourth bit is interpreted by *IR* as the acknowledgment of a value sent on h, so blame for its absence is properly assigned to the environment.) Putting the environment at fault means making *LSpec* true. We can achieve this by modifying (10.9) to define *LSpec* as follows:

$$(10.11) \quad LSpec \;\triangleq\; Liveness \;\Rightarrow\; \exists\, h \,:\, IR \wedge HSpec$$

where *Liveness* is a formula requiring that any bit sent on l, other than the last bit of a number, must eventually be acknowledged. However, if l is set to an illegal value, then we want the safety part of the specification to determine who is responsible. So, we want *Liveness* to be true in this case.

We define *Liveness* in terms of the inner variables h and *bitsSent*, which are related to l by formula *InnerIR* from the *Inner* submodule of module *ChannelRefinement*. (Remember that l should be the only free variable of *LSpec*.) The action that acknowledges receipt of one of the first three bits of the number is $RcvBit \wedge (bitsSent \neq \langle \rangle)$. Weak fairness of this action asserts that the required acknowledgments must eventually be sent. For the case of illegal values, recall that sending a bad value on l causes h to equal *ErrorVal*. We want *Liveness* to be true if this ever happens, which means if it eventually happens. We therefore add the following definition to the submodule *Inner* of the *ChannelRefinement* module:

$$
\begin{aligned}
InnerLiveness \;\triangleq\; &\wedge\, InnerIR \\
&\wedge\, \vee\, \mathrm{WF}_{\langle l,\, h,\, bitsSent \rangle}(RcvBit \wedge (bitsSent \neq \langle \rangle)) \\
&\quad\; \vee\, \Diamond(h = ErrorVal)
\end{aligned}
$$

To define *Liveness*, we have to hide h and *bitsSent* in *InnerLiveness*. We can do this, in a context in which l is declared, as follows:

$$
\begin{aligned}
ICR(h) \;&\triangleq\; \textsc{instance}\ ChannelRefinement \\
Liveness \;&\triangleq\; \exists\, h,\, bitsSent \,:\, ICR(h)!I(bitsSent)!InnerLiveness
\end{aligned}
$$

Now, suppose it is the environment that sends numbers over h and the system is supposed to acknowledge their receipt and then process them in some way. In this case, we want failure to acknowledge a bit to be a system error. So, *LSpec* should be false if *Liveness* is. The specification should then be

$$LSpec \;\triangleq\; Liveness \wedge (\exists\, h \,:\, IR \wedge HSpec)$$

Since h does not occur free in *Liveness*, this definition is equivalent to

$$LSpec \;\triangleq\; \exists\, h \,:\, Liveness \wedge IR \wedge HSpec$$

which has the form (10.9) if the interface refinement *IR* of (10.9) is taken to be *Liveness* $\wedge$ *IR*. In other words, we can we make the liveness condition part of the interface refinement. (In this case, we can simplify the definition by adding liveness directly to *InnerIR*.)

In general, if *HSpec* is an open-system specification that describes liveness as well as safety, then an interface refinement may have to take the form of formula (10.11). Both *Liveness* and the liveness condition of *IR* may depend on which changes to the lower-level interface variable *l* are attributed to the system and which to the environmet. For the channel refinement, this means that they will depend on whether the system or the environment is sending values on the channel.

10.9 Should You Compose?

When specifying a system, should we write a monolithic specification with a single next-state action, a closed-system composition that is the conjunction of specifications of individual components, or an open-system specification? The answer is: it usually makes little difference. For a real system, the definitions of the components' actions will take hundreds or thousands of lines. The different forms of specification differ only in the few lines where we assemble the initial predicates and next-state actions into the final formula.

If you are writing a specification from scratch, it's probably better to write a monolithic specification. It is usually easier to understand. Of course, there are exceptions. We write a real-time specification as the conjunction of an untimed specification and timing constraints; describing the changes to the system variables and the timers with a single next-state action usually makes the specification harder to understand.

Writing a composite specification may be sensible when you are starting from an existing specification. If you already have a specification of one component, you may want to write a separate specification of the other component and compose the two specifications. If you have a higher-level specification, you may want to write a lower-level version as an interface refinement. However, these are rather rare situations. Moreover, it's likely to be just as easy to modify the original specification or reuse it in another way. For example, instead of conjoining a new component to the specification of an existing one, you can simply include the definition of the existing component's next-state action, with an EXTENDS or INSTANCE statement, as part of the new specification.

Composition provides a new way of writing a complete-system specification; it doesn't change the specification. The choice between a composite specification and a monolithic one is therefore ultimately a matter of taste. Disjoint-state compositions are generally straightforward and present no problems. Shared-state compositions can be tricky and require care.

Open-system specifications introduce a mathematically different kind of specification. A closed-system specification $E \wedge M$ and its open-system counterpart $E \xrightarrow{+} M$ are not equivalent. If we really want a specification to serve as a legal contract between a user and an implementer, then we have to write an open-system specification. We also need open-system specifications if we want to specify and reason about systems built by composing off-the-shelf components with pre-existing specifications. All we can assume about such a component is that it satisfies a contract between the system builder and the supplier, and such a contract can be formalized only as an open-system specification. However, you are unlikely to encounter off-the-shelf component specifications during the early part of the twenty-first century. In the near future, open-system specifications are likely to be of theoretical interest only.

Chapter 11

Advanced Examples

It would be nice to provide an assortment of typical examples that cover most of the specification problems that arise in practice. However, there is no such thing as a typical specification. Every real specification seems to pose its own problems. But we can partition all specifications into two classes, depending on whether or not they contain VARIABLE declarations.

A specification with no variables defines data structures and operations on those structures. For example, the *Sequences* module defines various operations on sequences. When specifying a system, you may need some kind of data structure other than the ones provided by the standard modules like *Sequences* and *Bags*, described in Chapter 18. Section 11.1 gives some examples of data structure specifications.

A system specification contains variables that represent the system's state. We can further divide system specifications into two classes—high-level specifications that describe what it means for a system to be correct, and lower-level specifications that describe what the system actually does. In the memory example of Chapter 5, the linearizable memory specification of Section 5.3 is a high-level specification of correctness, while the write-through cache specification of Section 5.6 describes how a particular algorithm works. This distinction is not precise; whether a specification is high- or low-level is a matter of perspective. But it can be a useful way of categorizing system specifications.

Lower-level system specifications tend to be relatively straightforward. Once the level of abstraction has been chosen, writing the specification is usually just a matter of getting the details right when describing what the system does. Specifying high-level correctness can be much more subtle. Section 11.2 considers a high-level specification problem—formally specifying a multiprocessor memory.

11.1   Specifying Data Structures

Most of the data structures required for writing specifications are mathematically simple and are easy to define in terms of sets, functions, and records. Section 11.1.2 describes the specification of one such structure—a graph. On rare occasions, a specification will require sophisticated mathematical concepts. The only examples I know of are hybrid system specifications, discussed in Section 9.5. There, we used a module for describing the solutions to differential equations. That module is specified in Section 11.1.3 below. Section 11.1.4 considers the tricky problem of defining operators for specifying BNF grammars. Although not the kind of data structure you're likely to need for a system specification, specifying BNF grammars provides a nice little exercise in "mathematization". The module developed in that section is used in Chapter 15 for specifying the grammar of TLA⁺. But, before specifying data structures, you should know how to make local definitions.

11.1.1   Local Definitions

In the course of specifying a system, we write lots of auxiliary definitions. A system specification may consist of a single formula *Spec*, but we define dozens of other identifiers in terms of which we define *Spec*. These other identifiers often have fairly common names—for example, the identifier *Next* is defined in many specifications. The different definitions of *Next* don't conflict with one another because, if a module that defines *Next* is used as part of another specification, it is usually instantiated with renaming. For example, the *Channel* module is used in module *InnerFIFO* on page 38 with the statement

$$InChan \;\triangleq\; \text{INSTANCE } Channel \text{ WITH } \dots$$

The action *Next* of the *Channel* module is then instantiated as *InChan!Next*, so its definition doesn't conflict with the definition of *Next* in the *InnerFIFO* module.

A module that defines operations on a data structure is likely to be used in an EXTENDS statement, which does no renaming. The module might define some auxiliary operators that are used only to define the operators in which we're interested. For example, we need the *DifferentialEquations* module only to define the single operator *Integrate*. However, *Integrate* is defined in terms of other defined operators with names like *Nbhd* and *IsDeriv*. We don't want these definitions to conflict with other uses of those identifiers in a module that extends *DifferentialEquations*. So, we want the definitions of *Nbhd* and *IsDeriv* to be local to the *DifferentialEquations* module.[1]

---

[1] We could use the LET construct to put these auxiliary definitions inside the definition of *Integrate*, but that trick wouldn't work if the *DifferentialEquations* module exported other operators besides *Integrate* that were defined in terms of *Nbhd* and *IsDeriv*.

TLA$^+$ provides a LOCAL modifier for making definitions local to a module. If a module M contains the definition

LOCAL *Foo(x)* $\triangleq$ ...

then *Foo* can be used inside module M just like any ordinary defined identifier. However, a module that extends or instantiates M does not obtain the definition of *Foo*. That is, the statement EXTENDS M in another module does not define *Foo* in that module. Similarly, the statement

N $\triangleq$ INSTANCE M

does not define $N!Foo$. The LOCAL modifier can also be applied to an instantiation. The statement

LOCAL INSTANCE *Sequences*

in module M incorporates into M the definitions from the *Sequences* module. However, another module that extends or instantiates M does not obtain those definitions. Similarly, a statement like

LOCAL $P(x)$ $\triangleq$ INSTANCE N

makes all the instantiated definitions local to the current module.

The LOCAL modifier can be applied only to definitions and INSTANCE statements. It cannot be applied to a declaration or to an EXTENDS statement, so you *cannot* write either of the following:

LOCAL CONSTANT N These are not legal
LOCAL EXTENDS *Sequences*   statements.

If a module has no CONSTANT or VARIABLE declarations and no submodules, then extending it and instantiating it are equivalent. Thus, the two statements

EXTENDS *Sequences*   INSTANCE *Sequences*

are equivalent.

In a module that defines general mathematical operators, I like to make all definitions local except for the ones that users of the module would expect. For example, users expect the *Sequences* module to define operators on sequences, such as *Append*. They don't expect it to define operators on numbers, such as $+$. The *Sequences* module uses $+$ and other operators defined in the *Naturals* module. But instead of extending *Naturals*, it defines those operators with the statement

LOCAL INSTANCE *Naturals*

The definitions of the operators from *Naturals* are therefore local to *Sequences*. A module that extends the *Sequences* module could then define $+$ to mean something other than addition of numbers.

11.1.2 Graphs

A graph is an example of the kind of simple data structure often used in specifications. Let's now write a *Graphs* module for use in writing system specifications.

We must first decide how to represent a graph in terms of data structures that are already defined—either built-in TLA$^+$ data structures like functions, or ones defined in existing modules. Our decision depends on what kind of graphs we want to represent. Are we interested in directed graphs or undirected graphs? Finite or infinite graphs? Graphs with or without self-loops (edges from a node to itself)? If we are specifying graphs for a particular specification, the specification will tell us how to answer these questions. In the absence of such guidance, let's handle arbitrary graphs. My favorite way of representing both directed and undirected graphs is to specify arbitrary directed graphs, and to define an undirected graph as a directed graph that contains an edge iff it contains the opposite-pointing edge. Directed graphs have a pretty obvious representation: a directed graph consists of a set of nodes and a set of edges, where an edge from node m to node n is represented by the ordered pair $\langle m, n \rangle$.

In addition to deciding how to represent graphs, we must decide how to structure the *Graphs* module. The decision depends on how we expect the module to be used. For a specification that uses a single graph, it is most convenient to define operations on that specific graph. So, we want the *Graphs* module to have (constant) parameters *Node* and *Edge* that represent the sets of nodes and edges of a particular graph. A specification could use such a module with a statement

INSTANCE *Graphs* WITH *Node* ← . . . , *Edge* ← . . .

where the ". . . "s are the sets of nodes and edges of the particular graph appearing in the specification. On the other hand, a specification might use many different graphs. For example, it might include a formula that asserts the existence of a subgraph, satisfying certain properties, of some given graph G. Such a specification needs operators that take a graph as an argument—for example, a *Subgraph* operator defined so *Subgraph*(G) is the set of all subgraphs of a graph G. In this case, the *Graphs* module would have no parameters, and specifications would incorporate it with an EXTENDS statement. Let's write this kind of module.

An operator like *Subgraph* takes a graph as an argument, so we have to decide how to represent a graph as a single value. A graph G consists of a set N of nodes and a set E of edges. A mathematician would represent G as the ordered pair $\langle N, E \rangle$. However, $G.node$ is more perspicuous than $G[1]$, so we represent G as a record with *node* field N and *edge* field E.

Having made these decisions, it's easy to define any standard operator on graphs. We just have to decide what we should define. Here are some generally useful operators:

IsDirectedGraph(G)

> True iff G is an arbitrary directed graph—that is, a record with *node* field N and *edge* field E such that E is a subset of $N \times N$. This operator is useful because a specification might want to assert that something is a directed graph. (To understand how to assert that G is a record with *node* and *edge* fields, see the definition of *IsChannel* in Section 10.3 on page 140.)

DirectedSubgraph(G)

> The set of all subgraphs of a directed graph G. Alternatively, we could define *IsDirectedSubgraph(H, G)* to be true iff H is a subgraph of G. However, it's easy to express *IsDirectedSubgraph* in terms of *DirectedSubgraph*:

> $$IsDirectedSubgraph(H, G) \equiv H \in DirectedSubgraph(G)$$

> On the other hand, it's awkward to express *DirectedSubgraph* in terms of *IsDirectedSubgraph*:

> $$DirectedSubgraph(G) =$$
> $$\text{CHOOSE } S : \forall H : (H \in S) \equiv IsDirectedSubgraph(H, G)$$

> Section 6.1 explains why we can't define a set of all directed graphs, so we had to define the *IsDirectedGraph* operator.

IsUndirectedGraph(G)
UndirectedSubgraph(G)

> These are analogous to the operators for directed graphs. As mentioned above, an undirected graph is a directed graph G such that for every edge $\langle m, n \rangle$ in $G.edge$, the inverse edge $\langle n, m \rangle$ is also in $G.edge$. Note that *DirectedSubgraph(G)* contains directed graphs that are not undirected graphs—except for certain "degenerate" graphs G, such as graphs with no edges.

Path(G)

> The set of all paths in G, where a path is any sequence of nodes that can be obtained by following edges in the direction they point. This definition is useful because many properties of a graph can be expressed in terms of its set of paths. It is convenient to consider the one-element sequence $\langle n \rangle$ to be a path, for any node n.

AreConnectedIn(m, n, G)

> True iff there is a path from node m to node n in G. The utility of this operator becomes evident when you try defining various common graph properties, like connectivity.

There are any number of other graph properties and classes of graphs that we might define. Let's define these two:

IsStronglyConnected(G)
>   True iff G is strongly connected, meaning that there is a path from any
>   node to any other node. For an undirected graph, strongly connected is
>   equivalent to the ordinary definition of connected.

IsTreeWithRoot(G, r)
>   True iff G is a tree with root r, where we represent a tree as a graph
>   with an edge from each nonroot node to its parent. Thus, the parent of a
>   nonroot node n equals
>
>   > CHOOSE $m \in G.node$: $\langle n, m \rangle \in G.edge$

The *Graphs* module appears on the next page. By now, you should be able to
work out for yourself the meanings of all the definitions.

11.1.3   Solving Differential Equations

Section 9.5 on page 132 describes how to specify a hybrid system whose state includes a physical variable satisfying an ordinary differential equation. The specification uses an operator *Integrate* such that $Integrate(D, t_0, t_1, \langle x_0, \ldots, x_{n-1} \rangle)$
is the value at time t_1 of the n-tuple

$$\langle x, dx/dt, \ldots, d^{n-1}x/dt^{n-1} \rangle$$

where x is a solution to the differential equation

$$D[t, x, dx/dt, \ldots, d^n x/dt^n] = 0$$

whose 0th through $(n-1)^{\text{st}}$ derivatives at time t_0 are $x_0, \ldots, x_{n-1}$. We assume
that there is such a solution and that it is unique. Defining *Integrate* illustrates
how to express sophisticated mathematics in TLA$^+$.

   We start by defining some mathematical notation that we will use to define
the derivative. As usual, we obtain from the *Reals* module the definitions of the
set *Real* of real numbers and of the ordinary arithmetic operators. Let *PosReal*
be the set of all positive reals:

> $PosReal \triangleq \{r \in Real : r > 0\}$

and let *OpenInterval(a, b)* be the open interval from a to b (the set of numbers
greater than a and less than b):

> $OpenInterval(a, b) \triangleq \{s \in Real : (a < s) \land (s < b)\}$

(Mathematicians usually write this set as (a, b).) Let's also define $Nbhd(r, e)$
to be the open interval of width $2e$ centered at r:

> $Nbhd(r, e) \triangleq OpenInterval(r - e, r + e)$

┌─────────────────────── MODULE *Graphs* ───────────────────────┐

A module that defines operators on graphs. A directed graph is represented as a record whose *node* field is the set of nodes and whose *edge* field is the set of edges, where an edge is an ordered pair of nodes.

LOCAL INSTANCE *Naturals*

LOCAL INSTANCE *Sequences*

───

$IsDirectedGraph(G) \triangleq$ True iff G is a directed graph.
    $\wedge\ G = [node \mapsto G.node,\ edge \mapsto G.edge]$
    $\wedge\ G.edge \subseteq (G.node \times G.node)$

$DirectedSubgraph(G) \triangleq$ The set of all (directed) subgraphs of a directed graph.
    $\{H \in [node :$ SUBSET $G.node,\ edge :$ SUBSET $(G.node \times G.node)] :$
       $IsDirectedGraph(H) \wedge H.edge \subseteq G.edge\}$

───

$IsUndirectedGraph(G) \triangleq$ An undirected graph is a directed graph in which every
    $\wedge\ IsDirectedGraph(G)$ edge has an oppositely directed one.
    $\wedge\ \forall\, e \in G.edge : \langle e[2],\ e[1]\rangle \in G.edge$

$UndirectedSubgraph(G) \triangleq$ The set of (undirected) subgraphs of an undirected graph.
    $\{H \in DirectedSubgraph(G) : IsUndirectedGraph(H)\}$

───

$Path(G) \triangleq$ The set of paths in G, where a path is represented as a sequence of nodes.
    $\{p \in Seq(G.node) : \wedge\ p \neq \langle\,\rangle$
                  $\wedge\ \forall\, i \in 1\ ..\ (Len(p) - 1) : \langle p[i],\ p[i+1]\rangle \in G.edge\}$

$AreConnectedIn(m,\ n,\ G) \triangleq$ True iff there is a path from m to n in graph G.
    $\exists\, p \in Path(G) : (p[1] = m) \wedge (p[Len(p)] = n)$

$IsStronglyConnected(G) \triangleq$ True iff graph G is strongly connected.
    $\forall\, m,\ n \in G.node : AreConnectedIn(m,\ n,\ G)$

───

$IsTreeWithRoot(G,\ r) \triangleq$ True if G is a tree with root r, where edges point
    $\wedge\ IsDirectedGraph(G)$ from child to parent.
    $\wedge\ \forall\, e \in G.edge : \wedge\ e[1] \neq r$
                     $\wedge\ \forall\, f \in G.edge : (e[1] = f[1]) \Rightarrow (e = f)$
    $\wedge\ \forall\, n \in G.node : AreConnectedIn(n,\ r,\ G)$

└───┘

Figure 11.1: A module for specifying operators on graphs.

To explain the definitions, we need some notation for the derivative of a function. It's rather difficult to make mathematical sense of the usual notation df/dt for the derivative of f. (What exactly is t?) So, let's use a mathematically simpler notation and write the nth derivative of the function f as $f^{(n)}$. (We don't have to use TLA$^+$ notation because differentiation will not appear explicitly in our definitions.) Recall that $f^{(0)}$, the 0th derivative of f, equals f.

We can now start to define *Integrate*. If a and b are numbers, *InitVals* is an n-tuple of numbers, and D is a function from $(n+2)$-tuples of numbers to numbers, then

$$Integrate(D,\ a,\ b,\ InitVals)\ =\ \langle f^{(0)}[b],\ \ldots,\ f^{(n-1)}[b]\rangle$$

where f is the function satisfying the following two conditions:

- $D[r,\ f^{(0)}[r],\ f^{(1)}[r],\ \ldots, f^{(n)}[r]] = 0$, for all r in some open interval containing a and b.

- $\langle f^{(0)}[a],\ \ldots,\ f^{(n-1)}[a]\rangle\ =\ InitVals$

We want to define *Integrate*$(D,\ a,\ b,\ InitVals)$ in terms of this function f, which we can specify using the CHOOSE operator. It's easiest to choose not just f, but its first n derivatives as well. So, we choose a function g such that $g[i] = f^{(i)}$ for $i \in 0\ ..\ n$. The function g maps numbers in $0\ ..\ n$ into functions. More precisely, g is an element of

$$[0\ ..\ n \rightarrow [OpenInterval(a-e,\ b+e) \rightarrow Real]]$$

for some positive e. It is the function in this set that satisfies the following conditions:

1. $g[i]$ is the ith derivative of $g[0]$, for all $i \in 0\ ..\ n$.

2. $D[r,\ g[0][r],\ \ldots,\ g[n][r]] = 0$, for all r in $OpenInterval(a-e,\ b+e)$.

3. $\langle g[0][a],\ \ldots,\ g[n-1][a]\rangle = InitVals$

We now have to express these conditions formally.

To express the first condition, we will define *IsDeriv* so that *IsDeriv*$(i,\ df,\ f)$ is true iff df is the ith derivative of f. More precisely, this will be the case if f is a real-valued function on an open interval; we don't care what *IsDeriv*$(i,\ df,\ f)$ equals for other values of f. Condition 1 is then

$$\forall\, i \in 1\ ..\ n\ :\ IsDeriv(i,\ g[i],\ g[0])$$

To express the second condition formally, without the "$\ldots$", we reason as follows:

$$\begin{aligned}
&D[r,\ g[0][r],\ \ldots,\ g[n][r]] \\
&= D[\langle r,\ g[0][r],\ \ldots,\ g[n][r]\rangle] && \text{See page 50.} \\
&= D[\langle r\rangle \circ \langle g[0][r],\ \ldots,\ g[n][r]\rangle] && \text{Tuples are sequences} \\
&= D[\langle r\rangle \circ [i \in 1\ ..\ (n+1) \mapsto g[i-1][r]]] && \text{An } (n{+}1)\text{-tuple is a function with domain } 1\ ..\ n{+}1.
\end{aligned}$$

The third condition is simply

$$\forall\, i \in 1 \ldots n \,:\, g[i - 1][a] = InitVals[i]$$

We can therefore write the formula specifying g as

$$\exists\, e \in PosReal \,:\, \wedge\, g \in [0 \ldots n \rightarrow [OpenInterval(a - e,\, b + e) \rightarrow Real]]$$
$$\wedge\, \forall\, i \in 1 \ldots n \,:\, \wedge\, IsDeriv(i,\, g[i],\, g[0])$$
$$\wedge\, g[i - 1][a] = InitVals[i]$$
$$\wedge\, \forall\, r \in OpenInterval(a - e,\, b + e) \,:\,$$
$$D[\,\langle r \rangle \circ [i \in 1 \ldots (n + 1) \mapsto g[i - 1][r]]\,] = 0$$

where n is the length of *InitVals*. The value of *Integrate*$(D,\, a,\, b,\, InitVals)$ is the tuple $\langle g[0][b], \ldots, g[n - 1][b] \rangle$, which can be written formally as

$$[i \in 1 \ldots n \mapsto g[i - 1][b]]$$

To complete the definition of *Integrate*, we now define the operator *IsDeriv*. It's easy to define the ith derivative inductively in terms of the first derivative. So, we define *IsFirstDeriv*$(df,\, f)$ to be true iff df is the first derivative of f, assuming that f is a real-valued function whose domain is an open interval. Our definition actually works if the domain of f is any open set.[2] Elementary calculus tells us that $df[r]$ is the derivative of f at r iff

$$df[r] \;=\; \lim_{s \to r} \frac{f[s] - f[r]}{s - r}$$

The classical "δ-ϵ" definition of the limit states that this is true iff, for every $\epsilon > 0$, there is a $\delta > 0$ such that $0 < |s - r| < \delta$ implies

$$\left| df[r] - \frac{f[s] - f[r]}{s - r} \right| < \epsilon$$

Stated formally, this condition is

$$\forall\, \epsilon \in PosReal \,:$$
$$\exists\, \delta \in PosReal \,:$$
$$\forall\, s \in Nbhd(r,\, \delta) \setminus \{r\} \,:\, \frac{f[s] - f[r]}{s - r} \in Nbhd(df[r],\, \epsilon)$$

We define *IsFirstDeriv*$(df,\, f)$ to be true iff the domains of df and f are equal, and this condition holds for all r in their domain.

The definitions of *Integrate* and all the other operators introduced above appear in the *DifferentialEquations* module of Figure 11.2 on the next page. The LOCAL construct described in Section 11.1.1 above is used to make all these definitions local to the module, except for the definition of *Integrate*.

---

[2]A set S is open iff for every $r \in S$ there exists an $\epsilon > 0$ such that the interval from $r - \epsilon$ to $r + \epsilon$ is contained in S.

———— MODULE *DifferentialEquations* ————

This module defines the operator *Integrate* for specifying the solution to a differential equation. If a and b are reals with $a \leq b$; *InitVals* is an n-tuple of reals; and D is a function from $(n+1)$-tuples of reals to reals; then this is the n-tuple of values

$$\langle f[b], \frac{df}{dt}[b], \ldots, \frac{d^{n-1}f}{dt^{n-1}}[b] \rangle$$

where f is the solution to the differential equation

$$D[t, f, \frac{df}{dt}, \ldots, \frac{d^n f}{dt^n}] = 0$$

such that

$$\langle f[a], \frac{df}{dt}[a], \ldots, \frac{d^{n-1}f}{dt^{n-1}}[a] \rangle = \mathit{InitVals}$$

LOCAL INSTANCE *Reals*

LOCAL INSTANCE *Sequences*

LOCAL *PosReal* $\triangleq$ $\{r \in Real : r > 0\}$

LOCAL *OpenInterval*(a, b) $\triangleq$ $\{s \in Real : (a < s) \wedge (s < b)\}$

LOCAL *Nbhd*(r, e) $\triangleq$ *OpenInterval*$(r - e, r + e)$

> The INSTANCE statement and these definitions are local, so a module that extends this one obtains only the definition of *Integrate*.

LOCAL *IsFirstDeriv*(df, f) $\triangleq$
$\wedge\ df \in [\text{DOMAIN}\,f \rightarrow Real]$
$\wedge\ \forall\, r \in \text{DOMAIN}\,f :$
  $\forall\, e \in PosReal :$
   $\exists\, d \in PosReal :$
    $\forall\, s \in Nbhd(r, d) \setminus \{r\} : (f[s] - f[r])/(s - r) \in Nbhd(df[r], e)$

> Assuming DOMAIN f is an open subset of *Real*, this is true iff f is differentiable and df is its first derivative. Recall that the derivative of f at r is the number $df[r]$ satisfying the following condition: for every ϵ there exists a δ such that $0 < |s - r| < \delta$ implies $|df[r] - (f[s] - f[r])/(s - r)| < \epsilon$.

LOCAL *IsDeriv*(n, df, f) $\triangleq$ True iff f is n times differentiable and df is its nth derivative.

LET $IsD[k \in 0\,..\,n,\ g \in [\text{DOMAIN}\,f \rightarrow Real]]$ $\triangleq$ $IsD[k, g] = IsDeriv(k, g, f)$
  IF $k = 0$ THEN $g = f$
     ELSE $\exists\, gg \in [\text{DOMAIN}\,f \rightarrow Real] : \wedge\ IsFirstDeriv(g, gg)$
                                   $\wedge\ IsD[k - 1, gg]$

IN $IsD[n, df]$

Integrate$(D, a, b, InitVals)$ $\triangleq$
LET n $\triangleq$ $Len(InitVals)$
  gg $\triangleq$ CHOOSE $g : \exists\, e \in PosReal : \wedge\ g \in [0\,..\,n \rightarrow [OpenInterval(a - e, b + e) \rightarrow Real]]$
                        $\wedge\ \forall\, i \in 1\,..\,n : \wedge\ IsDeriv(i, g[i], g[0])$
                                       $\wedge\ g[i - 1][a] = InitVals[i]$
                        $\wedge\ \forall\, r \in OpenInterval(a - e, b + e) :$
                           $D[\langle r \rangle \circ [i \in 1\,..\,(n + 1) \mapsto g[i - 1][r]]] = 0$
IN $[i \in 1\,..\,n \mapsto gg[i - 1][b]]$

Figure 11.2: A module for specifying the solution to a differential equation.

11.1.4 BNF Grammars

BNF, which stands for Backus-Naur Form, is a standard way of describing the syntax of computer languages. This section develops the *BNFGrammars* module, which defines operators for writing BNF grammars. A BNF grammar isn't the kind of data structure that arises in system specification, and TLA⁺ is not particularly well suited to specifying one. Its syntax doesn't allow us to write BNF grammars exactly the way we'd like, but we can come reasonably close. Moreover, I think it's fun to use TLA⁺ to specify its own syntax. So, module *BNFGrammars* is used in Chapter 15 to specify part of the syntax of TLA⁺, as well as in Chapter 14 to specify the syntax of the TLC model checker's configuration file.

Let's start by reviewing BNF grammars. Consider the little language SE of simple expressions described by the BNF grammar

$$expr \ ::= \ \textbf{ident} \ | \ expr \ \textbf{op} \ expr \ | \ (\,expr\,) \ | \ \text{LET} \ def \ \text{IN} \ expr$$
$$def \ \ ::= \ \textbf{ident} \ \texttt{==} \ expr$$

where **op** is some class of infix operators like +, and **ident** is some class of identifiers such as *abc* and *x*. The language SE contains expressions like

$$abc \ + \ (\text{LET} \ x \ \texttt{==} \ y + abc \ \text{IN} \ x * x)$$

Let's represent this expression as the sequence

⟨ "abc", "+", "(", "LET", "x", "==",
  "y", "+", "abc", "IN", "x", "*", "x", ")" ⟩

of strings. The strings such as "abc" and "+" appearing in this sequence are usually called *lexemes*. In general, a sequence of lexemes is called a *sentence*; and a set of sentences is called a *language*. So, we want to define the language SE to consist of the set of all such sentences described by the BNF grammar.[3]

To represent a BNF grammar in TLA⁺, we must assign a mathematical meaning to nonterminal symbols like *def*, to terminal symbols like **op**, and to the grammar's two productions. The method that I find simplest is to let the meaning of a nonterminal symbol be the language that it generates. Thus, the meaning of *expr* is the language SE itself. I define a *grammar* to be a function G such that, for any string "str", the value of $G[\text{``str''}]$ is the language generated by the nonterminal *str*. Thus, if $G$ is the BNF grammar above, then $G[\text{``expr''}]$ is the complete language SE, and $G[\text{``def''}]$ is the language defined by the production for *def*, which contains sentences like

⟨ "y", "==", "qq", "*", "wxyz" ⟩

---

[3]BNF grammars are also used to specify how an expression is parsed—for example, that $a + b * c$ is parsed as $a + (b * c)$ rather than $(a + b) * c$. By considering the grammar to specify only a set of sentences, we are deliberately not capturing that use in our TLA⁺ representation of BNF grammars.

Instead of letting the domain of G consist of just the two strings "expr" and "def", it turns out to be more convenient to let its domain be the entire set STRING of strings, and to let $G[s]$ be the empty language (the empty set) for all strings s other than "expr" and "def". So, a grammar is a function from the set of all strings to the set of sequences of strings. We can therefore define the set *Grammar* of all grammars by

$$Grammar \;\triangleq\; [\text{STRING} \to \text{SUBSET } Seq(\text{STRING})]$$

In describing the mathematical meaning of records, Section 5.2 explained that $r.ack$ is an abbreviation for $r[\text{"ack"}]$. This is the case even if r isn't a record. So, we can write $G.op$ instead of $G[\text{"op"}]$. (A grammar isn't a record because its domain is the set of all strings rather than a finite set of strings.)

A terminal like **ident** can appear anywhere to the right of a "::=" that a nonterminal like *expr* can, so a terminal should also be a set of sentences. Let's represent a terminal as a set of sentences, each of which is a sequence consisting of a single lexeme. Let a *token* be a sentence consisting of a single lexeme, so a terminal is a set of tokens. For example, the terminal **ident** is a set containing tokens such as $\langle \text{"abc"} \rangle$, $\langle \text{"x"} \rangle$, and $\langle \text{"qq"} \rangle$. Any terminal appearing in the BNF grammar must be represented by a set of tokens, so the == in the grammar for SE is the set $\{\langle \text{"=="} \rangle\}$. Let's define the operator *tok* by

tok is short for *token*.

$$tok(s) \;\triangleq\; \{\langle s \rangle\}$$

so we can write this set of tokens as $tok(\text{"=="})$.

A production expresses a relation between the values of $G.str$ for some grammar G and some strings "str". For example, the production

$$def \quad ::= \quad \textbf{ident} == expr$$

asserts that a sentence s is in $G.def$ iff it has the form $i \circ \langle \text{"=="} \rangle \circ e$ for some token i in **ident** and some sentence e in $G.expr$. In mathematics, a formula about G must mention G (perhaps indirectly by using a symbol defined in terms of G). So, we can try writing this production in TLA$^+$ as

$$G.def \quad ::= \quad \textbf{ident} \; tok(\text{"=="}) \; G.expr$$

In the expression to the right of the ::=, adjacency is expressing some operation. Just as we have to make multiplication explicit by writing $2 * x$ instead of $2x$, we must express this operation by an explicit operator. Let's use &, so we can write the production as

$$(11.1) \quad G.def \quad ::= \quad \textbf{ident} \; \& \; tok(\text{"=="}) \; \& \; G.expr$$

This expresses the desired relation between the sets $G.def$ and $G.expr$ of sentences if ::= is defined to be equality and & is defined so that $L \; \& \; M$ is the set of all sentences obtained by concatenating a sentence in L with a sentence in M:

$$L \,\&\, M \;\triangleq\; \{s \circ t : s \in L, \, t \in M\}$$

The production

$$expr \;\; ::= \;\; \textbf{ident} \;\mid\; expr \; \textbf{op} \; expr \;\mid\; (\, expr \,) \;\mid\; \text{LET} \; def \; \text{IN} \; expr$$

can similarly be expressed as

The precedence rules of TLA$^+$ imply that $a \mid b \,\&\, c$ is interpreted as $a \mid (b \,\&\, c)$.

$$(11.2) \quad G.expr \;\; ::= \quad\; ident$$
$$\mid \quad G.expr \;\&\; op \;\&\; G.expr$$
$$\mid \quad tok(\text{``(''}) \;\&\; G.exp \;\&\; tok(\text{``)''})$$
$$\mid \quad tok(\text{``LET''}) \;\&\; G.def \;\&\; tok(\text{``IN''}) \;\&\; G.expr$$

This expresses the desired relation if $\mid$ (which means *or* in the BNF grammar) is defined to be set union ($\cup$).

We can also define the following operators that are sometimes used in BNF grammars:

- *Nil* is defined so that *Nil* $\&$ S equals S for any set S of sentences:

$$Nil \;\triangleq\; \{\langle\rangle\}$$

- L^+ equals $L \mid L \,\&\, L \mid L \,\&\, L \,\&\, L \mid \ldots$:

L^+ is typed L^+ and L^* is typed L^*.

$$L^+ \;\triangleq\; \text{LET} \;\; LL[n \in Nat] \;\triangleq\; \overbrace{LL[n] = L \mid \ldots \mid L \,\&\, \ldots L}^{n+1 \text{ copies}}.$$
$$\text{IF} \;\; n = 0 \;\; \text{THEN} \;\; L$$
$$\text{ELSE} \;\; LL[n-1] \mid LL[n-1] \,\&\, L$$
$$\text{IN} \quad \text{UNION} \; \{LL[n] : n \in Nat\}$$

- L^* equals *Nil* $\mid L \mid L \,\&\, L \mid L \,\&\, L \,\&\, L \mid \ldots$:

$$L^* \;\triangleq\; Nil \mid L^+$$

The BNF grammar for SE consists of two productions, expressed by the TLA$^+$ formulas (11.1) and (11.2). The entire grammar is the single formula that is the conjunction of these two formulas. We must turn this formula into a mathematical definition of a grammar GSE, which is a function from strings to languages. The formula is an assertion about a grammar G. We define GSE to be the smallest grammar G satisfying the conjunction of (11.1) and (11.2), where grammar G_1 smaller than G_2 means that $G_1[s] \subseteq G_2[s]$ for every string s. To express this in TLA$^+$, we define an operator *LeastGrammar* so that *LeastGrammar*(P) is the smallest grammar G satisfying $P(G)$:

$$LeastGrammar(P(\_)) \;\triangleq$$
$$\text{CHOOSE} \; G \in Grammar :$$
$$\quad \wedge P(G)$$
$$\quad \wedge \forall H \in Grammar : P(H) \Rightarrow (\forall s \in \text{STRING} : G[s] \subseteq H[s])$$

Letting $P(G)$ be the conjunction of (11.1) and (11.2), we can define the grammar
GSE to be $LeastGrammar(P)$. We can then define the language SE to equal
$GSE.expr$. The smallest grammar G satisfying a formula P must have $G[s]$
equal to the empty language for any string s that doesn't appear in P. Thus,
$GSE[s]$ equals the empty language $\{\}$ for any string s other than "expr" and
"def".

To complete our specification of GSE, we must define the sets $ident$ and
op of tokens. We can define the set op of operators by enumerating them—for
example:

$$op \;\triangleq\; tok(\text{"+"}) \mid tok(\text{"$-$"}) \mid tok(\text{"*"}) \mid tok(\text{"/"})$$

To express this a little more compactly, let's define $Tok(S)$ to be the set of all
tokens formed from elements in the set S of lexemes:

$$Tok(S) \;\triangleq\; \{\langle s \rangle : s \in S\}$$

We can then write

$$op \;\triangleq\; Tok(\{\text{"+"}, \text{"$-$"}, \text{"*"}, \text{"/"}\})$$

Let's define $ident$ to be the set of tokens whose lexemes are words made
entirely of lower-case letters, such as "abc", "qq", and "x". To learn how to do
that, we must first understand what strings in TLA$^+$ really are. In TLA$^+$, a
string is a sequence of characters. (We don't care, and the semantics of TLA$^+$
doesn't specify, what a character is.) We can therefore apply the usual sequence
operators on them. For example, $Tail(\text{"abc"})$ equals "bc", and "abc" $\circ$ "de"
equals "abcde".

See Section
16.1.10 on page
307 for more
about strings. Re-
member that we
take *sequence* and
tuple to be syn-
onymous.

The operators like & that we just defined for expressing BNF were applied
to sets of sentences, where a sentence is a sequence of lexemes. These operators
can be applied just as well to sets of sequences of any kind—including sets
of strings. For example, $\{\text{"one"}, \text{"two"}\}$ & $\{\text{"s"}\}$ equals $\{\text{"ones"}, \text{"twos"}\}$, and
$\{\text{"ab"}\}^+$ is the set consisting of all the strings "ab", "abab", "ababab", etc. So,
we can define $ident$ to equal $Tok(Letter^+)$, where $Letter$ is the set of all lexemes
consisting of a single lower-case letter:

$$Letter \;\triangleq\; \{\text{"a"}, \text{"b"}, \ldots, \text{"z"}\}$$

Writing this definition out in full (without the "...") is tedious. We can make
this a little easier as follows. We first define $OneOf(s)$ to be the set of all
one-character strings made from the characters of the string s:

$$OneOf(s) \;\triangleq\; \{\langle s[i] \rangle : i \in \text{DOMAIN } s\}$$

We can then define

$$Letter \;\triangleq\; OneOf(\text{"abcdefghijklmnopqrstuvwxyz"})$$

$$GSE \;\triangleq\; \text{LET} \;\; op \;\triangleq\; Tok(\,\{\,\text{``+''}, \text{``--''}, \text{``*''}, \text{``/''}\,\}\,)$$

$$ident \;\triangleq\; Tok(\, OneOf(\text{``abcdefghijklmnopqrstuvwxyz''})^{+}\,)$$

$$P(G) \;\triangleq\; \wedge\; G.expr \;::=\; \quad ident$$
$$|\;\; G.expr \;\&\; op \;\&\; G.expr$$
$$|\;\; tok(\text{``(''}) \;\&\; G.expr \;\&\; tok(\text{``)''})$$
$$|\;\; tok(\text{``LET''}) \;\&\; G.def \;\&\; tok(\text{``IN''}) \;\&\; G.expr$$
$$\wedge\; G.def \;::=\; ident \;\&\; tok(\text{``==''}) \;\&\; G.expr$$

$$\text{IN} \quad LeastGrammar(P)$$

Figure 11.3: The definition of the grammar *GSE* for the language SE.

The complete definition of the grammar *GSE* appears in Figure 11.3 on this page.

All the operators we've defined here for specifying grammars are grouped into module *BNFGrammars*, which appears in Figure 11.4 on the next page.

Using TLA^{+} to write ordinary BNF grammars is a bit silly. However, ordinary BNF grammars are not very convenient for describing the syntax of a complicated language like TLA^{+}. In fact, they can't describe the alignment rules for its bulleted lists of conjuncts and disjuncts. Using TLA^{+} to specify such a language is not so silly. In fact, a TLA^{+} specification of the complete syntax of TLA^{+} was written as part of the development of the Syntactic Analyzer, described in Chapter 12. Although valuable when writing a TLA^{+} parser, this specification isn't very helpful to an ordinary user of TLA^{+}, so it does not appear in this book.

11.2 Other Memory Specifications

Section 5.3 specifies a multiprocessor memory. The specification is unrealistically simple for three reasons: a processor can have only one outstanding request at a time, the basic correctness condition is too restrictive, and only simple read and write operations are provided. (Real memories provide many other operations, such as partial-word writes and cache prefetches.) We now specify a memory that allows multiple outstanding requests and has a realistic, weaker correctness condition. To keep the specification short, we still consider only the simple operations of reading and writing one word of memory.

11.2.1 The Interface

The first thing we must do to specify a memory is determine the interface. The interface we choose depends on the purpose of the specification. There are many different reasons why we might be specifying a multiprocessor memory. We could

—— MODULE $BNFGrammars$ ——

A sentence is a sequence of strings. (In standard terminology, the term "lexeme" is used instead of "string".) A token is a sentence of length one—that is, a one-element sequence whose single element is a string. A language is a set of sentences.

LOCAL INSTANCE $Naturals$

LOCAL INSTANCE $Sequences$

OPERATORS FOR DEFINING SETS OF TOKENS

$OneOf(s) \triangleq \{\langle s[i]\rangle : i \in \text{DOMAIN } s\}$

If s is a string, then $OneOf(s)$ is the set of strings formed from the individual characters of s. For example, $OneOf(\text{"abc"}) = \{\text{"a"}, \text{"b"}, \text{"c"}\}$.

$tok(s) \triangleq \{\langle s\rangle\}$

$Tok(S) \triangleq \{\langle s\rangle : s \in S\}$

If s is a string, then $tok(s)$ is the set containing only the token made from s. If S is a set of strings, then $Tok(S)$ is the set of tokens made from elements of S.

OPERATORS FOR DEFINING LANGUAGES

$Nil \triangleq \{\langle\rangle\}$ The language containing only the "empty" sentence.

$L \mathrel{\&} M \triangleq \{s \circ t : s \in L, t \in M\}$ All concatenations of sentences in L and M.

$L \mid M \triangleq L \cup M$

$L^{+} \triangleq\quad L \mid L \mathrel{\&} L \mid L \mathrel{\&} L \mathrel{\&} L \mid \ldots$

$\quad$ LET $LL[n \in Nat] \triangleq$ IF $n = 0$ THEN L

$\qquad\qquad\qquad\qquad\qquad\qquad\quad$ ELSE $\quad LL[n-1] \mid LL[n-1] \mathrel{\&} L$

$\quad$ IN $\quad$ UNION $\{LL[n] : n \in Nat\}$

$L^{*} \triangleq Nil \mid L^{+}$

$L ::= M \triangleq L = M$

$Grammar \triangleq [\text{STRING} \to \text{SUBSET } Seq(\text{STRING})]$

$LeastGrammar(P(\_)) \triangleq$ The smallest grammar G such that $P(G)$ is true.

$\quad$ CHOOSE $G \in Grammar :$

$\qquad \wedge P(G)$

$\qquad \wedge \forall H \in Grammar : P(H) \Rightarrow \forall s \in \text{STRING} : G[s] \subseteq H[s]$

Figure 11.4: The module $BNFGrammars$.

be specifying a computer architecture, or the semantics of a programming language. Let's suppose we are specifying the memory of an actual multiprocessor computer.

A modern processor performs multiple instructions concurrently. It can begin new memory operations before previous ones have been completed. The memory responds to a request as soon as it can; it need not respond to different requests in the order that they were issued.

A processor issues a request to a memory system by setting some register. We assume that each processor has a set of registers through which it communicates with the memory. Each register has three fields: an *adr* field that holds an address, a *val* field that holds a word of memory, and an *op* field that indicates what kind of operation, if any, is in progress. The processor can issue a command using a register whose *op* field equals "Free". It sets the *op* field to "Rd" or "Wr" to indicate the operation; it sets the *adr* field to the address of the memory word; and, for a write, it sets the *val* field to the value being written. (On a read, the processor can set the *val* field to any value.) The memory responds by setting the *op* field back to "Free" and, for a read, setting the *val* field to the value read. (The memory does not change the *val* field when responding to a write.)

Module *RegisterInterface* in Figure 11.5 on the next page contains some declarations and definitions for specifying the interface. It declares the constants *Adr*, *Val*, and *Proc*, which are the same as in the memory interface of Section 5.1, and the constant *Reg*, which is the set of registers. (More precisely, *Reg* is a set of register identifiers.) A processor has a separate register corresponding to each element of *Reg*. The variable *regFile* represents the processors' registers, $regFile[p][r]$ being register r of processor p. The module also defines the sets of requests and register values, as well as a type invariant for *regFile*.

11.2.2 The Correctness Condition

Section 5.3 specifies what is called a linearizable memory. In a linearizable memory, a processor never has more than one outstanding request. The correctness condition for the memory can be stated as

> The result of any execution is the same as if the operations of all the processors were executed in some sequential order, and each operation is executed between the request and the response.

The second clause, which requires the system to act as if each operation were executed between its request and its response, is both too weak and too strong for our specification. It's too weak because it says nothing about the execution order of two operations from the same processor unless one is issued after the other's response. For example, suppose a processor p issues a write and then a read to the same address. We want the read to obtain either the value p just wrote, or a value written by another processor—even if p issues the read before

$\overline{\quad\qquad\qquad\qquad\qquad\qquad}$ MODULE $RegisterInterface$ $\overline{\quad\qquad\qquad\qquad}$

CONSTANT $Adr,$ The set of memory addresses.

$\qquad\qquad Val,$ The set of memory-word values.

$\qquad\qquad Proc,$ The set of processors.

$\qquad\qquad Reg$ The set of registers used by a processor.

VARIABLE $regFile$ $regFile[p][r]$ represents the contents of register r of processor p.

$RdRequest \quad\triangleq\; [adr : Adr,\; val : Val,\; op : \{\,\text{"Rd"}\,\}]$

$WrRequest \quad\triangleq\; [adr : Adr,\; val : Val,\; op : \{\,\text{"Wr"}\,\}]$

$FreeRegValue \;\triangleq\; [adr : Adr,\; val : Val,\; op : \{\,\text{"Free"}\,\}]$

$Request \;\triangleq\; RdRequest \cup WrRequest$ The set of all possible requests.

$RegValue \;\triangleq\; Request \cup FreeRegValue$ The set of all possible register values.

$RegFileTypeInvariant \;\triangleq\;$ The type correctness invariant for $regFile$.

$\qquad regFile \in [Proc \to [Reg \to RegValue]]$

Figure 11.5: A module for specifying a register interface to a memory.

receiving the response for the write. This isn't guaranteed by the condition. The second clause is too strong because it places unnecessary ordering constraints on operations issued by different processors. If operations A and B are issued by two different processors, then we don't need to require that A precedes B in the execution order just because B was requested after A's response.

We modify the second clause to require that the system act as if operations of each individual processor were executed in the order that they were issued, obtaining the condition

> The result of any execution is the same as if the operations of all the processors were executed in some sequential order, and the operations of each individual processor appear in this sequence in the order in which the requests were issued.

In other words, we require that the values returned by the reads can be explained by some total ordering of the operation executions that is consistent with the order in which each processor issued its requests. There are a number of different ways of formalizing this condition; they differ in how bizarre the explanation may be. The differences can be described in terms of whether or not certain scenarios are permitted. In the scenario descriptions, $Wr_p(a, v)$ represents a write operation of value v to address a by processor p, and $Rd_p(a, v)$ represents a read of a by p that returns the value v.

The first decision we must make is whether all operations in an infinite behavior must be ordered, or if the ordering must exist only at each finite point during the behavior. Consider a scenario in which each of two processors writes its own value to the same address and then keeps reading that value forever:

In these scenarios, values and addresses with different names are assumed to be different.

Processor p: $Wr_p(a, v1)$, $Rd_p(a, v1)$, $Rd_p(a, v1)$, $Rd_p(a, v1)$, ...

Processor q: $Wr_q(a, v2)$, $Rd_q(a, v2)$, $Rd_q(a, v2)$, $Rd_q(a, v2)$, ...

At each point in the execution, we can explain the values returned by the reads with a total order in which all the operations of either processor precede all the operations of the other. However, there is no way to explain the entire infinite scenario with a single total order. In this scenario, neither processor ever sees the value written by the other. Since a multiprocessor memory is supposed to allow processors to communicate, we disallow this scenario.

The second decision we must make is whether the memory is allowed to predict the future. Consider this scenario:

Processor p: $Wr_p(a, v1)$, $Rd_p(a, v2)$

Processor q: $\qquad\qquad\qquad\qquad Wr_q(a, v2)$

Here, q issues its write of $v2$ after p has obtained the result of its read. The scenario is explained by the ordering $Wr_p(a, v1)$, $Wr_q(a, v2)$, $Rd_p(a, v2)$. However, this is a bizarre explanation because, to return the value $v2$ for p's read, the memory had to predict that another processor would write $v2$ some time in the future. Since a real memory can't predict what requests will be issued in the future, such a behavior cannot be produced by a correct implementation. We can therefore rule out the scenario as unreasonable. Alternatively, since no correct implementation can produce it, there's no need to outlaw the scenario.

If we don't allow the memory to predict the future, then it must always be able to explain the values read in terms of the writes that have been issued so far. In this case, we have to decide whether the explanations must be stable. For example, suppose a scenario begins as follows:

Processor p: $Wr_p(a1, v1)$, $Rd_p(a1, v3)$

Processor q: $Wr_q(a2, v2)$, $Wr_q(a1, v3)$

At this point, the only explanation for p's read $Rd_p(a1, v3)$ is that q's write $Wr_q(a1, v3)$ preceded it, which implies that q's other write $Wr_q(a2, v2)$ also preceded the read. Hence, if p now reads $a2$, it must obtain the value $v2$. But suppose the scenario continues as follows, with another processor r joining in:

Processor p: $Wr_p(a1, v1)$, $Rd_p(a1, v3)$, $\qquad\qquad\qquad Rd_p(a2, v0)$

Processor q: $Wr_q(a2, v2)$, $Wr_q(a1, v3)$

Processor r: $\qquad\qquad\qquad\qquad Wr_r(a1, v3)$

We can explain this scenario with the following ordering of the operations:

$Wr_p(a1, v1)$, $Wr_r(a1, v3)$, $Rd_p(a1, v3)$,

$\qquad Rd_p(a2, v0)$, $Wr_q(a2, v2)$, $Wr_q(a1, v3)$

In this explanation, processor r provided the value of $a1$ read by p, and p read the initial value $v0$ of memory address $a2$. The explanation is bizarre because the write that provided the value of $a1$ to p was actually issued after the completion of p's read operation. But, because the explanation of that value changed in mid-execution, the system never predicted the existence of a write that had not yet occurred. When writing a specification, we must decide whether or not to allow such changes of the explanation.

11.2.3   A Serial Memory

We first specify a memory that cannot predict the future and cannot change its explanations. There seems to be no standard name for such a memory; I'll call it a *serial* memory.

Our informal correctness condition is in terms of the sequence of all operations that have ever been issued. There is a general method of formalizing such a condition that works for specifying many different kinds of systems. We add an internal variable opQ that records the history of the execution. For each processor p, the value of $opQ[p]$ is a sequence whose i^{th} element, $opQ[p][i]$, describes the ith request issued by p, the response to that request (if it has been issued), and any other information about the operation needed to express the correctness condition. If necessary, we can also add other internal variables to record information not readily associated with individual requests.

For a system with the kind of register interface we are using, the next-state action has the form

$$(11.3) \quad \lor\ \exists\, proc \in Proc,\ reg \in Reg :$$
$$\lor\ \exists\, req \in Request :\ IssueRequest(proc,\ req,\ reg)$$
$$\lor\ RespondToRequest(proc,\ reg)$$
$$\lor\ Internal$$

where the component actions are

IssueRequest(proc, req, reg)
   The action with which processor *proc* issues a request *req* in register *reg*.

RespondToRequest(proc, reg)
   The action with which the system responds to a request in processor *proc*'s register *reg*.

Internal
   An action that changes only the internal state.

Liveness properties are asserted by fairness conditions on the *RespondToRequest* and *Internal* actions.

A general trick for writing the specification is to choose the internal state so the safety part of the correctness condition can be expressed by the formula $\Box P$ for some state predicate P. We guarantee that P is always true by letting P' be a conjunct of each action. I'll use this approach to specify the serial memory, taking for P a state predicate *Serializable*.

We want to require that the value returned by each read is explainable as the value written by some operation already issued, or as the initial value of the memory. Moreover, we don't want this explanation to change. We therefore add to the *opQ* entry for each completed read a *source* field that indicates where the value came from. This field is set by the *RespondToRequest* action.

We want all operations in an infinite behavior eventually to be ordered. This means that, for any two operations, the memory must eventually decide which one precedes the other—and it must stick to that decision. We introduce an internal variable *opOrder* that describes the ordering of operations to which the memory has already committed itself. An *Internal* step changes only *opOrder*, and it can only enlarge the ordering.

The predicate *Serializable* used to specify the safety part of the correctness condition describes what it means for *opOrder* to be a correct explanation. It asserts that there is some consistent total ordering of the operations that satisfies the following conditions:

- It extends *opOrder*.

- It orders all operations from the same processor in the order that they were issued.

- It orders operations so that the source of any read is the latest write to the same address that precedes the read, and is the initial value iff there is no such write.

We now translate this informal sketch of the specification into TLA⁺. We first choose the types of the variables *opQ* and *opOrder*. To do this, we define a set *opId* of values that identify the operations that have been issued. An operation is identified by a pair $\langle p,\ i \rangle$ where p is a processor and i is a position in the sequence *opQ[p]*. (The set of all such positions i is DOMAIN *opQ[p]*.) We let the corresponding element of *opId* be the record with *proc* field p and *idx* field i. Writing the set of all such records is a bit tricky because the possible values of the *idx* field depend on the *proc* field. We define *opId* to be a subset of the set of records whose *idx* field can be any value:

$$opId \;\triangleq\; \{\,oiv \in [proc\ :\ Proc,\ idx\ :\ Nat]\ :$$
$$oiv.idx \in \text{DOMAIN}\ opQ[oiv.proc]\}$$

For convenience, we define *opIdQ(oi)* to be the value of the *opQ* entry identified by an element *oi* of *opId*:

$$opIdQ(oi) \;\triangleq\; opQ[oi.proc][oi.idx]$$

The source of a value need not be an operation; it can also be the initial contents of the memory. The latter possibility is represented by letting the *source* field of the *opQ* entry have the special value *InitWr*. We then let *opQ* be an element of $[Proc \rightarrow Seq(opVal)]$, where *opVal* is the union of three sets:

The sets *Request*, *WrRequest*, and *RdRequest* are defined in module *RegisterInterface* on page 186.

$[req : Request,\ reg : Reg]$
Represents an active request in the register of the requesting processor indicated by the *reg* field.

$[req : WrRequest,\ reg : \{Done\}]$
Represents a completed write request, where *Done* is a special value that is not a register.

$[req : RdRequest,\ reg : \{Done\},\ source : opId \cup \{InitWr\}]$
Represents a completed read request whose value came from the operation indicated by the *source* field, or from the initial value of the memory location if the *source* field equals *InitWr*.

Note that *opId* and *opVal* are state functions whose values depend upon the value of the variable *opQ*.

We need to specify the initial contents of memory. A program generally cannot assume anything about the memory's initial contents, except that every address does contain a value in *Val*. So, the initial contents of memory can be any element of $[Adr \rightarrow Val]$. We declare an "internal" constant *InitMem*, whose value is the memory's initial contents. In the final specification, *InitMem* will be hidden along with the internal variables *opQ* and *opOrder*. We hide a constant with ordinary existential quantification $\exists$. The requirement that *InitMem* is a function from addresses to values could be made part of the initial predicate, but it's more natural to express it in the quantifier. The final specification will therefore have the form

$$\exists\, InitMem \in [Adr \rightarrow Val] :\ \exists\, opQ, opOrder :\ \ldots$$

For later use, we define *goodSource(oi)* to be the set of plausible values for the source of a read operation *oi* in *opId*. A plausible value is either *InitWr* or a write to the same address that *oi* reads. It will be an invariant of the specification that the source of any completed read operation *oi* is an element of *goodSource(oi)*. Moreover, the value returned by a completed read operation must come from its source. If the source is *InitWr*, then the value must come from *InitMem*; otherwise, it must come from the source request's *val* field. To express this formally, observe that the *opQ* entries only of completed reads have a *source* field. Since a record has a *source* field iff the string "source" is in its domain, we can write this invariant as

Section 5.2 on page 48 explains that a record is a function whose domain is a set of strings.

(11.4) $\forall\, oi \in opId$:

    ("source" $\in$ DOMAIN $opIdQ(oi))\ \Rightarrow$

       $\wedge\ opIdQ(oi).source \in goodSource(oi)$

       $\wedge\ opIdQ(oi).req.val =$ IF $\ opIdQ(oi).source = InitWr$

                              THEN $\ InitMem[opIdQ(oi).req.adr]$

                              ELSE $\ opIdQ(opIdQ(oi).source).req.val$

We now choose the type of *opOrder*. We usually denote an ordering relation by an operator such as $\prec$, writing $A \prec B$ to mean that A precedes B. However, the value of a variable cannot be an operator. So, we must represent an ordering relation as a set or a function. Mathematicians usually describe a relation $\prec$ on a set S as a set R of ordered pairs of elements in S, with $\langle A, B\rangle$ in R iff $A \prec B$. So, we let *opOrder* be a subset of $opId \times opId$, where $\langle oi, oj \rangle \in opOrder$ means that oi precedes oj.

The difference between operators and functions is discussed in Section 6.4 on page 69.

Our internal state is redundant because, if register r of processor p contains an uncompleted operation, then there is an *opQ* entry that points to the register and contains the same request. This redundancy means that the following relations among the variables are invariants of the specification:

- If an *opQ* entry's *reg* field is not equal to *Done*, then it denotes a register whose contents is the entry's *req* field.

- The number of *opQ* entries pointing to a register equals 1 if the register contains an active operation, otherwise it equals 0.

In the specification, we combine this condition, formula (11.4), and the type invariant into a single state predicate *DataInvariant*.

Having chosen the types of the variables, we can now define the initial predicate *Init* and the predicate *Serializable*. The definition of *Init* is easy. We define *Serializable* in terms of *totalOpOrder*, the set of all total orders of *opId*. A relation $\prec$ is a total order of *opId* iff it satisfies the following three conditions, for any oi, oj, and ok in *opId*:

Totality: Either $oi = oj$, $oi \prec oj$, or $oj \prec oi$.

Transitivity: $oi \prec oj$ and $oj \prec ok$ imply $oi \prec ok$.

Irreflexivity: $oi \not\prec oi$.

The predicate *Serializable* asserts that there is a total order of *opId* satisfying the three conditions on page 189. We can express this formally as the assertion that there exists an R in *totalOpOrder* satisfying

    $\wedge\ opOrder \subseteq R$ R extends *opOrder*

    $\wedge\ \forall\, oi, oj \in opId$: R correctly orders operations from the same processor.

       $(oi.proc = oj.proc) \wedge (oi.idx < oj.idx) \Rightarrow (\langle oi, oj \rangle \in R)$

$$\land \;\; \forall\, oi \in opId :$$
$$(\text{``source''} \in \text{DOMAIN } opIdQ(oi))$$
$$\Rightarrow \neg\,(\exists\, oj \in goodSource(oi) :$$
$$\land \langle oj, oi \rangle \in R$$
$$\land\, (opIdQ(oi).source \neq InitWr) \Rightarrow$$
$$(\langle opIdQ(oi).source, oj \rangle \in R))$$

> For every completed read oi in $opId$, there is no write oj to the same address that (i) precedes oi and (ii) follows the source if that source is not $InitWr$.

We allow each step to extend $opOrder$ to any relation on $opId$ that satisfies *Serializable*. We do this by letting every subaction of the next-state action specify $opOrder'$ with the conjunct *UpdateOpOrder*, defined by

$$UpdateOpOrder \;\;\triangleq\;\; \land\;\; opOrder' \subseteq (opId' \times opId')$$
$$\land\;\; opOrder \subseteq opOrder'$$
$$\land\;\; Serializable'$$

The next-state action has the generic form of formula (11.3) on page 188. We split the *RespondToRequest* action into the disjunction of separate *RespondToWr* and *RespondToRd* actions that represent responding to writes and reads, respectively. *RespondToRd* is the most complicated of the next-state action's subactions, so let's examine its definition. The definition has the form

$$RespondToRd(proc, reg) \;\;\triangleq$$
$$\text{LET } req \;\triangleq\; regFile[proc][reg]$$
$$idx \;\triangleq\; \text{CHOOSE } i \in \text{DOMAIN } opQ[proc] : opQ[proc][i].reg = reg$$
$$\text{IN } \;\;\ldots$$

This defines req to be the request in the register and idx to be an element in the domain of $opQ[proc]$ such that $opQ[proc][idx].reg$ equals reg. If the register is not free, then there is exactly one such value idx; and $opQ[proc][idx].req$, the idx^{th} request issued by $proc$, equals req. (We don't care what idx equals if the register is free.) The IN expression begins with the enabling condition

$$\land\;\; req.op = \text{``Rd''}$$

which asserts that the register is not free and it contains a read request. The next conjunct of the IN expression is

$$\land\;\; \exists\, src \in goodSource([proc \mapsto proc, idx \mapsto idx]) :$$
$$\text{LET } val \;\triangleq\; \text{IF } src = InitWr \text{ THEN } InitMem[req.adr]$$
$$\text{ELSE } opIdQ(src).req.val$$
$$\text{IN } \;\;\ldots$$

It asserts the existence of a value src, which will be the source of the value returned by the read; and it defines val to be that value. If the source is the initial contents of memory, then the value is obtained from $InitMem$; otherwise, it is obtained from the source request's *val* field. The inner IN expression has two conjuncts that specify the values of *regFile′* and *opQ′*. The first conjunct asserts that the register's *val* field is set to *val* and its *op* field is set to "Free", indicating that the register is made free.

$$\wedge\ regFile' = [regFile \text{ EXCEPT } ![proc][reg].val\ =\ val,$$
$$![proc][reg].op\ =\ \text{"Free"}]$$

The second conjunct of the inner IN expression describes the new value of *opQ*. Only the idx^{th} element of *opQ*[*proc*] is changed. It is set to a record whose *req* field is the same as the original request *req*, except that its *val* field is equal to *val*; whose *reg* field equals *Done*; and whose *source* field equals *src*.

$$\wedge\ opQ' = [opQ \text{ EXCEPT}$$
$$![proc][idx]\ =\ [req \quad \mapsto\ [req \text{ EXCEPT } !.val = val],$$
$$reg \quad \mapsto\ Done,$$
$$source\ \mapsto\ src]]$$

Finally, the outer IN clause ends with the conjunct

$$\wedge\ UpdateOpOrder$$

that determines the value of *opOrder′*. It also implicitly determines the possible choices of the source of the read—that is, the value of *opQ′*[*proc*][*idx*].*source*. For some choices of this value allowed by the second outer conjunct, there will be no value of *opOrder′* satisfying *UpdateOpOrder*. The conjunct *UpdateOpOrder* rules out those choices for the source.

The definitions of the other subactions *IssueRequest*, *RespondToWr*, and *Internal* of the next-state action are simpler, and I won't explain them.

Having finished the initial predicate and the next-state action, we must determine the liveness conditions. The first condition is that the memory must eventually respond to every operation. The response to a request in register *reg* of processor *proc* is produced by a *RespondToWr*(*proc*, *reg*) or *RespondToRd*(*proc*, *reg*) action. So, the obvious way to express this condition is

$$\forall\ proc \in Proc,\ reg \in Reg\ :$$
$$\text{WF}_{\langle...\rangle}(RespondToWr(proc,\ reg)\ \vee\ RespondToRd(proc,\ reg))$$

For this fairness condition to imply that the response is eventually issued, a *RespondToWr*(*proc*, *reg*) or *RespondToRd*(*proc*, *reg*) step must be enabled whenever there is an uncompleted operation in *proc*'s register *reg*. It isn't completely obvious that a *RespondToRd*(*proc*, *reg*) step is enabled when there is a read operation in the register, since the step is enabled only if there exist a source for the read and a value of *opOrder′* that satisfy *Serializable′*. The required source and value do exist because *Serializable*, which holds in the first state of the step, implies the existence of a correct total order of all the operations; this order can be used to choose a source and a relation $opOrder'$ that satisfy *Serializability'*.

The second liveness condition asserts that the memory must eventually commit to an ordering for every pair of operations. It is expressed as a fairness condition, for every pair of distinct operations oi and oj in $opId$, on an *Internal* action that makes oi either precede or follow oj in the order $opOrder'$. A first attempt at this condition is

$$
\text{(11.5)} \ \forall \, oi, \, oj \in opId :
$$
$$
(oi \neq oj) \Rightarrow \mathrm{WF}_{\langle \ldots \rangle}(\land \ \textit{Internal}
$$
$$
\land \ (\langle oi, oj \rangle \in opOrder') \lor (\langle oj, oi \rangle \in opOrder'))
$$

However, this isn't correct. In general, a formula $\forall \, x \in S : F$ is equivalent to $\forall \, x : (x \in S) \Rightarrow F$. Hence, (11.5) is equivalent to the assertion that the following formula holds, for all constant values oi and oj:

$$
(oi \in opId) \land (oj \in opId) \Rightarrow
$$
$$
\left(
\begin{array}{l}
(oi \neq oj) \Rightarrow \\
\quad \mathrm{WF}_{\langle \ldots \rangle}(\land \ \textit{Internal} \\
\qquad\qquad \land \ (\langle oi, oj \rangle \in opOrder') \lor (\langle oj, oi \rangle \in opOrder'))
\end{array}
\right)
$$

In a temporal formula, a predicate with no temporal operators is an assertion about the initial state. Hence, (11.5) asserts that the fairness condition is true for all pairs of distinct values oi and oj in the initial value of $opId$. But $opId$ is initially empty, so this condition is vacuously true. Hence, (11.5) is trivially implied by the initial predicate. We must instead assert fairness for the action

$$
\text{(11.6)} \ \land \ (oi \in opId) \land (oj \in opId)
$$
$$
\land \ \textit{Internal}
$$
$$
\land \ (\langle oi, oj \rangle \in opOrder') \lor (\langle oj, oi \rangle \in opOrder'))
$$

for all distinct values oi and oj. It suffices to assert this only for oi and oj of the right type. Since it's best to use bounded quantifiers whenever possible, let's write this condition as

$$
\forall \, oi, \, oj \in [proc : Proc, \ idx : Nat] : \quad \text{All operations are eventually ordered.}
$$
$$
(oi \neq oj) \Rightarrow \mathrm{WF}_{\langle \ldots \rangle}(\land \ (oi \in opId) \land (oj \in opId)
$$
$$
\land \ \textit{Internal}
$$
$$
\land \ (\langle oi, oj \rangle \in opOrder') \lor (\langle oj, oi \rangle \in opOrder'))
$$

For this formula to imply that any two operations are eventually ordered by $opOrder$, action (11.6) must be enabled if oi and oj are unordered operations in $opId$. It is, because *Serializable* is always enabled, so it is always possible to extend $opOrder$ to a total order of all issued operations.

The complete inner specification, with *InitMem*, *opQ*, and *opOrder* visible, is in module *InnerSerial* on pages 196–198. I have made two minor modifications to allow the specification to be checked by the TLC model checker. (Chapter 14 describes TLC and explains why these changes are needed.) Instead of the definition of *opId* given on page 189, the specification uses the equivalent definition

$$opId \;\triangleq\; \text{UNION}\; \{\, [proc : p,\; idx : \text{DOMAIN}\; opQ[p]\,] \,:\, p \in Proc\}$$

In the definition of *UpdateOpOrder*, the first conjunct is changed from

$$opOrder' \subseteq opId' \times opId'$$

to the equivalent

$$opOrder' \in \text{SUBSET}\; (opId' \times opId')$$

For TLC's benefit, I also ordered the conjuncts of all actions so *UpdateOpOrder* follows the "assignment of a value to" *opQ'*. This resulted in the UNCHANGED conjunct not being the last one in action *Internal*.

The complete specification is written, as usual, with a parametrized instantiation of *InnerSerial* to hide the constant *InitMem* and the variables *opQ* and *opOrder*:

────────────── MODULE *SerialMemory* ──────────────

EXTENDS *RegisterInterface*

$Inner(InitMem, opQ, opOrder) \;\triangleq\; \text{INSTANCE}\; InnerSerial$

$Spec \;\triangleq\; \exists\, InitMem \in [Adr \rightarrow Val] :$
$\qquad\qquad \exists\, opQ, opOrder : Inner(InitMem, opQ, opOrder)!Spec$

───

11.2.4   A Sequentially Consistent Memory

The serial memory specification does not allow the memory to predict future requests. We now remove this restriction and specify what is called a *sequentially consistent* memory. The freedom to predict the future can't be used by any real implementation,[4] so there's little practical difference between a serial and a sequentially consistent memory. However, the sequentially consistent memory has a simpler specification. This specification is surprising and instructive.

The next-state action of the sequential memory specification has the same structure as that of the serial memory specification, with actions *IssueRequest*,

---

[4]The freedom to change explanations, which a sequentially consistent memory allows, could conceivably be used to permit a more efficient implementation, but it's not easy to see how.

―――――――――――――――――――――― MODULE *InnerSerial* ――――――――――――――――――――――

EXTENDS *RegisterInterface, Naturals, Sequences, FiniteSets*

CONSTANT *InitMem*   The initial contents of memory, which will be an element of $[Proc \rightarrow Adr]$.

VARIABLE *opQ*, $opQ[p][i]$ is the ith operation issued by processor p.

  opOrder   The order of operations, which is a subset of $opId \times opId$. ($opId$ is defined below).

―――

$opId \triangleq$ UNION $\{ [proc : \{p\},\ idx : \text{DOMAIN } opQ[p]] : p \in Proc\}$ $[proc \mapsto p,\ idx \mapsto i]$ identifies
$opIdQ(oi) \triangleq opQ[oi.proc][oi.idx]$ operation i of processor p.

$InitWr \triangleq$ CHOOSE $v : v \notin [proc : Proc,\ idx : Nat]$ The source for an initial memory value.

$Done \triangleq$ CHOOSE $v : v \notin Reg$ The *reg* field value for a completed operation.

$opVal \triangleq$ Possible values of $opQ[p][i]$.

  $[req : Request,\ reg : Reg]$ An active request using register *reg*.
  $\cup\ [req : WrRequest,\ reg : \{Done\}]$ A completed write.
  $\cup\ [req : RdRequest,\ reg : \{Done\},\ source : opId \cup \{InitWr\}]$ A completed read of *source* value.

$goodSource(oi) \triangleq$
$\{InitWr\} \cup \{o \in opId : \wedge\ opIdQ(o).req.op = \text{"Wr"}$
$\wedge\ opIdQ(o).req.adr = opIdQ(oi).req.adr\}$

―――

$DataInvariant \triangleq$
  $\wedge\ RegFileTypeInvariant$ Simple type invariants for *regFile*,
  $\wedge\ opQ \in [Proc \rightarrow Seq(opVal)]$ *opQ*, and
  $\wedge\ opOrder \subseteq (opId \times opId)$ *opOrder*.

  $\wedge\ \forall\ oi \in opId :$
    $\wedge\ (\text{"source"} \in \text{DOMAIN } opIdQ(oi)) \Rightarrow$ The source of any completed read is either *InitWr*
      $\wedge\ opIdQ(oi).source \in goodSource(oi)$ or a write operation to the same address.
      $\wedge\ opIdQ(oi).req.val = \text{IF}\ opIdQ(oi).source = InitWr$ A read's value comes
        THEN $InitMem[opIdQ(oi).req.adr]$ from its source.
        ELSE $opIdQ(opIdQ(oi).source).req.val$

  $\wedge\ (opIdQ(oi).reg \neq Done) \Rightarrow$ *opQ* correctly describes the register contents.

    $(opIdQ(oi).req = regFile[oi.proc][opIdQ(oi).reg])$

  $\wedge\ \forall\ p \in Proc,\ r \in Reg :$ Only nonfree registers have corresponding *opQ* entries.

    $Cardinality(\{i \in \text{DOMAIN } opQ[p] : opQ[p][i].reg = r\}) =$
      IF $regFile[p][r].op = \text{"Free"}$ THEN 0 ELSE 1

Figure 11.6a: Module *InnerSerial* (beginning).

$Init \;\stackrel{\Delta}{=}\;$ The initial predicate.

$\quad \wedge\; regFile \in [Proc \to [Reg \to FreeRegValue]]$ Every register is free.

$\quad \wedge\; opQ = [p \in Proc \mapsto \langle\, \rangle]$ There are no operations in opQ.

$\quad \wedge\; opOrder = \{\}$ The order relation $opOrder$ is empty.

$totalOpOrder \;\stackrel{\Delta}{=}\;$ The set of all total orders on the set $opId$.

$\quad \{ R \in \textsc{subset}\,(opId \times opId) \; : $

$\qquad \wedge\; \forall\, oi,\, oj \in opId \;:\; (oi = oj) \;\vee\; (\langle oi,\, oj \rangle \in R) \;\vee\; (\langle oj,\, oi \rangle \in R)$

$\qquad \wedge\; \forall\, oi,\, oj,\, ok \in opId \;:\; (\langle oi,\, oj \rangle \in R) \wedge (\langle oj,\, ok \rangle \in R) \;\Rightarrow\; (\langle oi,\, ok \rangle \in R)$

$\qquad \wedge\; \forall\, oi \in opId \;:\; \langle oi,\, oi \rangle \notin R \, \}$

$Serializable \;\stackrel{\Delta}{=}$ Asserts that there exists a total order R of all operations that extends

$\quad \exists\, R \in totalOpOrder \;:$ $opOrder$, orders the operations of each processor correctly, and makes the

$\qquad \wedge\; opOrder \subseteq R$ source of each read the most recent write to the address.

$\qquad \wedge\; \forall\, oi,\, oj \in opId \;:\; (oi.proc = oj.proc) \wedge (oi.idx < oj.idx) \;\Rightarrow\; (\langle oi,\, oj \rangle \in R)$

$\qquad \wedge\; \forall\, oi \in opId \;:\; (\text{``source''} \in \textsc{domain}\; opIdQ(oi)) \;\Rightarrow$

$\qquad\qquad\qquad \neg\, (\exists\, oj \in goodSource(oi) \;:$

$\qquad\qquad\qquad\qquad \wedge\; \langle oj,\, oi \rangle \in R$

$\qquad\qquad\qquad\qquad \wedge\; (opIdQ(oi).source \neq InitWr) \;\Rightarrow\; (\langle opIdQ(oi).source,\, oj \rangle \in R))$

$UpdateOpOrder \;\stackrel{\Delta}{=}$ An action that chooses the new value of $opOrder$, allowing

$\quad \wedge\; opOrder' \in \textsc{subset}\,(opId' \times opId')$ it to be any relation that equals or extends the current value

$\quad \wedge\; opOrder \subseteq opOrder'$ of $opOrder$ and satisfies *Serializable*. This action is used in

   defining the subactions of the next-state action.

$\quad \wedge\; Serializable'$

$IssueRequest(proc,\, req,\, reg) \;\stackrel{\Delta}{=}$ Processor *proc* issues request *req* in register *reg*.

$\quad \wedge\; regFile[proc][reg].op = \text{``Free''}$ The register must be free.

$\quad \wedge\; regFile' = [regFile \;\textsc{except}\; ![proc][reg] = req]$ Put the request in the register.

$\quad \wedge\; opQ' = [opQ \;\textsc{except}\; ![proc] = Append(@,\, [req \mapsto req,\, reg \mapsto reg])]$ Add request to $opQ[proc]$.

$\quad \wedge\; UpdateOpOrder$

$RespondToWr(proc,\, reg) \;\stackrel{\Delta}{=}$ The memory responds to a write request in processor *proc*'s register *reg*.

$\quad \wedge\; regFile[proc][reg].op = \text{``Wr''}$ The register must contain an active write request.

$\quad \wedge\; regFile' = [regFile \;\textsc{except}\; ![proc][reg].op = \text{``Free''}]$ The register is freed.

$\quad \wedge\; \textsc{let}\; idx \;\stackrel{\Delta}{=}\; \textsc{choose}\; i \in \textsc{domain}\; opQ[proc] \;:\; opQ[proc][i].reg = reg$ The appropriate opQ

$\qquad \textsc{in}\quad opQ' = [opQ \;\textsc{except}\; ![proc][idx].reg = Done]$ entry is updated.

$\quad \wedge\; UpdateOpOrder$ $opOrder$ is updated.

Figure 11.6b: Module *InnerSerial* (middle).

$RespondToRd(proc,\ reg)\ \triangleq$ The memory responds to a read request in processor $proc$'s register reg.

> LET $req\ \triangleq\ regFile[proc][reg]$ $proc$'s register reg contains the request req, which is in $opQ[proc][idx]$.
>
> $idx\ \triangleq\ $ CHOOSE $i \in$ DOMAIN $opQ[proc]\ :\ opQ[proc][i].reg = reg$
>
> IN $\wedge\ req.op =$ "Rd"   The register must contain an active read request.
>
> $\wedge\ \exists\, src \in goodSource([proc \mapsto proc, idx \mapsto idx])\ :$ The read obtains its value from a source src.
>
>> LET $val\ \triangleq\ $ IF $src = InitWr$ THEN $InitMem[req.adr]$ The value returned by
>>   ELSE $opIdQ(src).req.val$ the read.
>>
>> IN $\wedge\ regFile' = [regFile$ EXCEPT $![proc][reg].val\ =\ val,$ Set register's val field,
>> $![proc][reg].op\ =\ $ "Free"]   and free the register.
>>
>> $\wedge\ opQ' = [opQ$ EXCEPT $opQ[proc][idx]$ is updated appropriately.
>> $![proc][idx] = [req\qquad \mapsto\ [req$ EXCEPT $!.val = val],$
>> $reg\qquad \mapsto\ Done,$
>> $source\ \mapsto\ src]\,]$
>
> $\wedge\ UpdateOpOrder$ $opOrder$ is updated.

$Internal\ \triangleq\ \wedge$ UNCHANGED $\langle regFile, opQ \rangle$
$\qquad\qquad\quad\ \wedge\ UpdateOpOrder$

$Next\ \triangleq$ The next-state action.
$\quad \vee\ \exists\, proc \in Proc,\ reg \in Reg\ :\ \vee\ \exists\, req \in Request\ :\ IssueRequest(proc,\ req,\ reg)$
$\qquad\qquad\qquad\qquad\qquad\qquad\qquad \vee\ RespondToRd(proc,\ reg)$
$\qquad\qquad\qquad\qquad\qquad\qquad\qquad \vee\ RespondToWr(proc,\ reg)$
$\quad \vee\ Internal$

$Spec\ \triangleq$ The complete internal specification.

$\quad \wedge\ Init$

$\quad \wedge\ \Box[Next]_{\langle regFile,\ opQ,\ opOrder \rangle}$

$\quad \wedge\ \forall\, proc \in Proc,\ reg \in Reg\ :$ The memory eventually responds to every request.
$\qquad \text{WF}_{\langle regFile,\ opQ,\ opOrder \rangle}(RespondToWr(proc,\ reg) \vee RespondToRd(proc,\ reg))$

$\quad \wedge\ \forall\, oi,\ oj \in [proc : Proc,\ idx : Nat]\ :$ All operations are eventually ordered.
$\qquad (oi \neq oj) \Rightarrow \text{WF}_{\langle regFile,\ opQ,\ opOrder \rangle}(\wedge\ (oi \in opId) \wedge (oj \in opId)$
$\qquad\qquad\qquad\qquad\qquad\qquad\qquad\qquad \wedge\ Internal$
$\qquad\qquad\qquad\qquad\qquad\qquad\qquad\qquad \wedge\ (\langle oi, oj \rangle \in opOrder') \vee (\langle oj, oi \rangle \in opOrder'))$

THEOREM $Spec \Rightarrow \Box(DataInvariant \wedge Serializable)$

Figure 11.6c: Module $InnerSerial$ (end).

RespondToRd, *RespondToWr*, and *Internal*. Like the serial memory specification, it has an internal variable *opQ* to which the *IssueRequest* operation appends an entry with *req* (request) and *reg* (register) fields. However, an operation does not remain forever in *opQ*. Instead, an *Internal* step removes it after it has been completed. The specification has a second internal variable *mem* that represents the contents of a memory—that is, the value of *mem* is a function from *Adr* to *Val*. The value of *mem* is changed only by an *Internal* action that removes a write from *opQ*.

Recall that the correctness condition has two requirements:

1. There is a sequential execution order of all the operations that explains the values returned by reads.

2. This execution order is consistent with the order in which operations are issued by each individual processor.

The order in which operations are removed from *opQ* is an explanatory execution order that satisfies requirement 1 if the *Internal* action satisfies these properties:

- When a write of value *val* to address *adr* is removed from *opQ*, the value of *mem*[*adr*] is set to *val*.

- A read of address *adr* that returned a value *val* can be removed from *opQ* only if *mem*[*adr*] = *val*.

Requirement 2 is satisfied if operations issued by processor *p* are appended by the *IssueRequest* action to the tail of *opQ*[*p*], and are removed by the *Internal* action only from the head of *opQ*[*p*].

We have now determined what the *IssueRequest* and *Internal* actions should do. The *RespondToWr* action is obvious; it's essentially the same as in the serial memory specification. The problem is the *RespondToRd* action. How can we define it so that the value returned by a read is one that *mem* will contain when the *Internal* action has to remove the read from *opQ*? The answer is surprisingly simple: we allow the read to return any value. If the read were to return a bad value—for example, one that is never written—then the *Internal* action would never be able to remove the read from *opQ*. We rule out that possibility with a liveness condition requiring that every operation in *opQ* eventually be removed. This makes it easy to write the *Internal* action. The only remaining problem is expressing the liveness condition.

To guarantee that every operation is eventually removed from *opQ*, it suffices to guarantee that, for every processor *proc*, the operation at the head of *opQ*[*proc*] is eventually removed. The desired liveness condition can therefore be expressed as

$$\forall \, proc \in Proc \; : \; \mathrm{WF}_{\langle \ldots \rangle}(RemoveOp(proc))$$

where $RemoveOp(proc)$ is an action that unconditionally removes the operation from the head of $opQ[proc]$. For convenience, we let the $RemoveOp(proc)$ action also update *mem*. We then define a separate action $Internal(proc)$ for each processor *proc*. It conjoins to $RemoveOp(proc)$ the following enabling condition, which asserts that if the operation being removed is a read, then it has returned the correct value:

$$(Head(opQ[proc]).req.op \; = \; \text{``Rd''}) \Rightarrow$$
$$(mem[Head(opQ[proc]).req.adr] \; = \; Head(opQ[proc]).req.val)$$

The complete internal specification, with the variables opQ and *mem* visible, appears in module *InnerSequential* on the following two pages. At this point, you should have no trouble understanding it. You should also have no trouble writing a module that instantiates *InnerSequential* and hides the internal variables opQ and *mem* to produce the final specification, so I won't bother doing it for you.

11.2.5   The Memory Specifications Considered

Almost every specification we write admits a direct implementation, based on the initial predicate and next-state action. Such an implementation may be completely impractical, but it is theoretically possible. It's easy to implement the linearizable memory with a single central memory. A direct implementation of the serial memory would require maintaining queues of all operations issued thus far, and a computationally infeasible search for possible total orderings. But, in theory, it's easy.

Our specification of a sequentially consistent memory cannot be implemented directly. A direct implementation would have to guess the correct value to return on a read, which is impossible. The specification is not directly implementable because it is not machine closed. As explained in Section 8.9.2 on page 111, a non-machine-closed specification is one in which a direct implementation can "paint itself into a corner," reaching a point at which it is no longer possible to satisfy the specification. Any finite scenario of memory operations can be produced by a behavior satisfying the sequentially consistent memory's initial predicate and next-state action—namely, a behavior that contains no *Internal* steps. However, not every finite scenario can be extended to one that is explainable by a sequential execution. For example, no scenario that begins as follows is possible in a two-processor system:

This notation for describing scenarios was introduced on page 186.

> Processor p: $Wr_p(a1, v1)$, $Rd_p(a1, v2)$, $Wr_p(a2, v2)$
> Processor q: $Wr_q(a2, v1)$, $Rd_q(a2, v2)$, $Wr_q(a1, v2)$

Here's why:

$$\text{— MODULE } InnerSequential \text{ —}$$

EXTENDS *RegisterInterface, Naturals, Sequences, FiniteSets*

VARIABLE opQ, $opQ[p][i]$ is the ith operation issued by processor p.

$\quad\quad mem$ An internal memory.

$Done \;\triangleq\;$ CHOOSE $v : v \notin Reg$ The *reg* field value for a completed operation.

$DataInvariant \;\triangleq\;$
$\quad \wedge\; RegFileTypeInvariant$ Simple type invariants for *regFile*,
$\quad \wedge\; opQ \in [Proc \to Seq([req : Request, reg : Reg \cup \{Done\}])]$ *opQ*, and
$\quad \wedge\; mem \in [Adr \to Val]$ *mem*.
$\quad \wedge\; \forall\, p \in Proc,\; r \in Reg :$ Only nonfree registers have corresponding *opQ* entries.
$\qquad Cardinality(\{i \in \text{DOMAIN } opQ[p] : opQ[p][i].reg = r\}) \;=\;$
$\qquad\quad$ IF $regFile[p][r].op =$ "Free" THEN 0 ELSE 1

$Init \;\triangleq\;$ The initial predicate.
$\quad \wedge\; regFile \in [Proc \to [Reg \to FreeRegValue]]$ Every register is free.
$\quad \wedge\; opQ = [p \in Proc \mapsto \langle\,\rangle]$ There are no operations in *opQ*.
$\quad \wedge\; mem \in [Adr \to Val]$ The internal memory can have any initial contents.

$IssueRequest(proc, req, reg) \;\triangleq\;$ Processor *proc* issues request *req* in register *reg*.
$\quad \wedge\; regFile[proc][reg].op =$ "Free" The register must be free.
$\quad \wedge\; regFile' = [regFile \text{ EXCEPT } ![proc][reg] = req]$ Put request in register.
$\quad \wedge\; opQ' = [opQ \text{ EXCEPT } ![proc] = Append(@, [req \mapsto req, reg \mapsto reg])]$ Add request to *opQ[proc]*.
$\quad \wedge\;$ UNCHANGED mem

$RespondToRd(proc, reg) \;\triangleq\;$ The memory responds to a read request in processor *proc*'s register *reg*.
$\quad \wedge\; regFile[proc][reg].op =$ "Rd" The register must contain an active read request.
$\quad \wedge\; \exists\, val \in Val :$ *val* is the value returned.
$\qquad \wedge\; regFile' = [regFile \text{ EXCEPT } ![proc][reg].val = val,$ Set the register's *val* field,
$\qquad\qquad\qquad\qquad\qquad\quad ![proc][reg].op \;=\;$ "Free"] and free the register.
$\qquad \wedge\; opQ' =$ LET $idx \;\triangleq\;$ *opQ[proc][idx]* contains the request in register *reg*.
$\qquad\qquad\qquad\qquad$ CHOOSE $i \in \text{DOMAIN } opQ[proc] : opQ[proc][i].reg = reg$
$\qquad\qquad\quad$ IN $\quad [opQ \text{ EXCEPT } ![proc][idx].req.val \;= val,$ Set *opQ[proc][idx]*'s *val* field to
$\qquad\qquad\qquad\qquad\qquad\quad ![proc][idx].reg \qquad = Done]$ *val* and its *reg* field to *Done*.
$\quad \wedge\;$ UNCHANGED mem

Figure 11.7a: Module *InnerSequential* (beginning).

$RespondToWr(proc, reg) \;\triangleq\;$ The memory responds to a write request in processor $proc$'s register reg.

$\;\;\;\wedge\; regFile[proc][reg].op = \text{``Wr''}$ The register must contain an active write request.

$\;\;\;\wedge\; regFile' = [regFile \text{ EXCEPT } ![proc][reg].op = \text{``Free''}]$ Free the register.

$\;\;\;\wedge\; \text{LET } idx \;\triangleq\; \text{CHOOSE } i \in \text{DOMAIN } opQ[proc] : opQ[proc][i].reg = reg$ Update the appropriate opQ entry.
$\;\;\;\;\;\;\;\; \text{IN}\;\;\;\; opQ' = [opQ \text{ EXCEPT } ![proc][idx].reg = Done]$

$\;\;\;\wedge\; \text{UNCHANGED } mem$

$RemoveOp(proc) \;\triangleq\;$ Unconditionally remove the operation at the head of $opQ[proc]$ and update mem.

$\;\;\;\wedge\; opQ[proc] \neq \langle\rangle$ $opQ[proc]$ must be nonempty.

$\;\;\;\wedge\; Head(opQ[proc]).reg = Done$ The operation must have been completed.

$\;\;\;\wedge\; mem' = \text{IF } Head(opQ[proc]).req.op = \text{``Rd''}$ Leave mem unchanged for a read operation, update it for a write operation.
$\;\;\;\;\;\;\;\;\;\;\;\;\;\;\;\; \text{THEN } mem$
$\;\;\;\;\;\;\;\;\;\;\;\;\;\;\;\; \text{ELSE } [mem \text{ EXCEPT } ![Head(opQ[proc]).req.adr] =$
$\; Head(opQ[proc]).req.val]$

$\;\;\;\wedge\; opQ' = [opQ \text{ EXCEPT } ![proc] = Tail(@)]$ Remove the operation from $opQ[proc]$.

$\;\;\;\wedge\; \text{UNCHANGED } regFile$ No register is changed.

$Internal(proc) \;\triangleq\;$ Remove the operation at the head of $opQ[proc]$. But if it's a read, only do so if it returned the value now in mem.
$\;\;\;\wedge\; RemoveOp(proc)$
$\;\;\;\wedge\; (Head(opQ[proc]).req.op = \text{``Rd''}) \Rightarrow$
$\;\;\;\;\;\;\;\; (mem[Head(opQ[proc]).req.adr] = Head(opQ[proc]).req.val)$

$Next \;\triangleq\;$ The next-state action.
$\;\;\;\exists\, proc \in Proc : \vee\; \exists\, reg \in Reg : \vee\; \exists\, req \in Request : IssueRequest(proc, req, reg)$
$\; \vee\; RespondToRd(proc, reg)$
$\; \vee\; RespondToWr(proc, reg)$
$\; \vee\; Internal(proc)$

$Spec \;\triangleq\; \wedge\; Init$
$\;\;\;\;\;\;\;\;\;\;\;\; \wedge\; \Box[Next]_{\langle regFile,\, opQ,\, mem\rangle}$
$\;\;\;\;\;\;\;\;\;\;\;\; \wedge\; \forall\, proc \in Proc,\, reg \in Reg :$ The memory eventually responds to every request.
$\;\;\;\;\;\;\;\;\;\;\;\;\;\;\;\;\;\;\; \text{WF}_{\langle regFile,\, opQ,\, mem\rangle}(RespondToWr(proc, reg) \vee RespondToRd(proc, reg))$
$\;\;\;\;\;\;\;\;\;\;\;\; \wedge\; \forall\, proc \in Proc :$ Every operation is eventually removed from opQ.
$\;\;\;\;\;\;\;\;\;\;\;\;\;\;\;\;\;\;\; \text{WF}_{\langle regFile,\, opQ,\, mem\rangle}(RemoveOp(proc))$

THEOREM $Spec \Rightarrow \Box\, DataInvariant$

Figure 11.7b: Module *InnerSequential* (end).

$Wr_q(a1, v2)$

precedes $Rd_p(a1, v2)$	This is the only explanation of the value read by p.
precedes $Wr_p(a2, v2)$	By the order in which operations are issued.
precedes $Rd_q(a2, v2)$	This is the only explanation of the value read by q.
precedes $Wr_q(a1, v2)$	By the order in which operations are issued.

Hence q's write of $a1$ must precede itself, which is impossible.

As mentioned in Section 8.9.2, a specification is machine closed if the liveness property is the conjunction of fairness properties for actions that imply the next-state action. The sequential memory specification asserts weak fairness of *RemoveOp(proc)*, for processors *proc*, and *RemoveOp(proc)* does not imply the next-state action. (The next-state action does not allow a *RemoveOp(proc)* step that removes from *opQ[proc]* a read that has returned the wrong value.)

Very high-level system specifications, such as our memory specifications, are subtle. It's easy to get them wrong. The approach we used in the serial memory specification—namely, writing conditions on the history of all operations—is dangerous. It's easy to forget some conditions. A non-machine-closed specification can occasionally be the simplest way to express what you want so say.

Part III

The Tools

Chapter 12

The Syntactic Analyzer

The Syntactic Analyzer is a Java program, written by Jean-Charles Grégoire and David Jefferson, that parses a TLA$^+$ specification and checks it for errors. The analyzer also serves as a front end for other tools, such as TLC (see Chapter 14). It is available from the TLA Web page.

You will probably run the analyzer by typing the command

   program_name   option   spec_file

where

program_name depends on your particular system. It might be

```
java tlasany.SANY
```

spec_file is the name of the file containing the TLA$^+$ specification. Each module named *M* that appears in the specification (except for submodules) must be in a separate file named *M*.tla. The extension .tla may be omitted from *spec_file*.

option is either empty or consists of one of the following two options:

-s Causes the analyzer to check only for syntactic errors and not for semantic errors. (These two classes of error are explained below.) You can use this option to find syntax errors when you begin writing a specification.

-d Causes the analyzer to enter debugging mode after checking the specification. In this mode, you can examine the specification's structure—for example, finding out where it thinks a particular identifier is defined or declared. The documentation that comes with the analyzer explains how to do this.

The rest of this brief chapter provides some hints for what to do when the Syntactic Analyzer reports an error.

The errors that the analyzer detects fall into two separate classes, which are usually called *syntactic* and *semantic* errors. A syntactic error is one that makes the specification grammatically incorrect, meaning that it violates either the BNF grammar or the precedence and alignment rules, described in Chapter 15. A semantic error is one that violates the legality conditions mentioned in Chapter 17. The term *semantic error* is misleading, because it suggests an error that makes a specification have the wrong meaning. All errors found by the analyzer are ones that make the specification illegal—that is, not syntactically well-formed—and hence make it have no meaning at all.

The analyzer reads the file sequentially, starting from the beginning, and it reports a syntax error if and when it reaches a point at which it becomes impossible for any continuation to produce a grammatically correct specification. For example, if we omitted the colon after $\exists\, req \in MReq$ in the definition of *Req* from module *InternalMemory* on page 52, we would get

$$
\begin{aligned}
Req(p) \;\triangleq\; &\land\; ctl[p] = \text{``rdy''} \\
&\land\; \exists\, req \in MReq \;\land\; Send(p,\, req,\, memInt,\, memInt') \\
&\qquad\quad \land\; buf' = [buf \;\text{EXCEPT}\; ![p] = req] \\
&\qquad\quad \land\; ctl' = [ctl \;\text{EXCEPT}\; ![p] = \text{``busy''}] \\
&\land\; \text{UNCHANGED}\; mem
\end{aligned}
$$

This would cause the analyzer to print something like

```
***Parse Error***
Encountered "/\" at line 19, column 11
```

Line 19, column 11 is the position of the $\land$ that begins the last line of the definition (right before the UNCHANGED). Until then, the analyzer thought it was parsing a quantified expression that began

$$\exists\, req \in (MReq \;\land\; Send(p,\, req,\, memInt,\, memInt') \;\land\; buf' = \ldots$$

(Such an expression is silly, having the form $\exists\, req \in p : \ldots$ where p is a Boolean, but it's legal.) The analyzer was interpreting each of these $\land$ symbols as an infix operator. However, interpreting the last $\land$ of this definition (at line 19, column 11) as an infix operator would violate the alignment rules for the outer conjunction list, so the analyzer reported an error.

As this example suggests, the analyzer may discover a syntax error far past the actual mistake. To help you locate the problem, it prints out a trace of where it was in the parse tree when it found the error. For this example, it prints

```
Residual stack trace follows:
Quantified form starting at line 16, column 14.
Junction Item starting at line 16, column 11.
AND-OR Junction starting at line 15, column 11.
Definition starting at line 15, column 1.
Module body starting at line 3, column 1.
```

If you can't find the source of an error, try the "divide and conquer" method: keep removing different parts of the module until you isolate the source of the problem.

Semantic errors are usually easy to find because the analyzer can locate them precisely. A typical semantic error is an undefined symbol that arises because you mistype an identifier. If, instead of leaving out the colon in the definition of $Req(p)$, we had left out the e in $MReq$, the analyzer would have reported

```
line 16, col 26 to line 16, col 28 of module InternalMemory
```

```
Could not resolve name 'MRq'.
```

The analyzer stops when it encounters the first syntactic error. It can detect multiple semantic errors in a single run.

Chapter 13

The TLATeX Typesetter

TLATeX is a Java program for typesetting TLA$^+$ modules, based on ideas by Dmitri Samborski. It can be obtained through the TLA Web page.

13.1 Introduction

TLATeX calls the LaTeX program to do the actual typesetting. LaTeX is a document-production system based on Donald Knuth's TeX typesetting program.[1] LaTeX normally produces as its output a *dvi file*—a file with extension `dvi` containing a device-independent description of the typeset output. TLATeX has options that allow it to call another program to translate the dvi file into a PostScript or PDF file. Some versions of LaTeX produce a PDF file directly.

You must have LaTeX installed on your computer to run TLATeX. LaTeX is public-domain software that can be downloaded from the Web; proprietary versions are also available. The TLA Web page points to the TLATeX Web page, which contains information about obtaining LaTeX and a PostScript or PDF converter.

You will probably run TLATeX by typing

```
java tlatex.TLA [options] fileName
```

where *fileName* is the name of the input file, and [*options*] is an optional sequence of options, each option name preceded by "-". Some options are followed by an argument, a multi-word argument being enclosed in double-quotes. If *fileName* does not contain an extension, then the input file is *fileName.tla*. For example, the command

---

[1] LaTeX is described in *LaTeX: A Document Preparation System, Second Edition*, by Leslie Lamport, published by Addison-Wesley, Reading, Massachusetts, 1994. TeX is described in *The TeXbook* by Donald E. Knuth, published by Addison-Wesley, Reading, Massachusetts, 1986.

```
java tlatex.TLA  -ptSize 12  -shade  MySpec
```

typesets the module in the file *MySpec.tla* using the *ptSize* option with argument 12 and the *shade* option. The input file must contain a complete TLA$^+$ module. Running TLATEX with the *help* option produces a list of all options. Running it with the *info* option produces most of the information contained in this chapter. (The *fileName* argument can be omitted when using the *help* or *info* option.)

All you probably need to know about using TLATEX is

- TLATEX can shade comments, as explained in the next section.

- The next section also explains how to get TLATEX to produce a PostScript or PDF file.

- The *number* option causes TLATEX to print line numbers in the left margin.

- You should use the *latexCommand* option if you run LATEX on your system by typing something other than `latex`. For example, if you run LATEX on file *f.tex* by typing

  ```
  locallatex  f.tex
  ```

  then you should run TLATEX by typing something like

  ```
  java tlatex.TLA -latexCommand locallatex  fileName
  ```

- If you happen to use any of these three two-character sequences in a comment:

  ```
  '~     '^     '.
  ```

  then you'd better read Section 13.4 on page 214 to learn about how TLATEX formats comments.

TLATEX's output should be good enough for most purposes. The following sections describe how you can get TLATEX to do a better job, and what to do in the unlikely case that it produces weird output.

13.2   Comment Shading

The *shade* option causes TLATEX to typeset comments in shaded boxes. A specification generally looks best when comments are shaded, as they are in this book. However, shading is not supported by some programs for viewing and printing dvi files. Hence, it may be necessary to create a PostScript or PDF file from the dvi file to view a specification with shaded comments. Here are all the options relevant to shading.

`-grayLevel` *num*

Determines the darkness of the shading, where *num* is a number between 0 and 1. The value 0 means completely black, and 1 means white; the default value is .85. The actual degree of shading depends on the output device and can vary from printer to printer and from screen to screen. You will have to experiment to find the right value for your system.

`-ps`
`-nops`

These options tell TLATEX to create or not to create a PostScript or PDF output file. The default is to create one if the *shade* option is specified, and otherwise not to.

`-psCommand` *cmd*

This is the command run by TLATEX to produce the PostScript or PDF output file. Its default value is `dvips`. TLATEX calls the operating system with the command

    cmd dviFile

where *dviFile* is the name of the dvi file produced by running LATEX. If a more sophisticated command is needed, you may want to use the *nops* option and run a separate program to create the PostScript or PDF file.

13.3   How It Typesets the Specification

TLATEX should typeset the specification itself pretty much the way you would want it to. It preserves most of the meaningful alignments in the specification—for example:

Input	**Output**

```
Action == /\ x'   = x - y
          /\ yy'  = 123
          /\ zzz' = zzz
```

$$Action \triangleq \land\ x'\ \ = x - y$$
$$\land\ yy'\ = 123$$
$$\land\ zzz' = zzz$$

Observe how the $\land$ and $=$ symbols are aligned in the output. Extra spaces in the input will be reflected in the output. However, TLATEX treats no space and one space between symbols the same:

Input	**Output**
`x+y`	$x + y$
`x + y`	$x + y$
`x   +   y`	$x\ +\ y$

TLATEX typesets the single TLA$^+$ module that must appear in the input file. It will also typeset any material that precedes and follows the module as if it were a comment. (However, that text won't be shaded.) The *noProlog* and *noEpilog* options suppress typesetting of material that precedes and follows the module, respectively.

TLATEX does not check that the specification is syntactically correct TLA$^+$ input. However, it will report an error if the specification contains an illegal lexeme, such as ";".

13.4   How It Typesets Comments

TLATEX distinguishes between one-line and multi-line comments. A one-line comment is any comment that is not a multi-line comment. Multi-line comments can be typed in any of the following three styles:

```
(*************)        \*************        (* This
(* This is    *)       \* This is            is a
(* a comment. *)       \* a comment.         comment. *)
(*************)        \*************
```

In the first two styles, the (* or * characters on the left must all be aligned, and the last line (containing the comment **···**) is optional. In the first style, nothing may appear to the right of the comment—otherwise, the input is considered to be a sequence of separate one-line comments. The third style works best when nothing appears on the same line to the left of the (* or to the right of the *).

TLATEX tries to do a sensible job of typesetting comments. In a multi-line comment, it usually considers a sequence of non-blank lines to be a single paragraph, in which case it typesets them as one paragraph and ignores line breaks in the input. But it does try to recognize tables and other kinds of multi-line formatting when deciding where to break lines. You can help it as follows:

- End each sentence with a period (".").

- Add blank lines to indicate the logical separation of items.

- Left-align the lines of each paragraph.

Below are some common ways in which TLATEX can mess up the typesetting of comments, and what you can do about it.

TLATEX can confuse parts of a specification with ordinary text. For example, identifiers should be italicized, and the minus in the expression $x - y$ should be typeset differently from the dash in *x-ray*. TLATEX gets this right most of the time, but it does make mistakes. You can tell TLATEX to treat something as part of a specification by putting single quotes (' and ') around it. You can tell it to treat something as ordinary text by putting '^ and ^' around it. For example:

Input

```
\*****************************
\* A better value of 'bar' is
\* now in '^http://foo/bar^'.
\*****************************
```

Output

A better value of *bar* is now in http://foo/bar.

But this is seldom necessary; TLATEX usually does the right thing.

Warning: Do not put any character between ' ^ and ^' except letters, numbers, and ordinary punctuation—unless you know what you're doing. In particular, the following characters have special meaning to LATEX and can have unexpected effects if used between ' ^ and ^':

$$ \_ \quad \texttt{\textasciitilde} \quad \# \quad \$ \quad \% \quad \texttt{\textasciicircum} \quad \& \quad < \quad > \quad \backslash \quad " \quad | \quad \{ \quad \} $$

See Section 13.8 on page 219 for further information about what can go between ' ^ and ^'.

TLATEX isn't very good at copying the way paragraphs are formatted in a comment. For example, note how it fails to align the two *A*s in

Input

```
\**********************
\* gnat: A tiny insect.
\*
\* gnu:  A short word.
\**********************
```

Output

gnat: A tiny insect.

gnu: A short word.

You can tell TLATEX to typeset a sequence of lines precisely the way they appear in the input, using a fixed-width font, by enclosing the lines with '. and .' , as in

Input

```
\**********************
\* This explains it all:
\*
\*    '. ---        ---
\*       | P |--->| M |
\*        ---      ---  .'
\**********************
```

Output

This explains it all:

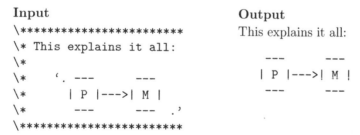

Using '. and .' is the only reasonable thing to do for a diagram. However, if you know (or want to learn) LATEX, Section 13.8 below on using LATEX commands in comments will explain how you can get TLATEX to do a good job of formatting things like lists and tables.

TLATEX will occasionally typeset a paragraph very loosely, with one or more lines containing lots of space between the words. This happens if there is no good way to typeset the paragraph. If it bothers you, the easiest solution is to rewrite the paragraph. You can also try to fix the problem with LATEX commands. (See Section 13.8 below.)

TLATEX usually handles pairs of double-quote characters (") the way it should:

Input

```
\********************
\* The string "ok" is
\* a "good" value.
\********************
```

Output

The string "ok" is a "good" value.

However, if it gets confused, you can use single quotes to identify string values and `` ` `` and `''` to produce the left and right double-quotes of ordinary text:

Input

```
\***********************
\* He asks ``Is `"good"'
\*  bad?''
\***********************
```

Output

He asks "Is "good" bad?"

TLATEX ignores any (* ... *) comment that appears within another comment. So, you can get it not to typeset part of a comment by enclosing that part between (* and *). But a better way to omit part of a comment is to enclose it between `` `~ `` and `~'`:

Input

```
\********************
\* x+y is always `~I
\* hope~' positive.
\********************
```

Output

$x + y$ is always positive.

13.5   Adjusting the Output Format

The following options allow you to adjust the font size, the dimensions of the printed area, and the position of the text on the page:

-ptSize *num*

> Specifies the size of the font. Legal values of *num* are 10, 11, or 12, which cause the specification to be typeset in a 10-, 11-, or 12-point font. The default value is 10.

-textwidth *num*
-textheight *num*

> The value of *num* specifies the width and height of the typeset output, in points. A point is 1/72 of an inch, or about 1/3 mm.

`-hoffset` *num*
`-voffset` *num*

> The value of *num* specifies the distance, in points, by which the text should be moved horizontally or vertically on the page. Exactly where on a page the text appears depends on the printer or screen-display program. You may have to adjust this value to get the output to appear centered on the printed page, or for the entire output to be visible when viewed on the screen.

13.6 Output Files

TLATEX itself writes either two or three files, depending on the options. The names of these files are normally determined from the name of the input file. However, options allow you to specify the name of each of these files. TLATEX also runs the separate LATEX program and possibly a program to produce a PostScript or PDF file. These programs produce additional files. Below are the file-related options. In their descriptions, the root of a file name is the name with any extension or path specifier removed; for example, the root of `c:\foo\bar.tla` is `bar`. All file names are interpreted relative to the directory in which TLATEX is run.

`-out` *fileName*

> If *f* is the root of *fileName*, then *f.tex* is the name of the LATEX input file that TLATEX writes to produce the final output. TLATEX then runs LATEX with *f.tex* as input, producing the following files:

> *f.dvi* The dvi output file.

> *f.log* A log file, containing LATEX's messages. In this file, an *overfull hbox* warning means that a specification line is too wide and extends into the right margin, and an *underfull hbox* warning means that LATEX could find no good line breaks in a comment paragraph. Unfortunately, the line numbers in the file refer to the *f.tex* file, not to the specification. But by examining the *f.tex* file, you can probably figure out where the corresponding part of the specification is.

> *f.aux* A LATEX auxiliary file that is of no interest.

> The default *out* file name is the root of the input file name.

`-alignOut` *fileName*

> This specifies the root name of the LATEX alignment file TLATEX writes—a file described in Section 13.7 below on trouble-shooting. If *f* is the root of *fileName*, then the alignment file is named *f.tex*, and running LATEX on it produces the files *f.dvi*, *f.log*, and *f.aux*. Only the *f.log* file is of interest. If

the *alignOut* option is not specified, the alignment file is given the same name as the *out* file. This option is used only for trouble-shooting, as described in the section below.

-tlaOut *fileName*

This option causes TLATEX to write to *fileName* a file that is almost the same as the input file. (The extension *tla* is added to *fileName* if it has no extension.) The *tlaOut* file differs from the input in that any portion of a comment enclosed by '^ and ^' is removed, and every occurrence of the two-character strings

      '~   ~'   '.   .'

is replaced by two blanks. As explained in Section 13.8 below, the *tlaOut* option allows you to maintain a version of the specification that is readable in ASCII while using LATEX commands to provide high-quality typesetting of comments. The default is not to write a *tlaOut* file.

-style *fileName*

This option is of interest only to LATEX users. Normally, TLATEX inserts a copy of the *tlatex* package file in the LATEX input files that it writes. The *style* option causes it instead to insert a \usepackage command to read the LATEX package named *fileName*. (LATEX package files have the extension *sty*. That extension is added to *fileName* if it's not already there.) The TLATEX style defines a number of special commands that are written by TLATEX in its LATEX input files. The package file specified by the *style* option must also define those commands. Any package file should therefore be created by modifying the standard *tlatex* package, which is the file *tlatex.sty* in the same directory as TLATEX's Java program files. You might want to create a new package to change the way TLATEX formats the specification, or to define additional commands for use in '^...^' text in comments.

13.7   Trouble-Shooting

TLATEX's error messages should be self-explanatory. However, it calls upon the operating system up to three times to execute other programs:

- It runs LATEX on the *alignOut* file that it wrote.

- It runs LATEX on the *out* file that it wrote.

- It may execute the *psCommand* to create the PostScript or PDF output file.

After each of the last two executions, TLAT_EX writes a message asserting that the appropriate output file was written. It might lie. Any of those executions might fail, possibly causing no output file to be written. Such a failure can even cause the operating system never to return control to TLAT_EX, so TLAT_EX never terminates. This type of failure is the likely problem if TLAT_EX does not produce a dvi file or a PostScript/PDF file, or if it never terminates. In that case, you should try rerunning TLAT_EX using the *alignOut* option to write a separate alignment file. Reading the two log files that LAT_EX produces, or any error file produced by executing *psCommand*, may shed light on the problem.

Normally, the LAT_EX input files written by TLAT_EX should not produce any LAT_EX errors. However, incorrect LAT_EX commands introduced in '`^`...`^`' regions can cause LAT_EX to fail.

13.8   Using LAT_EX Commands

TLAT_EX puts any text enclosed between '`^` and `^`' in a comment into the LAT_EX input file exactly as it appears. This allows you to insert LAT_EX formatting commands in comments. There are two ways to use this.

- You can enclose between '`^` and `^`' a short phrase appearing on a single line of input. LAT_EX typesets that phrase as part of the enclosing paragraph.

- You can enclose one or more complete lines of a multi-line comment between '`^` and `^`'. That text is typeset as one or more separate paragraphs whose prevailing left margin is determined by the position of the '`^`, as show here:

Input	Output
| ```
\*********************
\* The first paragraph.
\*
\*    The 2nd paragraph.
\*
\*    '^ Text formatted
\* by  \LaTeX. ^'
\*********************
``` | The first paragraph.<br><br>   The 2nd paragraph.<br><br>Text formatted by LAT<sub>E</sub>X. |

LAT_EX typesets the text between '`^` and `^`' in LR mode for a one-line comment and in paragraph mode for a multi-line comment. The LAT_EX file produced by TLAT_EX defines a `describe` environment that is useful for formatting text in a multi-line '`^`...`^`' region. This environment is the same as the standard LAT_EX `description` environment, except that it takes an argument, which should be the widest item label in the environment:
```

Input

```
\**************************
\* `^\begin{describe}{gnat:}
\*   \item[gnat:] Tiny insect.
\*   \item[gnu:] Short word.
\*   \end{describe} ^'
\**************************
```

Output

gnat: Tiny insect.

gnu: Short word.

As this example shows, putting LATEX commands in comments makes the comments in the input file rather unreadable. You can maintain both a typeset and an ASCII-readable version of the specification by enclosing text that should appear only in the ASCII version between '~ and ~'. You can then accompany each '^...^' region with its ASCII version enclosed by '~ and ~'. For example, the input file could contain

```
\************************************
\* `^ \begin{describe}{gnat:}
\*        \item[gnat:] A tiny insect.
\*        \item[gnu:]  A short word.
\*      \end{describe} ^'
\* `~ gnat: A tiny insect.
\*
\*      gnu:  A short word. ~'
\************************************
```

The *tlaOut* option causes TLATEX to write a version of the original specification with '^...^' regions deleted, and with '~ and ~' strings replaced by spaces. (The strings '. and .' are also replaced by spaces.) In the example above, the tlaOut file would contain the comment

```
\************************************
\*
\*      gnat: A tiny insect.
\*
\*      gnu:  A short word.
\************************************
```

The blank line at the top was produced by the end-of-line character that follows the ^'.

Warning: An error in a LATEX command inside '^...^' text can cause TLATEX not to produce any output. See Section 13.7 above on trouble-shooting.

Chapter 14

The TLC Model Checker

TLC is a program for finding errors in TLA$^+$ specifications. It was designed and implemented by Yuan Yu, with help from Leslie Lamport, Mark Hayden, and Mark Tuttle. It is available through the TLA Web page. This chapter describes TLC Version 2. At the time I am writing this, Version 2 is still being implemented and only Version 1 is available. Consult the documentation that accompanies the software to find out what version it is and how it differs from the version described here.

14.1 Introduction to TLC

TLC handles specifications that have the standard form

(14.1) $Init \wedge \Box[Next]_{vars} \wedge Temporal$

where *Init* is the initial predicate, *Next* is the next-state action, *vars* is the tuple of all variables, and *Temporal* is a temporal formula that usually specifies a liveness condition. Liveness and temporal formulas are explained in Chapter 8. If your specification contains no *Temporal* formula, so it has the form $Init \wedge \Box[Next]_{vars}$, then you can ignore the discussion of temporal checking. TLC does not handle the hiding operator $\exists$ (temporal existential quantification). You can check a specification with hidden variables by checking the internal specification, in which those variables are visible.

The most effective way to find errors in a specification is by trying to verify that it satisfies properties that it should. TLC can check that the specification satisfies (implies) a large class of TLA formulas—a class whose main restriction is that formulas may not contain $\exists$. You can also run TLC without having it check any property, in which case it will just look for two kinds of errors:

- "Silliness" errors. As explained in Section 6.2, a silly expression is one like $3 + \langle 1, 2 \rangle$, whose meaning is not determined by the semantics of TLA$^+$. A specification is incorrect if whether or not some particular behavior satisfies it depends on the meaning of a silly expression.

- Deadlock. The absence of deadlock is a particular property that we often want a specification to satisfy; it is expressed by the invariance property $\Box$(ENABLED *Next*). A counterexample to this property is a behavior exhibiting deadlock—that is, reaching a state in which *Next* is not enabled, so no further (nonstuttering) step is possible. TLC normally checks for deadlock, but this checking can be disabled since, for some systems, deadlock may just indicate successful termination.

The use of TLC will be illustrated with a simple example—a specification of the alternating bit protocol for sending data over a lossy FIFO transmission line. An algorithm designer might describe the protocol as a system that looks like this:

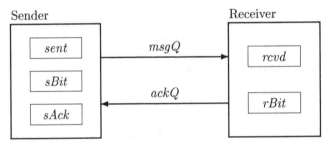

The sender can send a value when the one-bit values *sBit* and *sAck* are equal. It sets the variables *sent* to the value it is sending and complements *sBit*. This value is eventually delivered to the receiver by setting the variable *rcvd* and complementing the one-bit value *rBit*. Some time later, the sender's *sAck* value is complemented, permitting the next value to be sent. The protocol uses two lossy FIFO transmission lines: the sender sends data and control information on *msgQ*, and the receiver sends acknowledgments on *ackQ*.

The complete protocol specification appears in module *AlternatingBit* in Figure 14.1 on the following two pages. It is fairly straightforward, except for the liveness condition. Because messages can be repeatedly lost from the queues, strong fairness of the actions that receive messages is required to ensure that a message that keeps getting resent is eventually received. However, don't worry about the details of the specification. For now, all you need to know are the declarations

     CONSTANT *Data*   The set of data values that can be sent.

     VARIABLES *msgQ*, *ackQ*, *sBit*, *sAck*, *rBit*, *sent*, *rcvd*

and the types of the variables:

MODULE *AlternatingBit*

This specification describes a protocol for using lossy FIFO transmission lines to transmit a sequence of values from a sender to a receiver. The sender sends a data value d by sending a sequence of $\langle b, d \rangle$ messages on $msgQ$, where b is a control bit. It knows that the message has been received when it receives the ack b from the receiver on $ackQ$. It sends the next value with a different control bit. The receiver knows that a message on $msgQ$ contains a new value when the control bit differs from the last one it has received. The receiver keeps sending the last control bit it received on $ackQ$.

EXTENDS *Naturals, Sequences*

CONSTANTS *Data*   The set of data values that can be sent.

VARIABLES $msgQ$,   The sequence of $\langle$control bit, data value$\rangle$ messages in transit to the receiver.

$\qquad\quad ackQ$,   The sequence of one-bit acknowledgments in transit to the sender.

$\qquad\quad sBit$,   The last control bit sent by sender; it is complemented when sending a new data value.

$\qquad\quad sAck$,   The last acknowledgment bit received by the sender.

$\qquad\quad rBit$,   The last control bit received by the receiver.

$\qquad\quad sent$,   The last value sent by the sender.

$\qquad\quad rcvd$ The last value received by the receiver.

$$
\begin{aligned}
ABInit \;\triangleq\; &\land\; msgQ = \langle\rangle && \text{The initial condition:}\\
&\land\; ackQ = \langle\rangle && \quad \text{Both message queues are empty.}\\
&\land\; sBit \in \{0,1\} && \quad \text{All the bits equal 0 or 1}\\
&\land\; sAck = sBit && \quad\quad \text{and are equal to each other.}\\
&\land\; rBit = sBit &&\\
&\land\; sent \in Data && \quad \text{The initial values of } sent \text{ and } rcvd\\
&\land\; rcvd \in Data && \quad\quad \text{are arbitrary data values.}
\end{aligned}
$$

$$
\begin{aligned}
ABTypeInv \;\triangleq\; &\land\; msgQ \in Seq(\{0,1\} \times Data) && \text{The type-correctness invariant.}\\
&\land\; ackQ \in Seq(\{0,1\}) &&\\
&\land\; sBit \in \{0,1\} &&\\
&\land\; sAck \in \{0,1\} &&\\
&\land\; rBit \in \{0,1\} &&\\
&\land\; sent \in Data &&\\
&\land\; rcvd \in Data &&
\end{aligned}
$$

$SndNewValue(d) \;\triangleq\;$ The action in which the sender sends a new data value d.

$$
\begin{aligned}
&\land\; sAck = sBit && \text{Enabled iff } sAck \text{ equals } sBit.\\
&\land\; sent' = d && \text{Set } sent \text{ to } d.\\
&\land\; sBit' = 1 - sBit && \text{Complement control bit } sBit\\
&\land\; msgQ' = Append(msgQ, \langle sBit', d \rangle) && \text{Send value on } msgQ \text{ with new control bit.}\\
&\land\; \text{UNCHANGED } \langle ackQ, sAck, rBit, rcvd \rangle &&
\end{aligned}
$$

Figure 14.1a: The alternating bit protocol (beginning).

$ReSndMsg \triangleq$ The sender resends the last message it sent on $msgQ$.

$\quad \wedge sAck \neq sBit$ Enabled iff $sAck$ doesn't equal $sBit$.
$\quad \wedge msgQ' = Append(msgQ, \langle sBit, sent \rangle)$ Resend the last value in $send$.
$\quad \wedge$ UNCHANGED $\langle ackQ, sBit, sAck, rBit, sent, rcvd \rangle$

$RcvMsg \triangleq$ The receiver receives the message at the head of $msgQ$.

$\quad \wedge msgQ \neq \langle \rangle$ Enabled iff $msgQ$ not empty.
$\quad \wedge msgQ' = Tail(msgQ)$ Remove message from head of $msgQ$.
$\quad \wedge rBit' = Head(msgQ)[1]$ Set $rBit$ to message's control bit.
$\quad \wedge rcvd' = Head(msgQ)[2]$ Set $rcvd$ to message's data value.
$\quad \wedge$ UNCHANGED $\langle ackQ, sBit, sAck, sent \rangle$

$SndAck \triangleq \wedge ackQ' = Append(ackQ, rBit)$ The receiver sends $rBit$ on $ackQ$ at any time.
$\qquad\qquad \wedge$ UNCHANGED $\langle msgQ, sBit, sAck, rBit, sent, rcvd \rangle$

$RcvAck \triangleq \wedge ackQ \neq \langle \rangle$ The sender receives an ack on $ackQ$.
$\qquad\qquad \wedge ackQ' = Tail(ackQ)$ It removes the ack and sets $sAck$ to its
$\qquad\qquad \wedge sAck' = Head(ackQ)$ value.
$\qquad\qquad \wedge$ UNCHANGED $\langle msgQ, sBit, rBit, sent, rcvd \rangle$

$Lose(q) \triangleq$ The action of losing a message from queue q.

$\quad \wedge q \neq \langle \rangle$ Enabled iff q is not empty.
$\quad \wedge \exists i \in 1 .. Len(q) :$ For some i,
$\qquad q' = [j \in 1 .. (Len(q) - 1) \mapsto$ IF $j < i$ THEN $q[j]$ remove the ith message from q.
$\qquad\qquad\qquad\qquad\qquad\qquad\qquad$ ELSE $q[j + 1]]$ Leave every variable unchanged
$\quad \wedge$ UNCHANGED $\langle sBit, sAck, rBit, sent, rcvd \rangle$ except $msgQ$ and $ackQ$.

$LoseMsg \triangleq Lose(msgQ) \wedge$ UNCHANGED $ackQ$ Lose a message from $msgQ$.

$LoseAck \triangleq Lose(ackQ) \wedge$ UNCHANGED $msgQ$ Lose a message from $ackQ$.

$ABNext \triangleq \vee \exists d \in Data : SndNewValue(d)$ The next-state action.
$\qquad\qquad \vee ReSndMsg \vee RcvMsg \vee SndAck \vee RcvAck$
$\qquad\qquad \vee LoseMsg \vee LoseAck$

$abvars \triangleq \langle msgQ, ackQ, sBit, sAck, rBit, sent, rcvd \rangle$ The tuple of all variables.

$ABFairness \triangleq \wedge$ WF$_{abvars}(ReSndMsg) \wedge$ WF$_{abvars}(SndAck)$ The liveness condition.
$\qquad\qquad\qquad \wedge$ SF$_{abvars}(RcvMsg) \wedge$ SF$_{abvars}(RcvAck)$

$ABSpec \triangleq ABInit \wedge \square[ABNext]_{abvars} \wedge ABFairness$ The complete specification.

THEOREM $ABSpec \Rightarrow \square ABTypeInv$

Figure 14.1b: The alternating bit protocol (end).

- $msgQ$ is a sequence of elements in $\{0, 1\} \times Data$.

- $ackQ$ is a sequence of elements in $\{0, 1\}$.

- $sBit$, $sAck$, and $rBit$ are elements of $\{0, 1\}$.

- $sent$ and $rcvd$ are elements of $Data$.

The input to TLC consists of a TLA$^+$ module and a configuration file. TLC assumes the specification has the form of formula (14.1) on page 221. The configuration file tells TLC the names of the specification and of the properties to be checked. For example, the configuration file for the alternating bit protocol will contain the declaration

```
SPECIFICATION ABSpec
```

telling TLC to take *ABSpec* as the specification. If your specification has the form $Init \wedge \Box[Next]_{vars}$, with no liveness condition, then instead of using a `SPECIFICATION` statement, you can declare the initial predicate and next-state action by putting the following two statements in the configuration file:

```
INIT Init
NEXT Next
```

The property or properties to be checked are specified with a `PROPERTY` statement. For example, to check that *ABTypeInv* is actually an invariant, we could have TLC check that the specification implies $\Box ABTypeInv$ by adding the definition

$$InvProperty \;\triangleq\; \Box ABTypeInv$$

to module *AlternatingBit* and putting the statement

```
PROPERTY InvProperty
```

in the configuration file. Invariance checking is so common that TLC allows you instead to put the following statement in the configuration file:

```
INVARIANT ABTypeInv
```

The `INVARIANT` statement must specify a state predicate. To check invariance with a `PROPERTY` statement, the specified property has to be of the form $\Box P$. Specifying a state predicate P in a `PROPERTY` statement tells TLC to check that the specification implies P, meaning that P is true in the initial state of every behavior satisfying the specification.

TLC works by generating behaviors that satisfy the specification. To do this, it must be given what we call a *model* of the specification. To define a model, we must assign values to the specification's constant parameters. The only constant parameter of the alternating bit protocol specification is the set

Data of data values. We can tell TLC to let *Data* equal the set containing two arbitrary elements, named *d1* and *d2*, by putting the following declaration in the configuration file:

```
CONSTANT  Data = {d1, d2}
```

The keywords
CONSTANT and
CONSTANTS are
equivalent, as are
INVARIANT and
INVARIANTS.

(We can use any sequence of letters and digits containing at least one letter as the name of an element.)

There are two ways to use TLC. The default method is *model checking*, in which it tries to find all reachable states—that is, all states[1] that can occur in behaviors satisfying the formula $Init \land \Box[Next]_{vars}$. You can also run TLC in *simulation* mode, in which it randomly generates behaviors, without trying to check all reachable states. We now consider model checking; simulation mode is described in Section 14.3.2 on page 243.

Exhaustively checking all reachable states is impossible for the alternating bit protocol because the sequences of messages can get arbitrarily long, so there are infinitely many reachable states. We must further constrain the model to make it finite—that is, so it allows only a finite number of possible states. We do this by defining a state predicate called the *constraint* that asserts bounds on the lengths of the sequences. For example, the following constraint asserts that *msgQ* and *ackQ* have length at most 2:

Section 14.3 be-
low describes how
actions as well as
state predicates
can be used as
constraints.

$$\land\ Len(msgQ) \leq 2$$
$$\land\ Len(ackQ)\ \leq 2$$

Instead of specifying the bounds on the lengths of sequences in this way, I prefer to make them parameters and to assign them values in the configuration file. We don't want to put into the specification itself declarations and definitions that are just for TLC's benefit. So, we write a new module, called *MCAlternatingBit*, that extends the *AlternatingBit* module and can be used as input to TLC. This module appears in Figure 14.2 on the next page. A possible configuration file for the module appears in Figure 14.3 on the next page. Observe that the configuration file must specify values for all the constant parameters of the specification—in this case, the parameter *Data* from the *AlternatingBit* module and the two parameters declared in module *MCAlternatingBit* itself. You can put comments in the configuration file, using the TLA$^+$ comment syntax described in Section 3.5 (page 32).

When a constraint *Constr* is specified, TLC checks every state that appears in a behavior satisfying $Init \land \Box[Next]_{vars} \land \Box Constr$. In the rest of this chapter, these states will be called the *reachable* ones.

---

[1]As explained in Section 2.3 (page 18), a state is an assignment of values to all possible variables. However, when discussing a particular specification, we usually consider a state to be an assignment of values to that specification's variables. That's what I'm doing in this chapter.

```
┌──────────────────── MODULE MCAlternatingBit ────────────────────┐
  EXTENDS AlternatingBit

  CONSTANTS msgQLen, ackQLen

  SeqConstraint  ≜  ∧ Len(msgQ) ≤ msgQLen      A constraint on the lengths of
                    ∧ Len(ackQ) ≤ ackQLen      sequences for use by TLC.
└──────────────────────────────────────────────────────────────────┘
```

Figure 14.2: Module *MCAlternatingBit*.

Having TLC check the type invariant will catch many simple mistakes. When we've corrected all the errors we can find that way, we then want to look for less obvious ones. A common error is for an action not to be enabled when it should be, preventing some states from being reached. You can discover if an action is never enabled by using the *coverage* option, described on page 252. To discover if an action is just sometimes incorrectly disabled, try checking liveness properties. An obvious liveness property for the alternating bit protocol is that every message sent is eventually delivered. A message *d* has been sent when $sent = d$ and $sBit \neq sAck$. So, a naive way to state this property is

> The temporal operator $\rightsquigarrow$ is defined on page 91.

$$SentLeadsToRcvd \;\triangleq$$
$$\forall\, d \in Data \,:\, (sent = d) \wedge (sBit \neq sAck) \rightsquigarrow (rcvd = d)$$

Formula *SentLeadsToRcvd* asserts that, for any data value *d*, if *sent* ever equals *d* when *sBit* does not equal *sAck*, then *rcvd* must eventually equal *d*. This doesn't assert that every message sent is eventually delivered. For example, it is satisfied by a behavior in which a particular value *d* is sent twice, but received only once. However, the formula is good enough for our purposes because the protocol doesn't depend on the actual values being sent. If it were possible for the same value to be sent twice but received only once, then it would be possible for two different values to be sent and only one received, violating *SentLeadsToRcvd*. We therefore add the definition of *SentLeadsToRcvd* to module *MCAlternatingBit* and add the following statement to the configuration file:

```
PROPERTY SentLeadsToRcvd

CONSTANTS     Data   = {d1, d2}    (* Is this big enough?  *)
              msgQLen = 2
              ackQLen = 2    \* Try 3 next.
SPECIFICATION ABSpec
INVARIANT     ABTypeInv
CONSTRAINT    SeqConstraint
```

Figure 14.3: A configuration file for module *MCAlternatingBit*.

Checking liveness properties is a lot slower than other kinds of checking, so you should do it only after you've found all the errors you can by checking invariance properties.

Checking type correctness and property *SentLeadsToRcvd* is a good way to start looking for errors. But ultimately, we would like to see if the protocol meets its specification. However, we don't have its specification. In fact, it is typical in practice that we are called upon to check the correctness of a system design without any formal specification of what the system is supposed to do. In that case, we can write an *ex post facto* specification. Module *ABCorrectness* in Figure 14.4 on the next page is such a specification of correctness for the alternating bit protocol. It is actually a simplified version of the protocol's specification in which, instead of being read from messages, the variables *rcvd*, *rBit*, and *sAck* are obtained directly from the variables of the other process.

We want to check that the specification *ABSpec* of module *AlternatingBit* implies formula *ABCSpec* of module *ABCorrectness*. To do this, we modify module *MCAlternatingBit* by adding the statement

> INSTANCE *ABCorrectness*

and we modify the **PROPERTY** statement of the configuration file to

> **PROPERTIES ABCSpec SentLeadsToRcvd**

This example is atypical because the correctness specification *ABCSpec* does not involve variable hiding (temporal existential quantification). Let's now suppose module *ABCorrectness* did declare another variable h that appeared in *ABCSpec*, and that the correctness condition for the alternating bit protocol was *ABCSpec* with h hidden. The correctness condition would then be expressed formally in TLA$^+$ as follows:

> $AB(h) \;\triangleq\;$ INSTANCE *ABCorrectness*
> THEOREM $ABSpec \;\Rightarrow\; \exists\, h \,:\, AB(h)!ABCSpec$

TLC could not check this theorem directly because it cannot handle the temporal existential quantifier $\exists$. We would check this theorem with TLC the same way we would try to prove it—namely, by using a refinement mapping. As explained in Section 5.8 on page 62, we would define a state function *oh* in terms of the variables of module *AlternatingBit* and we would prove

(14.2) $ABSpec \;\Rightarrow\; AB(oh)!ABCSpec$

To get TLC to check this theorem, we would add the definition

> $ABCSpecBar \;\triangleq\; AB(oh)!ABCSpec$

and have TLC check the property *ABCSpecBar*.

The keywords PROPERTY and PROPERTIES are equivalent.

This use of INSTANCE is explained in Section 4.3 (page 41).

———— MODULE *ABCorrectness* ————

EXTENDS *Naturals*

CONSTANTS *Data*

VARIABLES *sBit, sAck, rBit, sent, rcvd*

$ABCInit \;\triangleq\; \wedge\; sBit\; \in \{0,1\}$
$\qquad\qquad \wedge\; sAck = sBit$
$\qquad\qquad \wedge\; rBit\; = sBit$
$\qquad\qquad \wedge\; sent\; \in\; Data$
$\qquad\qquad \wedge\; rcvd\; \in\; Data$

$CSndNewValue(d)\;\; \triangleq\; \wedge\; sAck = sBit$
$\qquad\qquad\qquad\qquad \wedge\; sent' = d$
$\qquad\qquad\qquad\qquad \wedge\; sBit' = 1 - sBit$
$\qquad\qquad\qquad\qquad \wedge\; \text{UNCHANGED}\; \langle sAck,\; rBit,\; rcvd \rangle$

$CRcvMsg \;\triangleq\; \wedge\; rBit \neq sBit$
$\qquad\qquad\quad \wedge\; rBit' = sBit$
$\qquad\qquad\quad \wedge\; rcvd' = sent$
$\qquad\qquad\quad \wedge\; \text{UNCHANGED}\; \langle sBit,\; sAck,\; sent \rangle$

$CRcvAck \;\triangleq\; \wedge\; rBit \neq sAck$
$\qquad\qquad\quad \wedge\; sAck' = rBit$
$\qquad\qquad\quad \wedge\; \text{UNCHANGED}\; \langle sBit,\; rBit,\; sent,\; rcvd \rangle$

$ABCNext \;\triangleq\; \vee\; \exists\, d \in Data\, :\, CSndNewValue(d)$
$\qquad\qquad\quad \vee\; CRcvMsg \vee CRcvAck$

$cvars \;\triangleq\; \langle sBit,\; sAck,\; rBit,\; sent,\; rcvd \rangle$

$ABCFairness \;\triangleq\; \text{WF}_{cvars}(CRcvMsg) \wedge \text{WF}_{cvars}(CRcvAck)$

$ABCSpec \;\triangleq\; ABCInit \wedge \square[ABCNext]_{cvars} \wedge ABCFairness$

Figure 14.4: A specification of correctness of the alternating bit protocol.

When TLC checks a property, it does not actually verify that the specification implies the property. Instead, it checks that (i) the safety part of the specification implies the safety part of the property and (ii) the specification implies the liveness part of the property. For example, suppose that the specification *Spec* and the property *Prop* are

$$Spec \;\triangleq\; Init \wedge \square[Next]_{vars} \wedge Temporal$$
$$Prop \;\triangleq\; ImpliedInit \wedge \square[ImpliedAction]_{pvars} \wedge ImpliedTemporal$$

where *Temporal* and *ImpliedTemporal* are liveness properties. In this case, TLC checks the two formulas

$$Init \land \Box[Next]_{vars} \Rightarrow ImpliedInit \land \Box[ImpliedAction]_{pvars}$$
$$Spec \Rightarrow ImpliedTemporal$$

This means that you cannot use TLC to check that a non-machine-closed specification satisfies a safety property. (Machine closure is discussed in Section 8.9.2 on page 111.) Section 14.3 below more precisely describes how TLC checks properties.

14.2   What TLC Can Cope With

No model checker can handle all the specifications that we can write in a language as expressive as TLA$^+$. However, TLC seems able to handle most TLA$^+$ specifications that people actually write. Getting TLC to handle a specification may require a bit of trickery, but it can usually be done without having to make any changes to the specification itself.

This section explains what TLC can and cannot cope with, and gives some ways to make it cope. The best way to understand TLC's limitations is to understand how it works. So, this section describes how TLC "executes" a specification.

14.2.1   TLC Values

A state is an assignment of values to variables. TLA$^+$ allows you to describe a wide variety of values—for example, the set of all sequences of prime numbers. TLC can compute only a restricted class of values, called TLC values. Those values are built from the following four types of primitive values:

Booleans	The values TRUE and FALSE.
Integers	Values like 3 and -1.
Strings	Values like "ab3".
Model Values	These are values introduced in the CONSTANT statement of the configuration file. For example, the configuration file shown in Figure 14.3 on page 227 introduces the model values d1 and d2. Model values with different names are assumed to be different.

A TLC value is defined inductively to be either

1. a primitive value, or

2. a finite set of comparable TLC values (*comparable* is defined below), or

3. a function f whose domain is a TLC value such that $f[x]$ is a TLC value, for all x in DOMAIN f.

For example, the first two rules imply that

(14.3) $\{\{$ "a", "b"$\}$, $\{$ "b", "c"$\}$, $\{$ "c", "d"$\}\}$

is a TLC value because rules 1 and 2 imply that $\{$ "a", "b"$\}$, $\{$ "b", "c"$\}$, and $\{$ "c", "d"$\}$ are TLC values, and the second rule then implies that (14.3) is a TLC value. Since tuples and records are functions, rule 3 implies that a record or tuple whose components are TLC values is a TLC value. For example, $\langle 1,$ "a", $2,$ "b" $\rangle$ is a TLC value.

To complete the definition of what a TLC value is, I must explain what *comparable* means in rule 2. The basic idea is that two values should be comparable iff the semantics of TLA$^+$ determines whether or not they are equal. For example, strings and numbers are not comparable because the semantics of TLA$^+$ doesn't tell us whether or not "abc" equals 42. The set $\{$ "abc", 42$\}$ is therefore not a TLC value; rule 2 doesn't apply because "abc" and 42 are not comparable. On the other hand, $\{$ "abc"$\}$ and $\{4, 2\}$ are comparable because sets having different numbers of elements must be unequal. Hence, the two-element set $\{\{$ "abc"$\}$, $\{4, 2\}\}$ is a TLC value. TLC considers a model value to be comparable to, and unequal to, any other value. The precise rules for comparability are given in Section 14.7.2.

14.2.2 How TLC Evaluates Expressions

Checking a specification requires evaluating expressions. For example, TLC does invariance checking by evaluating the invariant in each reachable state—that is, computing its TLC value, which should be TRUE. To understand what TLC can and cannot do, you have to know how it evaluates expressions.

TLC evaluates expressions in a straightforward way, generally evaluating subexpressions "from left to right". In particular:

- It evaluates $p \wedge q$ by first evaluating p and, if it equals TRUE, then evaluating q.

- It evaluates $p \vee q$ by first evaluating p and, if it equals FALSE, then evaluating q. It evaluates $p \Rightarrow q$ as $\neg p \vee q$.

- It evaluates IF p THEN e_1 ELSE e_2 by first evaluating p, then evaluating either e_1 or e_2.

To understand the significance of these rules, let's consider a simple example. TLC cannot evaluate the expression $x[1]$ if x equals $\langle \rangle$, since $\langle \rangle[1]$ is silly. (The empty sequence $\langle\rangle$ is a function whose domain is the empty set and hence does not contain 1.) The first rule implies that, if x equals $\langle\rangle$, then TLC can evaluate the formula

$$(x \neq \langle\rangle) \wedge (x[1] = 0)$$

but not the (logically equivalent) formula

$$(x[1] = 0) \wedge (x \neq \langle\rangle)$$

(When evaluating the latter formula, TLC first tries to compute $\langle\rangle[1] = 0$, reporting an error because it can't.) Fortunately, we naturally write the first formula rather than the second because it's easier to understand. People understand a formula by "mentally evaluating" it from left to right, much the way TLC does.

TLC evaluates $\exists x \in S : p$ by enumerating the elements $s_1, \ldots, s_n$ of S in some order and then evaluating p with s_i substituted for x, successively for $i = 1, \ldots, n$. It enumerates the elements of a set S in a very straightforward way, and it gives up and declares an error if the set is not obviously finite. For example, it can obviously enumerate the elements of $\{0, 1, 2, 3\}$ and $0 .. 3$. It enumerates a set of the form $\{x \in S : p\}$ by first enumerating S, so it can enumerate $\{i \in 0 .. 5 : i < 4\}$ but not $\{i \in Nat : i < 4\}$.

TLC evaluates the expressions $\forall x \in S : p$ and CHOOSE $x \in S : p$ by first enumerating the elements of S, much the same way as it evaluates $\exists x \in S : p$. The semantics of TLA$^+$ states that CHOOSE $x \in S : p$ is an arbitrary value if there is no x in S for which p is true. However, this case almost always arises because of a mistake, so TLC treats it as an error. Note that evaluating the expression

$$\text{IF } n > 5 \text{ THEN CHOOSE } i \in 1 .. n : i > 5 \text{ ELSE } 42$$

will not produce an error because TLC will not evaluate the CHOOSE expression if $n \leq 5$. (TLC would report an error if it tried to evaluate the CHOOSE expression when $n \leq 5$.)

TLC cannot evaluate "unbounded" quantifiers or CHOOSE expressions—that is, expressions having one of the forms

$$\exists x : p \qquad \forall x : p \qquad \text{CHOOSE } x : p$$

TLC cannot evaluate any expression whose value is not a TLC value, as defined in Section 14.2.1 above. In particular, TLC can evaluate a set-valued expression only if that expression equals a finite set, and it can evaluate a function-valued expression only if that expression equals a function whose domain is a finite set. TLC will evaluate expressions of the following forms only if it can enumerate the set S:

$$\exists x \in S : p \qquad \forall x \in S : p \qquad \text{CHOOSE } x \in S : p$$
$$\{x \in S : p\} \qquad \{e : x \in S\} \qquad [x \in S \mapsto e]$$
$$\text{SUBSET } S \qquad \text{UNION } S$$

TLC can often evaluate an expression even when it can't evaluate all subexpressions. For example, it can evaluate

$$[n \in Nat \mapsto n * (n+1)][3]$$

which equals the TLC value 12, even though it can't evaluate

$$[n \in Nat \mapsto n * (n+1)]$$

which equals a function whose domain is the set *Nat*. (A function can be a TLC value only if its domain is a finite set.)

TLC evaluates recursively defined functions with a simple recursive procedure. If f is defined by $f[x \in S] \triangleq e$, then TLC evaluates $f[c]$ by evaluating e with c substituted for x. This means that it can't handle some legal function definitions. For example, consider this definition from page 68:

$$
\begin{aligned}
mr[n \in Nat] \ \triangleq \ & \\
[f \ \mapsto \ & \text{IF} \ \ n = 0 \ \ \text{THEN} \ \ 17 \ \ \text{ELSE} \ \ mr[n-1].f * mr[n].g \,, \\
g \ \mapsto \ & \text{IF} \ \ n = 0 \ \ \text{THEN} \ \ 42 \ \ \text{ELSE} \ \ mr[n-1].f + mr[n-1].g\,]
\end{aligned}
$$

To evaluate $mr[3]$, TLC substitutes 3 for n and starts evaluating the right-hand side. But because $mr[n]$ appears in the right-hand side, TLC must evaluate the subexpression $mr[3]$, which it does by substituting 3 for n and starting to evaluate the right-hand side. And so on. TLC eventually detects that it's in an infinite loop and reports an error.

Legal recursive definitions that cause TLC to loop like this are rare, and they can be rewritten so TLC can handle them. Recall that we defined mr to express the mutual recursion:

$$
\begin{aligned}
f[n] \ &= \ \text{IF} \ \ n = 0 \ \ \text{THEN} \ \ 17 \ \ \text{ELSE} \ \ f[n-1] * g[n] \\
g[n] \ &= \ \text{IF} \ \ n = 0 \ \ \text{THEN} \ \ 42 \ \ \text{ELSE} \ \ f[n-1] + g[n-1]
\end{aligned}
$$

The subexpression $mr[n]$ appeared in the expression defining $mr[n]$ because $f[n]$ depends on $g[n]$. To eliminate it, we have to rewrite the mutual recursion so that $f[n]$ depends only on $f[n-1]$ and $g[n-1]$. We do this by expanding the definition of $g[n]$ in the expression for $f[n]$. Since the ELSE clause applies only to the case $n \neq 0$, we can rewrite the expression for $f[n]$ as

$$f[n] \ = \ \text{IF} \ \ n = 0 \ \ \text{THEN} \ \ 17 \ \ \text{ELSE} \ \ f[n-1] * (f[n-1] + g[n-1])$$

This leads to the following equivalent definition of mr:

$$
\begin{aligned}
mr[n \in Nat] \ \triangleq \ & \\
[f \ \mapsto \ & \text{IF} \ \ n = 0 \ \ \text{THEN} \ \ 17 \\
& \text{ELSE} \ \ mr[n-1].f * (mr[n-1].f + mr[n-1].g) \,, \\
g \ \mapsto \ & \text{IF} \ \ n = 0 \ \ \text{THEN} \ \ 42 \ \ \text{ELSE} \ \ mr[n-1].f + mr[n-1].g\,]
\end{aligned}
$$

With this definition, TLC has no trouble evaluating $mr[3]$.

The evaluation of ENABLED predicates and the action-composition operator "." are described on page 240 in Section 14.2.6. Section 14.3 explains how TLC evaluates temporal-logic formulas for temporal checking.

If you're not sure whether TLC can evaluate an expression, try it and see. But don't wait until TLC gets to the expression in the middle of checking the entire specification. Instead, make a small example in which TLC evaluates just that expression. See the explanation on page 14.5.3 of how to use TLC as a TLA$^+$ calculator.

14.2.3   Assignment and Replacement

As we saw in the alternating bit example, the configuration file must determine the value of each constant parameter. To assign a TLC value v to a constant parameter c of the specification, we write $c = v$ in the configuration file's CONSTANT statement. The value v may be a primitive TLC value or a finite set of primitive TLC values written in the form $\{v_1, \ldots, v_n\}$—for example, $\{1, -3, 2\}$. In v, any sequence of characters like a1 or foo that is not a number, a quoted string, or TRUE or FALSE is taken to be a model value.

In the assignment $c = v$, the symbol c need not be a constant parameter; it can also be a defined symbol. This assignment causes TLC to ignore the actual definition of c and to take v to be its value. Such an assignment is often used when TLC cannot compute the value of c from its definition. In particular, TLC cannot compute the value of *NotAnS* from the definition

$$NotAnS \;\triangleq\; \text{CHOOSE } n : n \notin S$$

because it cannot evaluate the unbounded CHOOSE expression. You can override this definition by assigning *NotAnS* a value in the CONSTANT statement of the configuration file. For example, the assignment

    NotAnS = NS

causes TLC to assign to *NotAnS* the model value NS. TLC ignores the actual definition of *NotAnS*. If you used the name *NotAnS* in the specification, you'd probably want TLC's error messages to call it NotAnS rather than NS. So, you'd probably use the assignment

    NotAnS = NotAnS

which assigns to the symbol *NotAnS* the model value NotAnS. Remember that, in the assignment $c = v$, the symbol c must be defined or declared in the TLA$^+$ module, and v must be a primitive TLC value or a finite set of such values.

The CONSTANT statement of the configuration file can also contain *replacements* of the form c <- d, where c and d are symbols defined in the TLA$^+$

Note that d is a defined symbol in the replacement c <- d, while v is a TLC value in the substitution $c = v$.

module. This causes TLC to replace c by d when performing its calculations. One use of replacement is to give a value to an operator parameter. For example, suppose we wanted to use TLC to check the write-through cache specification of Section 5.6 (page 54). The *WriteThroughCache* module extends the *MemoryInterface* module, which contains the declaration

CONSTANTS $Send(\_,\_,\_,\_)$, $Reply(\_,\_,\_,\_)$, ...

We have to tell TLC how to evaluate the operators *Send* and *Reply*. We do this by first writing a module *MCWriteThroughCache* that extends the *WriteThroughCache* module and defines two operators

$$MCSend(p, d, old, new) \triangleq \ldots$$
$$MCReply(p, d, old, new) \triangleq \ldots$$

We then add to the configuration file's CONSTANT statement the replacements

```
Send  <- MCSend
Reply <- MCReply
```

A replacement can also replace one defined symbol by another. In a specification, we usually write the simplest possible definitions. A simple definition is not always the easiest one for TLC to use. For example, suppose our specification requires an operator *Sort* such that $Sort(S)$ is a sequence containing the elements of S in increasing order, if S is a finite set of numbers. Our specification in module *SpecMod* might use the simple definition

$$Sort(S) \triangleq \text{CHOOSE } s \in [1 .. Cardinality(S) \rightarrow S] :$$
$$\forall i, j \in \text{DOMAIN } s : (i < j) \Rightarrow (s[i] < s[j])$$

To evaluate $Sort(S)$ for a set S containing n elements, TLC has to enumerate the n^n elements in the set $[1 .. n \rightarrow S]$ of functions. This may be unacceptably slow. We can write a module *MCSpecMod* that extends *SpecMod* and defines *FastSort* so it equals *Sort* when applied to finite sets of numbers, but can be evaluated more efficiently by TLC. We can then run TLC with a configuration file containing the replacement

```
Sort <- FastSort
```

One possible definition of *FastSort* is given in Section 14.4, on page 250.

14.2.4 Evaluating Temporal Formulas

Section 14.2.2 (page 231) explains what kind of ordinary expressions TLC can evaluate. The specification and properties that TLC checks are temporal formulas; this section describes the class of temporal formulas it can handle.

TLC can evaluate a TLA temporal formula iff (i) the formula is *nice*—a term defined in the next paragraph—and (ii) TLC can evaluate all the ordinary expressions of which the formula is composed. For example, a formula of the form $P \rightsquigarrow Q$ is nice, so TLC can evaluate it iff it can evaluate P and Q. (Section 14.3 below explains on what states and pairs of states TLC evaluates the component expressions of a temporal formula.)

A temporal formula is nice iff it is the conjunction of formulas that belong to one of the following four classes:

State Predicate

Invariance Formula A formula of the form $\Box P$, where P is a state predicate.

Box-Action Formula A formula of the form $\Box[A]_v$, where A is an action and v is a state function.

Simple Temporal Formula To define this class, we first make the following definitions:

> The terminology used here is not standard.

- The *simple Boolean operators* consist of the operators

$$\land \qquad \lor \qquad \neg \qquad \Rightarrow \qquad \equiv \qquad \text{TRUE} \qquad \text{FALSE}$$

  of propositional logic together with quantification over finite, constant sets.

- A *temporal state formula* is one obtained from state predicates by applying simple Boolean operators and the temporal operators $\Box$, $\Diamond$, and $\rightsquigarrow$. For example, if N is a constant, then

$$\forall\, i \in 1 \mathrel{..} N \,:\, \Box((x = i) \Rightarrow \exists\, j \in 1 \mathrel{..} i \,:\, \Diamond(y = j))$$

  is a temporal state formula.

- A *simple action formula* is one of the following, where A is an action and v a state function:

$$\text{WF}_v(A) \qquad \text{SF}_v(A) \qquad \Box\Diamond\langle A\rangle_v \qquad \Diamond\Box[A]_v$$

  The component expressions of $\text{WF}_v(A)$ and $\text{SF}_v(A)$ are $\langle A\rangle_v$ and ENABLED $\langle A\rangle_v$. (The evaluation of ENABLED formulas is described on page 240.)

A simple temporal formula is then defined to be one constructed from temporal state formulas and simple action formulas by applying simple Boolean operators.

For convenience, we exclude invariance formulas from the class of simple temporal formulas, so these four classes of nice temporal formulas are disjoint.

TLC can therefore evaluate the temporal formula

$$\forall\, i \in 1 \mathrel{..} N \,:\, \Diamond(y = i) \Rightarrow \text{WF}_y((y' = y + 1) \land (y \geq i))$$

if N is a constant, because this is a simple temporal formula (and hence nice) and TLC can evaluate all of its component expressions. TLC cannot evaluate $\Diamond \langle x' = 1 \rangle_x$, since this is not a nice formula. It cannot evaluate the formula $\mathrm{WF}_x(x'[1] = 0)$ if it must evaluate the action $\langle x'[1] = 0 \rangle_x$ on a step $s \to t$ in which $x = \langle \rangle$ in state t.

A PROPERTY statement can specify any formulas that TLC can evaluate. The formula of a SPECIFICATION statement must contain exactly one conjunct that is a box-action formula. That conjunct specifies the next-state action.

14.2.5   Overriding Modules

TLC cannot compute $2 + 2$ from the definition of $+$ contained in the standard *Naturals* module. Even if we did use a definition of $+$ from which TLC could compute sums, it would not do so very quickly. Arithmetic operators like $+$ are implemented directly in Java, the language in which TLC is written. This is achieved by a general mechanism of TLC that allows a module to be overridden by a Java class that implements the operators defined in the module. When TLC encounters an EXTENDS *Naturals* statement, it loads the Java class that overrides the *Naturals* module rather than reading the module itself. There are Java classes to override the following standard modules: *Naturals*, *Integers*, *Sequences*, *FiniteSets*, and *Bags*. (The *TLC* module described below in Section 14.4 is also overridden by a Java class.) Intrepid Java programmers will find that writing a Java class to override a module is not too hard.

14.2.6   How TLC Computes States

When TLC evaluates an invariant, it is calculating the invariant's value, which is either TRUE or FALSE. When TLC evaluates the initial predicate or the next-state action, it is computing a set of states—for the initial predicate, the set of all initial states, and for the next-state action, the set of possible *successor states* (primed states) reached from a given starting (unprimed) state. I will describe how TLC does this for the next-state action; the evaluation of the initial predicate is analogous.

Recall that a state is an assignment of values to variables. TLC computes the successors of a given state s by assigning to all unprimed variables their values in state s, assigning no values to the primed variables, and then evaluating the next-state action. TLC evaluates the next-state action as described in Section 14.2.2 (page 231), except for two differences, which I now describe. This description assumes that TLC has already performed all the assignments and replacements specified by the CONSTANT statement of the configuration file and has expanded all definitions. Thus, the next-state action is a formula containing only variables, primed variables, model values, and built-in TLA$^+$ operators and constants.

The first difference in evaluating the next-state action is that TLC does not evaluate disjunctions from left to right. Instead, when it evaluates a subformula $A_1 \vee \ldots \vee A_n$, it splits the computation into n separate evaluations, each taking the subformula to be one of the A_i. Similarly, when it evaluates $\exists x \in S : p$, it splits the computation into separate evaluations for each element of S. An implication $P \Rightarrow Q$ is treated as the disjunction $(\neg P) \vee Q$. For example, TLC splits the evaluation of

$$(A \Rightarrow B) \vee (C \wedge (\exists i \in S : D(i)) \wedge E)$$

into separate evaluations of the three disjuncts $\neg A$, B, and

$$C \wedge (\exists i \in S : D(i)) \wedge E$$

To evaluate the latter disjunct, it first evaluates C. If it obtains the value TRUE, then it splits this evaluation into the separate evaluations of $D(i) \wedge E$, for each i in S. It evaluates $D(i) \wedge E$ by first evaluating $D(i)$ and, if it obtains the value TRUE, then evaluating E.

The second difference in the way TLC evaluates the next-state action is that, for any variable x, if it evaluates an expression of the form $x' = e$ when x' has not yet been assigned a value, then the evaluation yields the value TRUE and TLC assigns to x' the value obtained by evaluating the expression e. TLC evaluates an expression of the form $x' \in S$ as if it were $\exists v \in S : x' = v$. It evaluates UNCHANGED x as $x' = x$ for any variable x, and UNCHANGED $\langle e_1, \ldots, e_n \rangle$ as

$$(\text{UNCHANGED } e_1) \wedge \ldots \wedge (\text{UNCHANGED } e_n)$$

for any expressions e_i. Hence, TLC evaluates UNCHANGED $\langle x, \langle y, z \rangle \rangle$ as if it were

$$(x' = x) \wedge (y' = y) \wedge (z' = z)$$

Except when evaluating an expression of the form $x' = e$, TLC reports an error if it encounters a primed variable that has not yet been assigned a value. An evaluation stops, finding no states, if a conjunct evaluates to FALSE. An evaluation that completes and obtains the value TRUE finds the state determined by the values assigned to the primed variables. In the latter case, TLC reports an error if some primed variable has not been assigned a value.

To illustrate how this works, let's consider how TLC evaluates the next-state action

$$
\begin{aligned}
(14.4) \quad &\vee \wedge x' \in 1 \ .. \ Len(y) \\
&\quad \wedge y' = Append(Tail(y), x') \\
&\vee \wedge x' = x + 1 \\
&\quad \wedge y' = Append(y, x')
\end{aligned}
$$

We first consider the starting state with $x = 1$ and $y = \langle 2, 3 \rangle$. TLC splits the computation into evaluating the two disjuncts separately. It begins evaluating the first disjunct of (14.4) by evaluating its first conjunct, which it treats as $\exists\, i \in 1 \,..\, Len(y) : x' = i$. Since $Len(y) = 2$, the evaluation splits into separate evaluations of

(14.5) $\wedge\ x' = 1$ $\wedge\ x' = 2$
         $\wedge\ y' = Append(Tail(y), x')$ $\wedge\ y' = Append(Tail(y), x')$

TLC evaluates the first of these actions as follows. It evaluates the first conjunct, obtaining the value TRUE and assigning to x' the value 1; it then evaluates the second conjunct, obtaining the value TRUE and assigning to y' the value $Append(Tail(\langle 2, 3\rangle), 1)$. So, evaluating the first action of (14.5) finds the successor state with $x = 1$ and $y = \langle 3, 1\rangle$. Similarly, evaluating the second action of (14.5) finds the successor state with $x = 2$ and $y = \langle 3, 2\rangle$. In a similar way, TLC evaluates the second disjunct of (14.4) to find the successor state with $x = 2$ and $y = \langle 2, 3, 2\rangle$. Hence, the evaluation of (14.4) finds three successor states.

Next, consider how TLC evaluates the next-state action (14.4) in a state with $x = 1$ and y equal to the empty sequence $\langle\rangle$. Since $Len(y) = 0$ and $1 \,..\, 0$ is the empty set $\{\,\}$, TLC evaluates the first disjunct as

      $\wedge\ \exists\, i \in \{\,\} : x' = i$
      $\wedge\ y' = Append(Tail(y), x')$

Evaluating the first conjunct yields FALSE, so the evaluation of the first disjunct of (14.4) stops, finding no successor states. Evaluating the second disjunct yields the successor state with $x = 2$ and $y = \langle 2\rangle$.

Since TLC evaluates conjuncts from left to right, their order can affect whether or not TLC can evaluate the next-state action. For example, suppose the two conjuncts in the first disjunct of (14.4) were reversed, like this:

      $\wedge\ y' = Append(Tail(y), x')$
      $\wedge\ x' \in 1 \,..\, Len(y)$

When TLC evaluates the first conjunct of this action, it encounters the expression $Append(Tail(y), x')$ before it has assigned a value to x', so it reports an error. Moreover, even if we were to change that x' to an x, TLC could still not evaluate the action starting in a state with $y = \langle\rangle$, since it would encounter the silly expression $Tail(\langle\rangle)$ when evaluating the first conjunct.

The description given above of how TLC evaluates an arbitrary next-state action is good enough to explain how it works in almost all cases that arise in practice. However, it is not completely accurate. For example, interpreted literally, it would imply that TLC can cope with the following two next-state actions, which are both logically equivalent to $(x' = \text{TRUE}) \wedge (y' = 1)$:

(14.6) $(x' = (y' = 1)) \wedge (x' = \text{TRUE})$ IF $x' = \text{TRUE}$ THEN $y' = 1$ ELSE FALSE

In fact, TLC will produce error messages when presented with either of these bizarre next-state actions.

Remember that TLC computes initial states by using a similar procedure to evaluate the initial predicate. Instead of starting from given values of the unprimed variables and assigning values to the primed variables, it assigns values to the unprimed variables.

TLC evaluates ENABLED formulas essentially the same way it evaluates a next-state action. More precisely, to evaluate a formula ENABLED A, TLC computes successor states as if A were the next-state action. The formula evaluates to TRUE iff there exists a successor state. To check if a step $s \to t$ satisfies the composition $A \cdot B$ of actions A and B, TLC first computes all states u such that $s \to u$ is an A step and then checks if $u \to t$ is a B step for some such u.

Action composition is discussed on page 77.

TLC may also have to evaluate an action when checking a property. In that case, it evaluates the action as it would any expression, and it has no trouble evaluating even the bizarre actions (14.6).

14.3   How TLC Checks Properties

Section 14.2 above explains how TLC evaluates expressions and computes initial states and successor states. This section describes how TLC uses evaluation to check properties—first for model-checking mode (its default), and then for simulation mode.

First, let's define some formulas that are obtained from the configuration file. In these definitions, a *specification conjunct* is a conjunct of the formula named by the SPECIFICATION statement (if there is one), a *property conjunct* is a conjunct of a formula named by a PROPERTY statement, and the conjunction of an empty set of formulas is defined to be TRUE. The definitions use the four classes of nice temporal formulas defined above in Section 14.2.4 on page 235.

Init The specification's initial state predicate. It is specified by an INIT or SPECIFICATION statement. In the latter case, it is the conjunction of all specification conjuncts that are state predicates.

Next The specification's next-state action. It is specified by a NEXT statement or a SPECIFICATION statement. In the latter case, it is the action N such that there is a specification conjunct of the form $\Box[N]_v$. There must not be more than one such conjunct.

Temporal The conjunction of every specification conjunct that is neither a state predicate nor a box-action formula. It is usually the specification's liveness condition.

Invariant The conjunction of every state predicate I that is either named by an INVARIANT statement or for which some property conjunct equals $\Box I$.

ImpliedInit The conjunction of every property conjunct that is a state predicate.

ImpliedAction The conjunction of every action $[A]_v$ such that some property conjunct equals $\Box[A]_v$.

ImpliedTemporal The conjunction of every property conjunct that is a simple temporal formula, but is not of the form $\Box I$, where I is a state predicate.

Constraint The conjunction of all state predicates named by CONSTRAINT statements.

ActionConstraint The conjunction of all actions named by ACTION-CONSTRAINT statements. An action constraint is similar to an ordinary constraint, except it eliminates possible transitions rather than states. An ordinary constraint P is equivalent to the action constraint P'.

14.3.1 Model-Checking Mode

TLC keeps two data structures: a directed graph $\mathcal{G}$ whose nodes are states, and a queue (a sequence) $\mathcal{U}$ of states. A *state in* $\mathcal{G}$ means a state that is a node of the graph $\mathcal{G}$. The graph $\mathcal{G}$ is the part of the state reachability graph that TLC has found so far, and $\mathcal{U}$ contains all states in $\mathcal{G}$ whose successors TLC has not yet computed. TLC's computation maintains the following invariants:

- The states of $\mathcal{G}$ satisfy the *Constraint* predicate.

- For every state s in $\mathcal{G}$, the edge from s to s is in $\mathcal{G}$.

- If there is an edge in $\mathcal{G}$ from state s to a different state t, then t is a successor state of s that satisfies the action constraint. In other words, the step $s \rightarrow t$ satisfies *Next* $\wedge$ *ActionConstraint*.

- Each state s of $\mathcal{G}$ is reachable from an initial state (one that satisfies the *Init* predicate) by a path in $\mathcal{G}$.

- $\mathcal{U}$ is a sequence of distinct states that are nodes in $\mathcal{G}$.

- For every state s in $\mathcal{G}$ that is not in $\mathcal{U}$, and for every state t satisfying *Constraint* such that the step $s \rightarrow t$ satisfies *Next* $\wedge$ *ActionConstraint*, the state t and the edge from s to t are in $\mathcal{G}$.

TLC executes the following algorithm, starting with $\mathcal{G}$ and $\mathcal{U}$ empty:

1. Check that every ASSUME in the specification is satisfied by the values assigned to the constant parameters.

2. Compute the set of initial states by evaluating the initial predicate *Init*, as described above in Section 14.2.6. For each initial state s found:

(a) Evaluate the predicates *Invariant* and *ImpliedInit* in state s; report an error and stop if either is false.

(b) If the predicate *Constraint* is true in state s, then add s to the queue $\mathcal{U}$ and add node s and edge $s \rightarrow s$ to the graph $\mathcal{G}$.

3. While $\mathcal{U}$ is nonempty, do the following:

   (a) Remove the first state from $\mathcal{U}$ and let s be that state.

   (b) Find the set T of all successor states of s by evaluating the next-state action starting from s, as described above in Section 14.2.6.

   (c) If T is empty and the *deadlock* option is *not* selected, then report a deadlock error and stop.

   (d) For each state t in T, do the following:

       i. If *Invariant* is false in state t or *ImpliedAction* is false for the step $s \rightarrow t$, then report an error and stop.

       ii. If the predicate *Constraint* is true in state t and the step $s \rightarrow t$ satisfies *ActionConstraint*, then

           A. If t is not in $\mathcal{G}$, then add it to the tail of $\mathcal{U}$ and add the node t and the edge $t \rightarrow t$ to $\mathcal{G}$.

           B. Add the edge $s \rightarrow t$ to $\mathcal{G}$.

TLC can use multiple threads, and steps 3(b)–(d) may be performed concurrently by different threads for different states s. See the description of the *workers* option on page 253 below.

If formula *ImpliedTemporal* is not equal to TRUE, then whenever it adds an edge $s \rightarrow t$ in the procedure above, TLC evaluates all the predicates and actions that appear in formulas *Temporal* and *ImpliedTemporal* for the step $s \rightarrow t$. (It does this when adding any edge, including the self-loops $s \rightarrow s$ and $t \rightarrow t$ in steps 2(b) and 3(d)ii.A.)

Periodically during the computation of $\mathcal{G}$, and when it has finished computing $\mathcal{G}$, TLC checks the *ImpliedTemporal* property as follows. Let $\mathcal{T}$ be the set consisting of every behavior τ that is the sequence of states in an infinite path in $\mathcal{G}$ starting with an initial state. (For example, $\mathcal{T}$ contains the path $s \rightarrow s \rightarrow s \rightarrow \dots$ for every initial state s in $\mathcal{G}$.) Note that every behavior in $\mathcal{T}$ satisfies $Init \wedge \square[Next]_{vars}$. TLC checks that every behavior in $\mathcal{T}$ also satisfies the formula $Temporal \Rightarrow ImpliedTemporal$. (This is conceptually what happens; TLC does not actually check each behavior separately.) See Section 14.3.5 on page 247 below for a discussion of why TLC's checking of the *ImpliedTemporal* property may not do what you expect.

The computation of $\mathcal{G}$ terminates only if the set of reachable states is finite. Otherwise, TLC will run forever—that is, until it runs out of resources or is stopped.

See page 226 for the definition of *reachable state*.

TLC does not always perform all three of the steps described above. It does step 2 only for a non-constant module, in which case the configuration file must specify an *Init* formula. TLC does step 3 only if the configuration file specifies a *Next* formula, which it must do if it specifies an *Invariant*, *ImpliedAction*, or *ImpliedTemporal* formula.

14.3.2 Simulation Mode

In simulation mode, TLC repeatedly constructs and checks individual behaviors of a fixed maximum length. The maximum length can be specified with the *depth* option, as described on page 251 below. (Its default value is 100 states.) In simulation mode, TLC runs until you stop it.

To create and check a behavior, TLC uses the procedure described above for constructing the graph $\mathcal{G}$—except with the following difference. After computing the set of initial states, and after computing the set T of successors for a state s, TLC randomly chooses an element of that set. If the element does not satisfy the constraint, then the computation of $\mathcal{G}$ stops. Otherwise, TLC puts only that state in $\mathcal{G}$ and $\mathcal{U}$, and checks the *Invariant* and the *ImpliedInit* or the *ImpliedAction* formula for it. (The queue $\mathcal{U}$ isn't actually maintained, since it would never contain more than a single element.) The construction of $\mathcal{G}$ stops, and the formula *Temporal* $\Rightarrow$ *ImpliedTemporal* is checked, when the specified maximum number of states have been generated. TLC then repeats the procedure, starting with $\mathcal{G}$ and $\mathcal{U}$ empty.

TLC's choices are not strictly random, but are generated using a pseudo-random number generator from a randomly chosen seed. The seed and another value called the *aril* are printed if TLC finds an error. As described in Section 14.5.1 below, using the *key* and *aril* options, you can get TLC to generate the behavior that displayed the error.

14.3.3 Views and Fingerprints

In the description above of how TLC checks properties, I wrote that the nodes of the graph $\mathcal{G}$ are states. That is not quite correct. The nodes of $\mathcal{G}$ are values of a state function called the *view*. TLC's default view is the tuple of all declared variables, whose value determines the state. However, you can specify that the view should be some other state function *myview* by putting the statement

        VIEW myview

in the configuration file, where *myview* is an identifier that is either defined or else declared to be a variable.

When TLC computes initial states, it puts their views rather than the states themselves in $\mathcal{G}$. (The view of a state s is the value of the VIEW state function in

Remember that we are using the term *state* informally to mean an assignment of values to declared variables, rather than to all variables.

state s.) If there are multiple initial states with the same view, only one of them is put in the queue $\mathcal{U}$. Instead of inserting an edge from a state s to a state t, TLC inserts the edge from the view of s to the view of t. In step 3(d)ii.A in the algorithm above, TLC checks if the view of t is in $\mathcal{G}$.

When using a view other than the default one, TLC may stop before it has found all reachable states. For the states it does find, it correctly performs safety checks—that is, the *Invariant*, *ImpliedInit*, and *ImpliedAction* checks. Moreover, it prints out a correct counterexample (a finite sequence of states) if it finds an error in one of those properties. However, it may incorrectly check the *ImpliedTemporal* property. Because the graph $\mathcal{G}$ that TLC is constructing is not the actual reachability graph, it may report an error in the *ImpliedTemporal* property when none exists, printing out a bogus counterexample.

Specifying a nonstandard view can cause TLC not to check many states. You should do it when there is no need to check different states that have the same view. The most likely alternate view is a tuple consisting of some, but not all, declared variables. For example, you may have added one or more variables to help debug the specification. Using the tuple of the original variables as the view lets you add debugging variables without increasing the number of states that TLC must explore. If the properties being checked do not mention the debugging variables, then TLC will find all reachable states of the original specification and will correctly check all properties.

In the actual implementation, the nodes of the graph $\mathcal{G}$ are not the views of states, but *fingerprints* of those views. A TLC fingerprint is a 64-bit number generated by a "hashing" function. Ideally, the probability that two different views have the same fingerprint is 2^{-64}, which is a very small number. However, it is possible for a *collision* to occur, meaning that TLC mistakenly thinks that two different views are the same because they have the same fingerprint. If this happens, TLC will not explore all the states that it should. In particular, with the default view, TLC will report that it has checked all reachable states when it hasn't.

When it terminates, TLC prints out two estimates of the probability that a fingerprint collision occurred. The first is based on the assumption that the probability of two different views having the same fingerprint is 2^{-64}. (Under this assumption, if TLC generated n views with m distinct fingerprints, then the probability of a collision is about $m*(n-m)*2^{-64}$.) However, the process of generating states is highly nonrandom, and no known fingerprinting scheme can guarantee that the probability of any two distinct states generated by TLC having the same fingerprint is actually 2^{-64}. So, TLC also prints an empirical estimate of the probability that a collision occurred. It is based on the observation that, if there was a collision, then it is likely that there was also a "near miss". The estimate is the maximum value of $1/|f_1 - f_2|$ over all pairs $\langle f_1, f_2 \rangle$ of distinct fingerprints generated by TLC. In practice, the probability of collision turns out to be very small unless TLC is generating billions of distinct states.

Views and fingerprinting apply only to model-checking mode. In simulation mode, TLC ignores any `VIEW` statement.

14.3.4 Taking Advantage of Symmetry

The memory specifications of Chapter 5 are symmetric in the set *Proc* of processors. Intuitively, this means that permuting the processors doesn't change whether or not a behavior satisfies a specification. To define symmetry more precisely, we first need some definitions.

A *permutation of* a finite set S is a function whose domain and range both equal S. In other words, π is a permutation of S iff

$$(S = \text{DOMAIN } \pi) \,\wedge\, (\forall\, w \in S : \exists\, v \in S : \pi[v] = w)$$

A *permutation* is a function that is a permutation of its (finite) domain. If π is a permutation of a set S of values and s is a state, let s^π be the state obtained from s by replacing each value v in S with $\pi[v]$. To see what s^π means, let's take as an example the permutation π of { "a", "b", "c" } such that $\pi[\text{"a"}] = \text{"b"}$, $\pi[\text{"b"}] = \text{"c"}$, and $\pi[\text{"c"}] = \text{"a"}$. Suppose that, in state s, the values of the variables x and y are

$$
\begin{aligned}
x \;&=\; \langle\, \text{"b"}, \text{"c"}, \text{"d"} \,\rangle \\
y \;&=\; [i \in \{ \text{"a"}, \text{"b"} \} \mapsto \text{IF } i = \text{"a"} \text{ THEN } 7 \text{ ELSE } 42]
\end{aligned}
$$

Then in state s^π, the values of the variables x and y are

$$
\begin{aligned}
x \;&=\; \langle\, \text{"c"}, \text{"a"}, \text{"d"} \,\rangle \\
y \;&=\; [i \in \{ \text{"b"}, \text{"c"} \} \mapsto \text{IF } i = \text{"b"} \text{ THEN } 7 \text{ ELSE } 42]
\end{aligned}
$$

This example should give you an intuitive idea of what s^π means; I won't try to define it rigorously. If σ is the behavior $s_1, s_2, \ldots$, let σ^π be the behavior $s_1^\pi, s_2^\pi, \ldots$.

We can now define what symmetry means. A specification *Spec* is *symmetric with respect to* a permutation π iff the following condition holds: for any behavior σ, formula *Spec* is satisfied by σ iff it is satisfied by σ^π.

The memory specifications of Chapter 5 are symmetric with respect to any permutation of *Proc*. This means that there is no need for TLC to check a behavior σ if it has already checked the behavior σ^π for some permutation π of *Proc*. (Any error revealed by σ would also be revealed by σ^π.) We can tell TLC to take advantage of this symmetry by putting the following statement in the configuration file:

```
SYMMETRY Perms
```

where *Perms* is defined in the module to equal *Permutations(Proc)*, the set of all permutations of *Proc*. (The *Permutations* operator is defined in the *TLC* module, described in Section 14.4 below.) This SYMMETRY statement causes TLC to modify the algorithm described on pages 241–242 so that it never adds a state s to its queue $\mathcal{U}$ of unexamined states and to its state graph $\mathcal{G}$ if $\mathcal{G}$ already contains the state s^π, for some permutation π of *Proc*. If there are n processes, this reduces the number of states that TLC examines by a factor of $n!$.

The memory specifications of Chapter 5 are also symmetric with respect to any permutation of the set *Adr* of memory addresses. To take advantage of this symmetry as well as the symmetry with respect to permutations of processors, we define the symmetry set (the set specified by the SYMMETRY statement) to equal

$$Permutations(Proc) \cup Permutations(Adr)$$

In general, the SYMMETRY statement can specify an arbitrary symmetry set Π, each element of which is a permutation of a set of model values. More precisely, each element π in Π must be a permutation such that all the elements of DOMAIN π are assigned model values by the configuration file's CONSTANT statement. (If the configuration has no SYMMETRY statement, we take the symmetry set Π to be the empty set.)

To explain what TLC does when given an arbitrary symmetry set Π, I need a few more definitions. If τ is a sequence $\langle \pi 1, \ldots, \pi n \rangle$ of permutations in Π, let s^τ equal $(\ldots((s^{\pi 1})^{\pi 2})\ldots)^{\pi n}$. (If τ is the empty sequence, then s^τ is defined to equal s.) Define the *equivalence class* $\widehat{s}$ of a state s to be the set of states s^τ for all sequences τ of permutations in Π. For any state s, TLC keeps only a single element of $\widehat{s}$ in $\mathcal{U}$ and $\mathcal{G}$. This is accomplished by the following modifications to the algorithm on pages 241–242. In step 2(b), TLC adds the state s to $\mathcal{U}$ and $\mathcal{G}$ only if $\mathcal{U}$ and $\mathcal{G}$ do not already contain a state in $\widehat{s}$. Step 3(d)ii is changed to

A. If no element in $\widehat{t}$ is in $\mathcal{G}$, then add t to the tail of $\mathcal{U}$ and add the node t and the edge $t \rightarrow t$ to $\mathcal{G}$.

B. Add the edge $s \rightarrow tt$ to $\mathcal{G}$, where tt is the unique element of $\widehat{t}$ that is (now) in $\mathcal{G}$.

When a VIEW statement appears in the configuration file, these changes are modified as described in Section 14.3.3 above so that views rather than states are put in $\mathcal{G}$.

If the specification and the properties being checked are, indeed, symmetric with respect to all permutations in the symmetry set, then TLC's *Invariant*, *ImpliedInit*, and *ImpliedAction* checking will find and correctly report any error that they would have found had the SYMMETRY statement been omitted. However, TLC may perform *ImpliedTemporal* checking incorrectly—it may miss errors, report an error that doesn't exist, or report a real error with an incorrect counterexample. So, you should do *ImpliedTemporal* checking when using a SYMMETRY statement only if you understand exactly what TLC is doing.

If the specification and properties are not symmetric with respect to all permutations in the symmetry set, then TLC may be unable to print an error trace if it does find an error. In that case, it will print the error message

```
Failed to recover the state from its fingerprint.
```

The symmetry set is used only in model-checking mode. TLC ignores it in simulation mode.

14.3.5 Limitations of Liveness Checking

If a specification violates a safety property, then there is a finite behavior that displays the violation. That behavior can be generated with a finite model. It is therefore, in principle, possible to discover the violation with TLC. It may be impossible to discover a violation of a liveness property with any finite model. To see why, consider the following simple specification *EvenSpec* that starts with x equal to zero and repeatedly increments it by 2:

Safety properties were defined on page 87.

$$EvenSpec \;\triangleq\; (x = 0) \wedge \Box[x' = x + 2]_x \wedge \mathrm{WF}_x(x' = x + 2)$$

Obviously, x never equals 1 in any behavior satisfying *EvenSpec*. So, *EvenSpec* does not satisfy the liveness property $\Diamond(x = 1)$. Suppose we ask TLC to check if *EvenSpec* implies $\Diamond(x = 1)$. To get TLC to terminate, we must provide a constraint that limits it to generating a finite number of reachable states. All the infinite behaviors satisfying $(x = 0) \wedge \Box[x' = x+2]_x$ that TLC generates will then end in an infinite number of stuttering steps. In any such behavior, action $x' = x + 2$ is always enabled, but only a finite number of $x' = x + 2$ steps occur, so $\mathrm{WF}_x(x' = x + 2)$ is false. TLC will therefore not report an error because the formula

$$\mathrm{WF}_x(x' = x + 2) \;\Rightarrow\; \Diamond(x = 1)$$

is satisfied by all the infinite behaviors it generates.

When doing temporal checking, make sure that your model will permit infinite behaviors that satisfy the specification's liveness condition. For example, consider the finite model of the alternating bit protocol specification defined by the configuration file of Figure 14.3 on page 227. You should convince yourself that it allows infinite behaviors that satisfy formula *ABFairness*.

It's a good idea to verify that TLC is performing the liveness checking you expect. Have it check a liveness property that the specification does not satisfy and make sure it reports an error.

──────────────────── MODULE TLC ────────────────────

LOCAL INSTANCE *Naturals*   The keyword LOCAL means that definitions from the instantiated
                                 module are not obtained by a module that extends TLC.
LOCAL INSTANCE *Sequences*

OPERATORS FOR DEBUGGING

$Print(out, val) \triangleq val$ Causes TLC to print the values *out* and *val*.

$Assert(val, out) \triangleq$ IF $val =$ TRUE THEN TRUE   Causes TLC to report an error
                                  ELSE CHOOSE v : TRUE   and print *out* if *val* is not true.

$JavaTime \triangleq$ CHOOSE n : $n \in Nat$ Causes TLC to print the current time, in milliseconds elapsed
                                                 since 00:00 on 1 Jan 1970 UT, modulo 2^{31}.

OPERATORS FOR REPRESENTING FUNCTIONS AND SETS OF PERMUTATIONS

$d :> e \triangleq [x \in \{d\} \mapsto e]$ The function f with domain $\{d_1, \ldots, d_n\}$
$f @@ g \triangleq [x \in (\text{DOMAIN } f) \cup (\text{DOMAIN } g) \mapsto$ such that $f[d_i] = e_i$, for $i = 1, \ldots, n$ can be
                                                                     written
                   IF $x \in$ DOMAIN f THEN $f[x]$ ELSE $g[x]]$ $d_1 :> e_1$ @@ $\ldots$ @@ $d_n :> e_n$

$Permutations(S) \triangleq \{f \in [S \to S] : \forall w \in S : \exists v \in S : f[v] = w\}$ The set of permutations of S.

AN OPERATOR FOR SORTING

$SortSeq(s, \_ \prec \_) \triangleq$ The result of sorting sequence s according to the ordering $\prec$.

LET $Perm \triangleq$ CHOOSE $p \in Permutations(1 .. Len(s))$:
                     $\forall i, j \in 1 .. Len(s) : (i < j) \Rightarrow (s[p[i]] \prec s[p[j]]) \vee (s[p[i]] = s[p[j]])$
IN $[i \in 1 .. Len(s) \mapsto s[Perm[i]]]$

Figure 14.5: The standard module TLC.

14.4   The *TLC* Module

The standard *TLC* module, in Figure 14.5 on this page, defines operators that
are handy when using TLC. The module on which you run TLC usually EX-
TENDS the *TLC* module, which is overridden by its Java implementation.   Module overriding
   Module *TLC* begins with the statement   is explained above
                                                                                 in Section 14.2.5.

        LOCAL INSTANCE *Naturals*

As explained on page 171, this is like an EXTENDS statement, except that the
definitions included from the *Naturals* module are not obtained by any other
module that extends or instantiates module *TLC*. Similarly, the next statement
locally instantiates the *Sequences* module.

Module *TLC* next defines three operators *Print*, *Assert*, and *JavaTime*. They are of no use except in running TLC, when they can help you track down problems.

The operator *Print* is defined so that *Print*(*out*, *val*) equals *val*. But when TLC evaluates this expression, it prints the values of *out* and *val*. You can add *Print* expressions to a specification to help locate an error. For example, if your specification contains

$\land$ *Print*("a", TRUE)
$\land$ *P*
$\land$ *Print*("b", TRUE)

and TLC prints the "a" but not the "b" before reporting an error, then the error happened while TLC was evaluating *P*. If you know where the error is but don't know why it's occurring, you can add *Print* expressions to give you more information about what values TLC has computed.

To understand what gets printed when, you must know how TLC evaluates expressions, which is explained above in Sections 14.2 and 14.3. TLC usually evaluates an expression many times, so inserting a *Print* expression in the specification can produce a lot of output. One way to limit the amount of output is to put the *Print* expression inside an IF/THEN expression, so it is executed only in interesting cases.

The *TLC* module next defines the operator *Assert* so *Assert*(*val*, *out*) equals TRUE if *val* equals TRUE. If *val* does not equal TRUE, evaluating *Assert*(*val*, *out*) causes TLC to print the value of *out* and to halt. (In this case, the value of *Assert*(*val*, *out*) is irrelevant.)

Next, the operator *JavaTime* is defined to equal an arbitrary natural number. However, TLC does not obey the definition of *JavaTime* when evaluating it. Instead, evaluating *JavaTime* yields the time at which the evaluation takes place, measured in milliseconds elapsed since 00:00 Universal Time on 1 January 1970, modulo 2^{31}. If TLC is generating states slowly, using the *JavaTime* operator in conjunction with *Print* expressions can help you understand why. If TLC is spending too much time evaluating an operator, you may be able to replace the operator's definition with an equivalent one that TLC can evaluate more efficiently. (See Section 14.2.3 on page 234.)

The TLC module next defines the operators :> and @@ so that the expression

$d_1 :> e_1 \ @@ \ \ldots \ @@ \ d_n :> e_n$

is the function f with domain $\{d_1, \ldots, d_n\}$ such that $f[d_i] = e_i$, for $i = 1, \ldots, n$. For example, the sequence $\langle$ "ab", "cd" $\rangle$, which is a function with domain $\{1, 2\}$, can be written as

```
1 :> "ab"   @@   2 :> "cd"
```

TLC uses these operators to represent function values that it prints when evaluating a *Print* expression or reporting an error. However, it usually prints values the way they appear in the specification, so it usually prints a sequence as a sequence, not in terms of the :> and @@ operators.

Next comes the definition of *Permutations(S)* to be the set of all permutations of S, if S is a finite set. The *Permutations* operator can be used to specify a set of permutations for the SYMMETRY statement described in Section 14.3.4 above. More complicated symmetries can be expressed by defining a set $\{\pi_1, \dots, \pi_n\}$ of permutations, where each π_i is written as an explicit function using the :> and @@ operators. For example, consider a specification of a memory system in which each address is in some way associated with a processor. The specification would be symmetric under two kinds of permutations: ones that permute addresses associated with the same processor, and ones that permute the processors along with their associated addresses. Suppose we tell TLC to use two processors and four addresses, where addresses $a11$ and $a12$ are associated with processor $p1$ and addresses $a21$ and $a22$ are associated with processor $p2$. We can get TLC to take advantage of the symmetries by giving it the following set of permutations as the symmetry set:

$$Permutations(\{a11, a12\}) \ \cup \ \{p1\!:\!> p2 \ @@ \ p2\!:\!> p1$$
$$@@ \ a11\!:\!> a21 \ @@ \ a21\!:\!> a11$$
$$@@ \ a12\!:\!> a22 \ @@ \ a22\!:\!> a12\}$$

The permutation $p1\!:\!> p2 \ @@ \ \dots \ @@ \ a22\!:\!> a12$ interchanges the processors and their associated addresses. The permutation that just interchanges $a21$ and $a22$ need not be specified explicitly because it is obtained by interchanging the processors, interchanging $a11$ and $a12$, and interchanging the processors again.

The TLC module ends by defining the operator *SortSeq*, which can be used to replace operator definitions with ones that TLC can evaluate more efficiently. If s is a finite sequence and $\prec$ is a total ordering relation on its elements, then $SortSeq(s, \prec)$ is the sequence obtained from s by sorting its elements according to $\prec$. For example, $SortSeq(\langle 3, 1, 3, 8 \rangle, >)$ equals $\langle 8, 3, 3, 1 \rangle$. The Java implementation of *SortSeq* allows TLC to evaluate it more efficiently than a user-defined sorting operator. For example, here's how we can use *SortSeq* to define an operator *FastSort* to replace the *Sort* operator defined on page 235.

$$FastSort(S) \ \triangleq$$
$$\text{LET } \ MakeSeq[SS \in \text{SUBSET } S] \ \triangleq$$
$$\text{IF } \ SS = \{\} \text{ THEN } \langle \rangle$$
$$\text{ELSE LET } \ ss \ \triangleq \ \text{CHOOSE } ss \in SS : \text{TRUE}$$
$$\text{IN} \quad Append(MakeSeq[SS \setminus \{ss\}], ss)$$
$$\text{IN} \quad SortSeq(MakeSeq[S], <)$$

14.5 How to Use TLC

14.5.1 Running TLC

Exactly how you run TLC depends on what operating system you are using and how it is configured. You will probably type a command of the form

> *program_name options spec_file*

where

program_name is specific to your system. It might be `java tlatk.TLC`.

spec_file is the name of the file containing the TLA$^+$ specification. Each TLA$^+$ module named M that appears in the specification must be in a separate file named M`.tla`. The extension `.tla` may be omitted from *spec_file*.

options is a sequence consisting of zero or more of the following options:

-deadlock
: Tells TLC not to check for deadlock. Unless this option is specified, TLC will stop if it finds a deadlock—that is, a reachable state with no successor state.

-simulate
: Tells TLC to run in simulation mode, generating randomly chosen behaviors, instead of generating all reachable states. (See Section 14.3.2 above.)

-depth *num*
: This option causes TLC to generate behaviors of length at most *num* in simulation mode. Without this option, TLC will generate runs of length at most 100. This option is meaningful only when the *simulate* option is used.

-seed *num*
: In simulation mode, the behaviors generated by TLC are determined by the initial seed given to a pseudorandom number generator. Normally, the seed is generated randomly. This option causes TLC to let the seed be *num*, which must be an integer from -2^{63} to $2^{63} - 1$. Running TLC twice in simulation mode with the same seed and aril (see the *aril* option below) will produce identical results. This option is meaningful only when using the *simulate* option.

-aril *num*
: This option causes TLC to use *num* as the *aril* in simulation mode. The aril is a modifier of the initial seed. When TLC finds an error in simulation mode, it prints out both the initial seed and an aril

number. Using this initial seed and aril will cause the first trace generated to be that error trace. Adding *Print* expressions will usually not change the order in which TLC generates traces. So, if the trace doesn't tell you what went wrong, you can try running TLC again on just that trace to print out additional information.

-coverage *num*

This option causes TLC to print "coverage" information every *num* minutes and at the end of its execution. For every action conjunct that "assigns a value" to a variable, TLC prints the number of times that conjunct has actually been used in constructing a new state. The values it prints may not be accurate, but their magnitude can provide useful information. In particular, a value of 0 indicates part of the next-state action that was never "executed". This might indicate an error in the specification, or it might mean that the model TLC is checking is too small to exercise that part of the action.

-recover *run_id*

This option causes TLC to start executing the specification not from the beginning, but from where it left off at the last checkpoint. When TLC takes a checkpoint, it prints the run identifier. (That identifier is the same throughout an execution of TLC.) The value of *run_id* should be that run identifier.

-cleanup

TLC creates a number of files when it runs. When it completes, it erases all of them. If TLC finds an error, or if you stop it before it finishes, TLC can leave some large files around. The *cleanup* option causes TLC to delete all files created by previous runs. Do not use this option if you are currently running another copy of TLC in the same directory; if you do, it can cause the other copy to fail.

-difftrace *num*

When TLC finds an error, it prints an error trace. Normally, that trace is printed as a sequence of complete states, where a state lists the values of all declared variables. The *difftrace* option causes TLC to print an abridged version of each state, listing only the variables whose values are different than in the preceding state. This makes it easier to see what is happening in each step, but harder to find the complete state.

-terse

Normally, TLC completely expands values that appear in error messages or in the output from evaluating *Print* expressions. The *terse* option causes TLC instead to print partially evaluated, shorter versions of these values.

-workers *num*

> Steps 3(b)–(d) of the TLC execution algorithm described on pages 241–242 can be speeded up on a multiprocessor computer by the use of multiple threads. This option causes TLC to use *num* threads when finding reachable states. There is no reason to use more threads than there are actual processors on your computer. If the option is omitted, TLC uses a single thread.

-config *config_file*

> Specifies that the configuration file is named *config_file*, which must be a file with extension .cfg. The extension .cfg may be omitted from *config_file*. If this option is omitted, the configuration file is assumed to have the same name as *spec_file*, except with the extension .cfg.

-nowarning

> There are TLA$^+$ expressions that are legal but are sufficiently unlikely that their presence probably indicates an error. For example, the expression $[f \text{ EXCEPT } ![v] = e]$ is probably incorrect if v is not an element of the domain of f. (In this case, the expression just equals f.) TLC normally issues a warning when it encounters such an unlikely expression; this option suppresses these warnings.

14.5.2 Debugging a Specification

When you write a specification, it usually contains errors. The purpose of running TLC is to find as many of those errors as possible. We hope an error in the specification will cause TLC to report an error. The challenge of debugging is to find the error in the specification that caused the error reported by TLC. Before addressing this challenge, let's first examine TLC's output when it finds no error.

TLC's Normal Output

When you run TLC, the first thing it prints is the version number and creation date:

> TLC's messages may differ in format from the ones described here.

```
TLC Version 2.12 of 26 May 2003
```

Always include this information when reporting any problems with TLC. Next, TLC describes the mode in which it's being run. The possibilities are

```
Model-checking
```

in which it is exhaustively checking all reachable states, or

```
Running Random Simulation with seed 1901803014088851111.
```

in which it is running in simulation mode, using the indicated seed. (Seeds are described on pages 251–252.) Let's suppose it's running in model-checking mode. If you asked TLC to do liveness checking, it will now print something like

```
Implied-temporal checking--relative complexity = 8.
```

The time TLC takes for liveness checking is approximately proportional to the relative complexity. Even with a relative complexity of 1, checking liveness takes longer than checking safety. So, if the relative complexity is not small, TLC will probably take a very long time to complete, unless the model is very small. In simulation mode, a large complexity means that TLC will not be able to simulate very many behaviors. The relative complexity depends on the number of terms and the size of sets being quantified over in the temporal formulas.

TLC next prints a message like

```
Finished computing initial states:
4 states generated, with 2 of them distinct.
```

This indicates that, when evaluating the initial predicate, TLC generated 4 states, among which there were 2 distinct ones. TLC then prints one or more messages such as

```
Progress(9):  2846 states generated, 984 distinct states
found.  856 states left on queue.
```

This message indicates that TLC has thus far constructed a state graph $\mathcal{G}$ of diameter[2] 9, that it has generated and examined 2846 states, finding 984 distinct ones, and that the queue $\mathcal{U}$ of unexplored states contains 856 states. After running for a while, TLC generates these progress reports about once every five minutes. For most specifications, the number of states on the queue increases monotonically at the beginning of the execution and decreases monotonically at the end. The progress reports therefore provide a useful guide to how much longer the execution is likely to take.

$\mathcal{G}$ and $\mathcal{U}$ are described in Section 14.3.1 on page 241.

When TLC successfully completes, it prints

```
Model checking completed.  No error has been found.
```

It then prints something like

---

[2] The diameter of $\mathcal{G}$ is the smallest number d such that every state in $\mathcal{G}$ can be reached from an initial state by a path containing at most d states. It is the depth TLC has reached in its breadth-first exploration of the set of states. When using multiple threads (specified with the *workers* option), the diameter TLC reports may not be quite correct.

```
Estimates of the probability that TLC did not check all
reachable states because two distinct states had the same
fingerprint:
   calculated (optimistic):  .000003
   based on the actual fingerprints:  .00007
```

As explained on page 244, these are TLC's two estimates of the probability of a fingerprint collision. Finally, TLC prints a message like

```
2846 states generated, 984 distinct states found,
0 states left on queue.
The state graph has diameter 15.
```

with the total number of states and the diameter of the state graph.

While TLC is running, it may also print a message such as

```
-- Checkpointing run states/99-05-20-15-47-55 completed
```

This indicates that it has written a checkpoint that you can use to restart TLC in the event of a computer failure. (As explained in Section 14.5.3 on page 260, checkpoints have other uses as well.) The run identifier

```
states/99-05-20-15-47-55
```

is used with the *recover* option to restart TLC from where the checkpoint was taken. If only part of this message was printed—for example, because your computer crashed while TLC was taking the checkpoint—there is a slight chance that all the checkpoints are corrupted and you must start TLC again from the beginning.

Error Reports

The first problems you find in your specification will probably be syntax errors. TLC reports them with

```
ParseException in parseSpec:
```

followed by the error message generated by the Syntactic Analyzer. Chapter 12 explains how to interpret the analyzer's error messages. Running your specification through the analyzer as you write it will catch a lot of simple errors quickly.

As explained in Section 14.3.1 above, TLC executes three basic phases. In the first phase, it checks assumptions; in the second, it computes the initial states; and in the third, it generates the successor states of states on the queue $\mathcal{U}$ of unexplored states. You can tell if it has entered the third phase by whether or not it has printed the "initial states computed" message.

TLC's most straightforward error report occurs when it finds that one of the properties it is checking does not hold. Suppose we introduce an error into our alternating bit specification (Figure 14.1 on pages 223 and 224) by replacing the first conjunct of the invariant *ABTypeInv* with

$$\land\ msgQ \in Seq(Data)$$

TLC quickly finds the error and prints

```
Invariant ABTypeInv is violated
```

It next prints a minimal-length[3] behavior that leads to the state not satisfying the invariant:

```
The behavior up to this point is:
STATE 1: <Initial predicate>
/\ rBit = 0
/\ sBit = 0
/\ ackQ = <<  >>
/\ rcvd = d1
/\ sent = d1
/\ sAck = 0
/\ msgQ = <<  >>

STATE 2: <Action at line 66 in AlternatingBit>
/\ rBit = 0
/\ sBit = 1
/\ ackQ = <<  >>
/\ rcvd = d1
/\ sent = d1
/\ sAck = 0
/\ msgQ = << << 1, d1 >> >>
```

Note that TLC indicates which part of the next-state action allows the step that produces each state.

TLC prints each state as a TLA⁺ predicate that determines the state. When printing a state, TLC describes functions using the operators :> and @@ defined in the TLC module. (See Section 14.4 on page 248.)

The hardest errors to locate are usually the ones detected when TLC is forced to evaluate an expression that it can't handle, or one that is "silly" because its value is not specified by the semantics of TLA⁺. As an example, let's introduce a typical "off-by-one" error into the alternating bit protocol by replacing the second conjunct in the definition of *Lose* with

$$\exists\, i \in 1 \,..\, Len(q) :$$
$$q' = [j \in 1 \,..\, (Len(q) - 1) \mapsto \text{IF}\ \ j < i\ \text{THEN}\ \ q[j-1]$$
$$\text{ELSE}\ \ q[j]]$$

---

[3]When using multiple threads, it is possible, though unlikely, for there to be a shorter behavior that also violates the invariant.

If q has length greater than 1, then this defines $Lose(q)[1]$ to equal $q[0]$, which is a nonsensical value if q is a sequence. (The domain of a sequence q is the set $1 \ldots Len(q)$, which does not contain 0.) Running TLC produces the error message

```
Error: Applying tuple
<< << 1, d1 >>, << 1, d1 >> >>
to integer 0 which is out of domain.
```

It then prints a behavior leading to the error. TLC finds the error when evaluating the next-state action to compute the successor states for some state s, and s is the last state in that behavior. Had the error occurred when evaluating the invariant or the implied-action, TLC would have been evaluating it on the last state or step of the behavior.

Finally, TLC prints the location of the error:

```
The error occurred when TLC was evaluating the nested
expressions at the following positions:
0. Line 57, column 7 to line 59, column 60 in AlternatingBit
1. Line 58, column 55 to line 58, column 60 in AlternatingBit
```

The first position identifies the second conjunct of the definition of *Lose*; the second identifies the expression $q[j-1]$. This tells you that the error occurred while TLC was evaluating $q[j-1]$, which it was doing as part of the evaluation of the second conjunct of the definition of *Lose*. You must infer from the printed trace that it was evaluating the definition of *Lose* while evaluating the action *LoseMsg*. In general, TLC prints a tree of nested expressions—higher-level ones first. It seldom locates the error as precisely as you would like; often it just narrows it down to a conjunct or disjunct of a formula. You may need to insert *Print* expressions to locate the problem. See the discussion on page 259 for further advice on locating errors.

14.5.3 Hints on Using TLC Effectively

Start Small

The constraint and the assignment of values to the constant parameters define a model of the specification. How long it takes TLC to check a specification depends on the specification and the size of the model. Running on a 600MHz work station, TLC finds about 700 distinct reachable states per second for the alternating bit protocol specification. For some specifications, the time it takes TLC to generate a state grows with the size of the model; it can also increase as the generated states become more complicated. For some specifications run on larger models, TLC finds fewer than one reachable state per second.

You should always begin testing a specification with a tiny model, which TLC can check quickly. Let sets of processes and of data values have only one element; let queues be of length one. A specification that has not been tested probably has lots of errors. A small model will quickly catch most of the simple ones. When a very small model reveals no more errors, you can then run TLC with larger models to try to catch more subtle errors.

One way to figure out how large a model TLC can handle is to estimate the approximate number of reachable states as a function of the parameters. However, this can be hard. If you can't do it, increase the model size very gradually. The number of reachable states is typically an exponential function of the model's parameters; and the value of a^b grows very fast with increasing values of b.

Many systems have errors that will show up only on models too large for TLC to check exhaustively. After having TLC model check your specification on as large a model as your patience allows, you can run it in simulation mode on larger models. Random simulation is not an effective way to catch subtle errors, but it's worth trying; you might get lucky.

Be Suspicious of Success

Section 14.3.5 on page 247 explains why you should be suspicious if TLC does not find a violation of a liveness property; the finite model may mask errors. You should also be suspicious if TLC finds no error when checking safety properties. It's very easy to satisfy a safety property by simply doing nothing. For example, suppose we forgot to include the *SndNewValue* action in the alternating bit protocol specification's next-state action. The sender would then never try to send any values. But the resulting specification would still satisfy the protocol's correctness condition, formula *ABCSpec* of module *ABCorrectness*. (The specification doesn't require that values must be sent.)

The *coverage* option described on page 252 provides one way to catch such problems. Another way is to make sure that TLC finds errors in properties that should be violated. For example, if the alternating bit protocol is sending messages, then the value of *sent* should change. You can verify that it does change by checking that TLC reports a violation of the property

$$\forall\, d \in Data : (sent = d) \Rightarrow \Box(sent = d)$$

A good sanity check is to verify that TLC finds states that are reached only by performing a number of operations. For example, the caching memory specification of Section 5.6 should have reachable states in which a particular processor has both a read and two write operations in the *memQ* queue. Reaching such a state requires a processor to perform two writes followed by a read to an uncached address. We can verify that such a state is reachable by having TLC find a violation of an invariant declaring that there aren't a read and two writes for the same processor in *memQ*. (Of course, this requires a model in which *memQ* can be large enough.) Another way to check that certain states are reached is by using the *Print* operator inside an IF/THEN expression in an invariant to print a message when a suitable state is reached.

Let TLC Help You Figure Out What Went Wrong

When TLC reports that an invariant is violated, it may not be obvious what part of the invariant is false. If you give separate names to the conjuncts of your invariant and list them separately in the configuration file's INVARIANT statement, TLC will tell you which conjunct is false. However, it may be hard to see why even an individual conjunct is false. Instead of spending a lot of time trying to figure it out by yourself, it's easier to add *Print* expressions and let TLC tell you what's going wrong.

If you rerun TLC from the beginning with a lot of *Print* expressions, it will print output for every state it checks. Instead, you should start TLC from the state in which the invariant is false. Define a predicate, say *ErrorState*, that describes this state, and modify the configuration file to use *ErrorState* as the initial predicate. Writing the definition of *ErrorState* is easy—just copy the last state in TLC's error trace.[4]

You can use the same trick if any safety property is violated, or if TLC reports an error when evaluating the next-state action. For an error in a property of the form $\Box[A]_v$, rerun TLC using the next-to-last state in the error trace as the initial predicate, and using the last state in the trace, with the variable names primed, as the next-state action. To find an error that occurs when evaluating the next-state action, use the last state in the error trace as the initial predicate. (In this case, TLC may find several successor states before reporting the error.)

If you have introduced model values in the configuration file, they will undoubtedly appear in the states printed by TLC. So, if you are to copy those states into the module, you will have to declare the model values as constant parameters and then assign to each of these parameters the model value of the same name. For example, the configuration file we used for the alternating bit protocol introduces model values $d1$ and $d2$. So, we would add to module *MCAlternatingBit* the declaration

    CONSTANTS $d1$, $d2$

and add to the CONSTANT statement of the configuration file the assignments

    `d1 = d1    d2 = d2`

which assign to the constant parameters $d1$ and $d2$ the model values d1 and d2, respectively.

---

[4]Defining *ErrorState* is not so easy if you use the *difftrace* option, which is a reason for not using that option.

Don't Start Over After Every Error

After you've eliminated the errors that are easy to find, TLC may have to run for a long time before finding an error. Very often, it takes more than one try to fix an error properly. If you start TLC from the beginning after correcting an error, it may run for a long time only to report that you made a silly mistake in the correction. If the error was discovered when taking a step from a correct state, then it's a good idea to check your correction by starting TLC from that state. As explained above, you do this by defining a new initial predicate that equals the state printed by TLC.

Another way to avoid starting from scratch after an error is by using checkpoints. A checkpoint saves the current state graph $\mathcal{G}$ and queue $\mathcal{U}$ of unexplored states. It does not save any other information about the specification. You can restart TLC from a checkpoint even if you have changed the specification, as long as the specification's variables and the values that they can assume haven't changed. More precisely, you can restart from a checkpoint iff the view of any state computed before the checkpoint has not changed and the symmetry set is the same. When you correct an error that TLC found after running for a long time, you may want to use the *recover* option (page 252) to continue TLC from the last checkpoint instead of having it recheck all the states it has already checked.[5]

> The view and symmetry set are defined in Sections 14.3.3 and 14.3.4, respectively.

Check Everything You Can

Verify that your specification satisfies all the properties you think it should. For example, you shouldn't be content to check that the alternating bit protocol specification satisfies the higher-level specification *ABCSpec* of module *ABCorrectness*. You should also check lower-level properties that you expect it to satisfy. One such property, revealed by studying the algorithm, is that there should never be more than two different messages in the *msgQ* queue. So, we can check that the following predicate is an invariant:

$$Cardinality(\{msgQ[i] : i \in 1 \mathinner{\ldotp\ldotp} Len(msgQ)\}) \leq 2$$

(We must add the definition of *Cardinality* to module *MCAlternatingBit* by adding *FiniteSets* to its EXTENDS statement.)

It's a good idea to check as many invariance properties as you can. If you think that some state predicate should be an invariant, let TLC test if it is. Discovering that the predicate isn't an invariant may not reveal an error, but it will probably teach you something about your specification.

---

[5]Some states in the graph $\mathcal{G}$ may not be saved by a checkpoint; they will be rechecked when restarting from the checkpoint.

Be Creative

Even if a specification seems to lie outside the realm of what it can handle, TLC may be able to help check it. For example, suppose a specification's next-state action has the form $\exists\, n \in Nat : A(n)$. TLC cannot evaluate quantification over an infinite set, so it apparently can't deal with this specification. However, we can enable TLC to evaluate the quantified formula by using the configuration file's CONSTANT statement to replace Nat with the finite set $0 \mathinner{\ldotp\ldotp} n$, for some n. This replacement profoundly changes the specification's meaning. However, it might nonetheless allow TLC to reveal errors in the specification. Never forget that your objective in using TLC is not to verify that a specification is correct; it's to find errors.

Replacement is explained in Section 14.2.3.

Use TLC as a TLA$^+$ Calculator

Misunderstanding some aspect of TLA$^+$ can lead to errors in your specification. Use TLC to check your understanding of TLA$^+$ by running it on small examples. TLC checks assumptions, so you can turn it into a TLA$^+$ calculator by having it check a module with no specification, only ASSUME statements. For example, if g equals

$$[f \text{ EXCEPT } ![d] = e_1,\ ![d] = e_2]$$

what is the value of $g[d]$? You can ask TLC by letting it check a module containing

$$\begin{aligned}
\text{ASSUME} \quad \text{LET } f \;&\triangleq\; [i \in 1 \mathinner{\ldotp\ldotp} 10 \mapsto 1] \\
g \;&\triangleq\; [f \text{ EXCEPT } ![2] = 3,\ ![2] = 4] \\
\text{IN} \quad & Print(g[2],\ \text{TRUE})
\end{aligned}$$

You can have it verify that $(F \Rightarrow G) \equiv (\neg F \vee G)$ is a tautology by checking

$$\text{ASSUME} \quad \forall\, F,\, G \in \text{BOOLEAN} : (F \Rightarrow G) \equiv (\neg F \vee G)$$

TLC can even look for counterexamples to a conjecture. Can every set be written as the disjunction of two different sets? Check it for all subsets of $1 \mathinner{\ldotp\ldotp} 4$ with

$$\begin{aligned}
\text{ASSUME} \quad \forall\, S \in\ &\text{SUBSET } (1 \mathinner{\ldotp\ldotp} 4) : \\
\text{IF} \quad &\exists\, T,\, U \in \text{SUBSET } (1 \mathinner{\ldotp\ldotp} 4) : (T \neq U) \wedge (S = T \cup U) \\
\text{THEN} \quad &\text{TRUE} \\
\text{ELSE} \quad &Print(S,\ \text{TRUE})
\end{aligned}$$

When TLC is run just to check assumptions, it may need no information from the configuration file. But you must provide a configuration file, even if that file is empty.

14.6 What TLC Doesn't Do

We would like TLC to generate all the behaviors that satisfy a specification. But no program can do this for an arbitrary specification. I have already mentioned some limitations of TLC. There are other limitations that you may stumble on. One of them is that the Java classes that override the *Naturals* and *Integers* modules handle only numbers in the interval $-2^{31} \mathinner{\ldotp\ldotp} (2^{31} - 1)$; TLC reports an error if any computation generates a value outside this interval.

TLC can't generate all behaviors satisfying an arbitrary specification, but it might achieve the easier goal of ensuring that every behavior it does generate satisfies the specification. However, for reasons of efficiency, TLC doesn't always meet this goal. It deviates from the semantics of TLA$^+$ in two ways.

The first deviation is that TLC doesn't preserve the precise semantics of CHOOSE. As explained in Section 16.1, if S equals T, then CHOOSE $x \in S : P$ should equal CHOOSE $x \in T : P$. However, TLC guarantees this only if S and T are syntactically the same. For example, TLC might compute different values for the two expressions

$$\text{CHOOSE } x \in \{1, 2, 3\} : x < 3 \qquad \text{CHOOSE } x \in \{3, 2, 1\} : x < 3$$

A similar violation of the semantics of TLA$^+$ exists with CASE expressions, whose semantics are defined (in Section 16.1.4) in terms of CHOOSE.

The second part of the semantics of TLA$^+$ that TLC does not preserve is the representation of strings. In TLA$^+$, the string "abc" is a three-element sequence—that is, a function with domain $\{1, 2, 3\}$. TLC treats strings as primitive values, not as functions. It thus considers the legal TLA$^+$ expression "abc"[2] to be an error.

14.7 The Fine Print

This section describes in detail two aspects of TLC that were sketched above: the grammar of the configuration file, and the precise definition of TLC values.

14.7.1 The Grammar of the Configuration File

The grammar of TLC's configuration file is described in the TLA$^+$ module *ConfigFileGrammar* in Figure 14.6 on the next page. More precisely, the set of sentences *ConfigGrammar.File*, where *ConfigGrammar* is defined in the module, describes all syntactically correct configuration files from which comments have been removed. The *ConfigFileGrammar* module extends the *BNFGrammars* module, which is explained above in Section 11.1.4 (page 179).

Here are some additional restrictions on the configuration file that are not specified by module *ConfigFileGrammar*. There can be at most one INIT and

─────────────── MODULE *ConfigFileGrammar* ───────────────
EXTENDS *BNFGrammars*

───
                          LEXEMES
Letter $\triangleq$ *OneOf*("abcdefghijklmnopqrstuvwxyz_ABCDEFGHIJKLMNOPQRSTUVWXYZ")

Num $\triangleq$ *OneOf*("0123456789")

LetterOrNum $\triangleq$ *Letter* $\cup$ *Num*

AnyChar $\triangleq$ *LetterOrNum* $\cup$ *OneOf*("~ ! @ # \ $ % ^ & * − + = | ( ) { } [ ] , : ; ' ' < > . ? / ")

SingularKW $\triangleq$ {"SPECIFICATION", "INIT", "NEXT", "VIEW", "SYMMETRY"}

PluralKW $\triangleq$
   {"CONSTRAINT", "CONSTRAINTS", "ACTION−CONSTRAINT", "ACTION−CONSTRAINTS",
    "INVARIANT", "INVARIANTS", "PROPERTY", "PROPERTIES"}

Keyword $\triangleq$ *SingularKW* $\cup$ *PluralKW* $\cup$ {"CONSTANT", "CONSTANTS"}

AnyIdent $\triangleq$ *LetterOrNum** & *Letter* & *LetterOrNum**

Ident $\triangleq$ *AnyIdent* \ *Keyword*
───
ConfigGrammar $\triangleq$ THE BNF GRAMMAR
  LET $P(G) \triangleq$
          $\wedge$ *G.File* ::= *G.Statement*$^+$
          $\wedge$ *G.Statement* ::=   *Tok*(*SingularKW*) & *Tok*(*Ident*)
                               | *Tok*(*PluralKW*) & *Tok*(*Ident*)*
                               | *Tok*({"CONSTANT", "CONSTANTS"})
                                   & (*G.Replacement* | *G.Assignment*)*
          $\wedge$ *G.Replacement* ::= *Tok*(*Ident*) & *tok*("<−") & *Tok*(*AnyIdent*)
          $\wedge$ *G.Assignment* ::= *Tok*(*Ident*) & *tok*("=") & *G.IdentValue*
          $\wedge$ *G.IdentValue* ::=   *Tok*(*AnyIdent*) | *G.Number* | *G.String*
                               | *tok*("{")
                                   & (*Nil* | *G.IdentValue* & (*tok*(",") & *G.IdentValue*)*)
                                   & *tok*("}")
          $\wedge$ *G.Number* ::= (*Nil* | *tok*("−")) & *Tok*(*Num*$^+$)
          $\wedge$ *G.String* ::= *tok*(" " ") & *Tok*(*AnyChar**) & *tok*(" " ")
  IN   *LeastGrammar*(*P*)
───

Figure 14.6: The BNF grammar of the configuration file.

one `NEXT` statement. There can be one `SPECIFICATION` statement, but only if there is no `INIT` or `NEXT` statement. (See page 243 in Section 14.3.1 for conditions on when these statements must appear.) There can be at most one `VIEW` statement and at most one `SYMMETRY` statement. Multiple instances of other statements are allowed. For example, the two statements

```
INVARIANT Inv1
INVARIANT Inv2 Inv3
```

specify that TLC is to check the three invariants *Inv*1, *Inv*2, and *Inv*3. These statements are equivalent to the single statement

```
INVARIANT Inv1 Inv2 Inv3
```

14.7.2   Comparable TLC Values

Section 14.2.1 (page 230) describes TLC values. That description is incomplete because it does not define exactly when values are comparable. The precise definition is that two TLC values are comparable iff the following rules imply that they are:

1. Two primitive values are comparable iff they have the same value type.

   This rule implies that "abc" and "123" are comparable, but "abc" and 123 are not.

2. A model value is comparable with any value. (It is equal only to itself.)

3. Two sets are comparable if they have different numbers of elements, or if they have the same numbers of elements and all the elements in one set are comparable with all the elements in the other.

   This rule implies that $\{1\}$ and $\{\text{"a"}, \text{"b"}\}$ are comparable and that $\{1, 2\}$ and $\{2, 3\}$ are comparable. However, $\{1, 2\}$ and $\{\text{"a"}, \text{"b"}\}$ are not comparable.

4. Two functions f and g are comparable iff (i) their domains are comparable and (ii) if their domains are equal, then $f[x]$ and $g[x]$ are comparable for every element x in their domain.

   This rule implies that $\langle 1, 2 \rangle$ and $\langle \text{"a"}, \text{"b"}, \text{"c"} \rangle$ are comparable, and that $\langle 1, \text{"a"} \rangle$ and $\langle 2, \text{"bc"} \rangle$ are comparable. However, $\langle 1, 2 \rangle$ and $\langle \text{"a"}, \text{"b"} \rangle$ are not comparable.

Part IV

The TLA$^+$ Language

This part of the book describes TLA$^+$ in detail. Chapter 15 explains the syntax; Chapters 16 and 17 explain the semantics; and Chapter 18 contains the standard modules. Almost all of the TLA$^+$ language has already been described—mainly through examples. In fact, most of the language was described in Chapters 1–6. This part gives complete specification of the language.

A completely formal specification of TLA$^+$ would consist of a formal definition of the set of legal (syntactically well-formed) modules, and a precisely defined meaning operator that assigns to every legal module M its mathematical meaning $[\![M]\!]$. Such a specification would be quite long and of limited interest. Instead, I have tried to provide a fairly informal specification that is detailed enough to show mathematically sophisticated readers how they could write a completely formal one.

These chapters are heavy going, and few people will want to read them completely. However, I hope they can serve as a reference manual for anyone who reads or writes TLA$^+$ specifications. If you have a question about the finer details of the syntax or the meaning of some part of the language, you should be able to find the answer here.

Tables 1–8 on the next page through page 273 provide a tiny reference manual. Tables 1–4 very briefly describe all the built-in operators of TLA$^+$. Table 5 lists all the user-definable operator symbols and indicates which ones are already used by the standard modules. It's a good place to look when choosing notation for your specification. Table 6 gives the precedence of the operators; it is explained in Section 15.2.1 on page 283. Table 7 lists all operators defined by the standard modules. Finally, Table 8 shows how to type any symbol that doesn't have an obvious ASCII equivalent.

Logic

$\land \quad \lor \quad \lnot \quad \Rightarrow \quad \equiv$

TRUE   FALSE   BOOLEAN [the set {TRUE, FALSE}]

$\forall x : p \quad \exists x : p \quad \forall x \in S : p \ ^{(1)} \quad \exists x \in S : p \ ^{(1)}$

CHOOSE $x : p$ [An x satisfying p]   CHOOSE $x \in S : p$ [An x in S satisfying p]

Sets

$= \quad \neq \quad \in \quad \notin \quad \cup \quad \cap \quad \subseteq \quad \setminus$ [set difference]

$\{e_1, \ldots, e_n\}$ [Set consisting of elements e_i]

$\{x \in S : p\} \ ^{(2)}$ [Set of elements x in S satisfying p]

$\{e : x \in S\} \ ^{(1)}$ [Set of elements e such that x in S]

SUBSET S [Set of subsets of S]

UNION S [Union of all elements of S]

Functions

$f[e]$ [Function application]

DOMAIN f [Domain of function f]

$[x \in S \mapsto e] \ ^{(1)}$ [Function f such that $f[x] = e$ for $x \in S$]

$[S \to T]$ [Set of functions f with $f[x] \in T$ for $x \in S$]

$[f \text{ EXCEPT } ![e_1] = e_2] \ ^{(3)}$ [Function $\widehat{f}$ equal to f except $\widehat{f}[e_1] = e_2$]

Records

$e.h$ [The h-field of record e]

$[h_1 \mapsto e_1, \ldots, h_n \mapsto e_n]$ [The record whose h_i field is e_i]

$[h_1 : S_1, \ldots, h_n : S_n]$ [Set of all records with h_i field in S_i]

$[r \text{ EXCEPT } !.h = e] \ ^{(3)}$ [Record $\widehat{r}$ equal to r except $\widehat{r}.h = e$]

Tuples

$e[i]$ [The ith component of tuple e]

$\langle e_1, \ldots, e_n \rangle$ [The n-tuple whose i^{th} component is e_i]

$S_1 \times \ldots \times S_n$ [The set of all n-tuples with i^{th} component in S_i]

Strings and Numbers

"$c_1 \ldots c_n$" [A literal string of n characters]

STRING [The set of all strings]

$d_1 \ldots d_n \quad d_1 \ldots d_n . d_{n+1} \ldots d_m$ [Numbers (where the d_i are digits)]

---

(1) $x \in S$ may be replaced by a comma-separated list of items $v \in S$, where v is either a comma-separated list or a tuple of identifiers.

(2) x may be an identifier or tuple of identifiers.

(3) $![e_1]$ or $!.h$ may be replaced by a comma separated list of items $!a_1 \cdots a_n$, where each a_i is $[e_i]$ or $.h_i$.

Table 1: The constant operators.

IF p THEN e_1 ELSE e_2 [e_1 if p true, else e_2]

CASE $p_1 \to e_1 \ \Box \ \ldots \ \Box \ p_n \to e_n$ [Some e_i such that p_i true]

CASE $p_1 \to e_1 \ \Box \ \ldots \ \Box \ p_n \to e_n \ \Box$ OTHER $\to e$ [Some e_i such that p_i true, or e if all p_i are false]

LET $d_1 \stackrel{\Delta}{=} e_1 \ \ldots \ d_n \stackrel{\Delta}{=} e_n$ IN e [e in the context of the definitions]

$\land \ p_1$ [the conjunction $p_1 \land \ldots \land p_n$] $\lor \ p_1$ [the disjunction $p_1 \lor \ldots \lor p_n$]

$\vdots$ $\vdots$

$\land \ p_n$ $\lor \ p_n$

Table 2: Miscellaneous constructs.

e' [The value of e in the final state of a step]

$[A]_e$ [$A \lor (e' = e)$]

$\langle A \rangle_e$ [$A \land (e' \neq e)$]

ENABLED A [An A step is possible]

UNCHANGED e [$e' = e$]

$A \cdot B$ [Composition of actions]

Table 3: Action operators.

$\Box F$ [F is always true]

$\Diamond F$ [F is eventually true]

$\mathrm{WF}_e(A)$ [Weak fairness for action A]

$\mathrm{SF}_e(A)$ [Strong fairness for action A]

$F \rightsquigarrow G$ [F leads to G]

$F \stackrel{+}{\rhd} G$ [F guarantees G]

$\boldsymbol{\exists} x : F$ [Temporal existential quantification (hiding)]

$\boldsymbol{\forall} x : F$ [Temporal universal quantification]

Table 4: Temporal operators.

Infix Operators

$+$ [1]	$-$ [1]	$*$ [1]	$/$ [2]	$\circ$ [3]	$++$
$\div$ [1]	$\%$ [1]	$\char`\^$ [1,4]	$..$ [1]	$...$	$--$
$\oplus$ [5]	$\ominus$ [5]	$\otimes$	$\oslash$	$\odot$	$**$
$<$ [1]	$>$ [1]	$\leq$ [1]	$\geq$ [1]	$\sqcap$	$//$
$\prec$	$\succ$	$\preceq$	$\succeq$	$\sqcup$	$\char`\^\char`\^$
$\ll$	$\gg$	$<:$	$:>$ [6]	$\&$	$\&\&$
$\sqsubset$	$\sqsupset$	$\sqsubseteq$ [5]	$\sqsupseteq$	$\mid$	$\parallel$
$\subset$	$\supset$		$\supseteq$	$\star$	$\%\%$
$\vdash$	$\dashv$	$\vDash$	$\Dashv$	$\bullet$	$\#\#$
$\sim$	$\simeq$	$\approx$	$\cong$	$\$$	$\$\$$
$:=$	$::=$	$\asymp$	$\doteq$	$??$	$!!$
$\propto$	$\wr$	$\uplus$	$\bigcirc$	$@@$ [6]	

Postfix Operators [7]

$\char`\^+$  $\char`\^*$ $\char`\^\#$

Prefix Operator

$-$ [8]

---

(1) Defined by the *Naturals*, *Integers*, and *Reals* modules.
(2) Defined by the *Reals* module.
(3) Defined by the *Sequences* module.
(4) $x\char`\^y$ is printed as x^y.
(5) Defined by the *Bags* module.
(6) Defined by the *TLC* module.
(7) $e\char`\^+$ is printed as $e^+$, and similarly for $\char`\^*$ and $\char`\^\#$.
(8) Defined by the *Integers* and *Reals* modules.

Table 5: User-definable operator symbols.

Prefix Operators

¬	4–4	□	4–15	UNION	8–8
ENABLED	4–15	◇	4–15	DOMAIN	9–9
UNCHANGED	4–15	SUBSET	8–8	−	12–12

Infix Operators

$\Rightarrow$	1–1	$\leq$	5–5	$<:$	7–7		$\ominus$	11–11 (a)	
$\overset{+}{\rightarrow}$	2–2	$\ll$	5–5	$\backslash$	8–8		$-$	11–11 (a)	
$\equiv$	2–2	$\prec$	5–5	$\cap$	8–8 (a)		$--$	11–11 (a)	
$\rightsquigarrow$	2–2	$\preceq$	5–5	$\cup$	8–8 (a)		$\&$	13–13 (a)	
$\wedge$	3–3 (a)	$\propto$	5–5	$..$	9–9		$\&\&$	13–13 (a)	
$\vee$	3–3 (a)	$\sim$	5–5	$...$	9–9		$\odot$	13–13 (a)	
$\neq$	5–5	$\simeq$	5–5	$!!$	9–13		$\oslash$	13–13	
$\dashv$	5–5	$\sqsubset$	5–5	$\#\#$	9–13 (a)		$\otimes$	13–13 (a)	
$::=$	5–5	$\sqsubseteq$	5–5	$\$$	9–13 (a)		$*$	13–13 (a)	
$:=$	5–5	$\sqsupset$	5–5	$\$\$$	9–13 (a)		$**$	13–13 (a)	
$<$	5–5	$\sqsupseteq$	5–5	$??$	9–13 (a)		$/$	13–13	
$=$	5–5	$\subset$	5–5	$\sqcap$	9–13 (a)		$//$	13–13	
$\rightvdash$	5–5	$\subseteq$	5–5	$\sqcup$	9–13 (a)		$\bigcirc$	13–13 (a)	
$>$	5–5	$\succ$	5–5	$\uplus$	9–13 (a)		$\bullet$	13–13 (a)	
$\approx$	5–5	$\succeq$	5–5	$\wr$	9–14		$\div$	13–13	
$\asymp$	5–5	$\supset$	5–5	$\oplus$	10–10 (a)		$\circ$	13–13 (a)	
$\cong$	5–5	$\supseteq$	5–5	$+$	10–10 (a)		$\star$	13–13 (a)	
$\doteq$	5–5	$\vdash$	5–5	$++$	10–10 (a)		$\hat{}$	14–14	
$\geq$	5–5	$\models$	5–5	$\%$	10–11		$\hat{}\hat{}$	14–14	
$\gg$	5–5	$.^{(1)}$	5–14 (a)	$\%\%$	10–11 (a)		$.^{(2)}$	17–17 (a)	
$\in$	5–5	$@@$	6–6 (a)	$\mid$	10–11 (a)				
$\notin$	5–5	$:>$	7–7	$\|$	10–11 (a)				

Postfix Operators

$\hat{}+$	15–15	$\hat{}*$	15–15	$\hat{}\#$	15–15	$'$	15–15

---

(1) Action composition (\cdot).
(2) Record field (period).

Table 6: The precedence ranges of operators. The relative precedence of two operators is unspecified if their ranges overlap. Left-associative operators are indicated by (a).

Modules *Naturals, Integers, Reals*

$+$	$-$ [1]	$*$	$/$ [2]	$\char`\^$ [3]	$..$	*Nat*	*Real* [2]
$\div$	$\%$	$\leq$	$\geq$	$<$	$>$	*Int* [4]	*Infinity* [2]

(1) Only infix $-$ is defined in *Naturals*.
(2) Defined only in *Reals* module.
(3) Exponentiation.
(4) Not defined in *Naturals* module.

Module *Sequences*

$\circ$	*Head*	*SelectSeq*	*SubSeq*
Append	*Len*	*Seq*	*Tail*

Module *FiniteSets*

IsFiniteSet   *Cardinality*

Module *Bags*

$\oplus$	*BagIn*	*CopiesIn*	*SubBag*
$\ominus$	*BagOfAll*	*EmptyBag*	
$\sqsubseteq$	*BagToSet*	*IsABag*	
BagCardinality	*BagUnion*	*SetToBag*	

Module *RealTime*

RTBound   *RTnow*   *now* (declared to be a variable)

Module *TLC*

$:>$	$@@$	*Print*	*Assert*	*JavaTime*	*Permutations*
SortSeq					

Table 7: Operators defined in the standard modules.

Symbol	ASCII	Symbol	ASCII	Symbol	ASCII
$\wedge$	`/\` or `\land`	$\vee$	`\/` or `\lor`	$\Rightarrow$	`=>`
$\neg$	`~` or `\lnot` or `\neg`	$\equiv$	`<=>` or `\equiv`	$\triangleq$	`==`
$\in$	`\in`	$\notin$	`\notin`	$\neq$	`#` or `/=`
$\langle$	`<<`	$\rangle$	`>>`	$\Box$	`[]`
$<$	`<`	$>$	`>`	$\Diamond$	`<>`
$\leq$	`\leq` or `=<` or `<=`	$\geq$	`\geq` or `>=`	$\rightsquigarrow$	`~>`
$\ll$	`\ll`	$\gg$	`\gg`	$\overset{+}{\rightarrow}$	`-+->`
$\prec$	`\prec`	$\succ$	`\succ`	$\mapsto$	`|->`
$\preceq$	`\preceq`	$\succeq$	`\succeq`	$\div$	`\div`
$\subseteq$	`\subseteq`	$\supseteq$	`\supseteq`	$\cdot$	`\cdot`
$\subset$	`\subset`	$\supset$	`\supset`	$\circ$	`\o` or `\circ`
$\sqsubset$	`\sqsubset`	$\sqsupset$	`\sqsupset`	$\bullet$	`\bullet`
$\sqsubseteq$	`\sqsubseteq`	$\sqsupseteq$	`\sqsupseteq`	$\star$	`\star`
$\vdash$	`|-`	$\dashv$	`-|`	$\bigcirc$	`\bigcirc`
$\vDash$	`|=`	$=\!|$	`=|`	$\sim$	`\sim`
$\rightarrow$	`->`	$\leftarrow$	`<-`	$\simeq$	`\simeq`
$\cap$	`\cap` or `\intersect`	$\cup$	`\cup` or `\union`	$\asymp$	`\asymp`
$\sqcap$	`\sqcap`	$\sqcup$	`\sqcup`	$\approx$	`\approx`
$\oplus$	`(+)` or `\oplus`	$\uplus$	`\uplus`	$\cong$	`\cong`
$\ominus$	`(-)` or `\ominus`	$\times$	`\X` or `\times`	$\doteq$	`\doteq`
$\odot$	`(.)` or `\odot`	$\wr$	`\wr`	x^y	`x^y` (2)
$\otimes$	`(\X)` or `\otimes`	$\propto$	`\propto`	x^+	`x^+` (2)
$\oslash$	`(/)` or `\oslash`	"s"	`"s"` (1)	x^*	`x^*` (2)
$\exists$	`\E`	$\forall$	`\A`	$x^\#$	`x^#` (2)
$\exists\!\exists$	`\EE`	$\forall\!\forall$	`\AA`	$'$	`,`
$]_v$	`]_v`	$\rangle_v$	`>>_v`		
WF_v	`WF_v`	SF_v	`SF_v`		

⌐‾‾⌐	`--------` (3)	⌐‾‾⌐	`--------` (3)
⊢—⊣	`--------` (3)	⌐__⌐	`========` (3)

(1) *s* is a sequence of characters. See Section 16.1.10 on page 307.
(2) *x* and *y* are any expressions.
(3) a sequence of four or more `-` or `=` characters.

Table 8: The ASCII representations of typeset symbols.

Chapter 15

The Syntax of TLA⁺

This book uses the ASCII version of TLA⁺—the version based on the ASCII character set. One can define other versions of TLA⁺ that use different sets of characters. A different version might allow an expression like $\Omega[\mathring{a}] \circ \langle\,\text{"ça"}\,\rangle$. Since mathematical formulas look pretty much the same in most languages, the basic syntax of all versions of TLA⁺ should be the same. Different versions would differ in their lexemes and in the identifiers and strings they allow. This chapter describes the syntax of the ASCII version, the only one that now exists.

The term *syntax* has two different usages, which I will somewhat arbitrarily attribute to mathematicians and computer scientists. A computer scientist would say that $\langle a,\ a \rangle$ is a syntactically correct TLA⁺ expression. A mathematician would say that the expression is syntactically correct iff it appears in a context in which a is defined or declared. A computer scientist would call this requirement a *semantic* rather than a syntactic condition. A mathematician would say that $\langle a,\ a \rangle$ is meaningless if a isn't defined or declared, and one can't talk about the semantics of a meaningless expression. This chapter describes the syntax of TLA⁺, in the computer scientist's sense of syntax. The "semantic" part of the syntax is specified in Chapters 16 and 17.

TLA⁺ is designed to be easy for humans to read and write. In particular, its syntax for expressions tries to capture some of the richness of ordinary mathematical notation. This makes a precise specification of the syntax rather complicated. Such a specification has been written in TLA⁺, but it's quite detailed and you probably don't want to look at it unless you are writing a parser for the language. This chapter gives a less formal description of the syntax that should answer any questions likely to arise in practice. Section 15.1 specifies precisely a simple grammar that ignores some aspects of the syntax such as operator precedence, indentation rules for ∧ and ∨ lists, and comments. These other aspects are explained informally in Section 15.2. Sections 15.1 and 15.2 describe the grammar of a TLA⁺ module viewed as a sequence of lexemes, where a lexeme is a sequence of characters such as |-> that forms an atomic unit of the grammar. Section 15.3 describes how the sequence of characters that you actually type are turned into a sequence of lexemes. It includes the precise syntax for comments.

This chapter describes the ASCII syntax for TLA$^+$ specifications. Typeset versions of specifications appear in this book. For example, the infix operator typeset as $\prec$ is represented in ASCII as \prec. Table 8 on page 273 gives the correspondence between the ASCII and typeset versions of all TLA$^+$ symbols for which the correspondence may not be obvious.

15.1 The Simple Grammar

The simple grammar of TLA$^+$ is described in BNF. More precisely, it is specified below in the TLA$^+$ module *TLAPlusGrammar*. This module uses the operators for representing BNF grammars defined in the *BNFGrammars* module of Section 11.1.4 (page 179). Module *TLAPlusGrammar* contains comments describing how to read the specification as an ordinary BNF grammar. So, if you are familiar with BNF grammars and just want to learn the syntax of TLA$^+$, you don't have to understand how the TLA$^+$ operators for writing grammars are defined. Otherwise, you should read Section 11.1.4 before trying to read the following module.

───────────────── MODULE *TLAPlusGrammar* ─────────────────

EXTENDS *Naturals, Sequences, BNFGrammars*

This module defines a simple grammar for TLA$^+$ that ignores many aspects of the language, such as operator precedence and indentation rules. I use the term *sentence* to mean a sequence of lexemes, where a lexeme is just a string. The *BNFGrammars* module defines the following standard conventions for writing sets of sentences: $L \mid M$ means an L or an M, L^* means the concatenation of zero or more Ls, and L^+ means the concatenation of one or more Ls. The concatenation of an L and an M is denoted by L & M rather than the customary juxtaposition $L\,M$. *Nil* is the null sentence, so *Nil* & L equals L for any L.

A *token* is a one-lexeme sentence. There are two operators for defining sets of tokens: if s is a lexeme, then $tok(s)$ is the set containing the single token $\langle s \rangle$; and if S is a set of lexemes, then $Tok(S)$ is the set containing all tokens $\langle s \rangle$ for $s \in S$. In comments, I will not distinguish between the token $\langle s \rangle$ and the string s.

We begin by defining two useful operators. First, a *CommaList(L)* is defined to be an L or a sequence of Ls separated by commas.

$CommaList(L) \;\triangleq\; L$ & $(tok(\text{","})$ & $L)^*$

Next, if c is a character, then we define *AtLeast4("c")* to be the set of tokens consisting of 4 or more c's.

$AtLeast4(s) \;\triangleq\; Tok(\{s \circ s \circ s\}$ & $\{s\}^+)$

We now define some sets of lexemes. First is *ReservedWord*, the set of words that can't be used as identifiers. (Note that BOOLEAN, TRUE, FALSE, and STRING are identifiers that are predefined.)

ReservedWord $\triangleq$

{ "ASSUME",	"ELSE",	"LOCAL",	"UNION",
"ASSUMPTION",	"ENABLED",	"MODULE",	"VARIABLE",
"AXIOM",	"EXCEPT",	"OTHER",	"VARIABLES",
"CASE",	"EXTENDS",	"SF_",	"WF_",
"CHOOSE",	"IF",	"SUBSET",	"WITH",
"CONSTANT",	"IN",	"THEN",	
"CONSTANTS",	"INSTANCE",	"THEOREM",	
"DOMAIN",	"LET",	"UNCHANGED"	}

Next are three sets of characters—more precisely, sets of 1-character lexemes. They are the sets of letters, numbers, and characters that can appear in an identifier.

Letter $\triangleq$
$\qquad OneOf(\text{"abcdefghijklmnopqrstuvwxyzABCDEFGHIJKLMNOPQRSTUVWXYZ"})$

Numeral $\triangleq OneOf(\text{"0123456789"})$

NameChar $\triangleq Letter \cup Numeral \cup \{\text{"\_"}\}$

We now define some sets of tokens. A *Name* is a token composed of letters, numbers, and _ characters that contains at least one letter, but does not begin with "WF_" or "SF_" (see page 290 for an explanation of this restriction). It can be used as the name of a record field or a module. An *Identifier* is a *Name* that isn't a reserved word.

Name $\triangleq Tok((NameChar^* \& Letter \& NameChar^*)$
$\qquad\qquad\qquad \setminus (\{\text{"WF\_"}, \text{"SF\_"}\} \& NameChar^+))$

Identifier $\triangleq Name \setminus Tok(ReservedWord)$

An *IdentifierOrTuple* is either an identifier or a tuple of identifiers. Note that $\langle \rangle$ is typed as << >>.

IdentifierOrTuple $\triangleq$
$\qquad Identifier \mid tok(\text{"<<"}) \& CommaList(Identifier) \& tok(\text{">>"})$

A *Number* is a token representing a number. You can write the integer 63 in the following ways: 63, 63.00, \b111111 or \B111111 (binary), \o77 or \O77 (octal), or \h3f, \H3f, \h3F, or \H3F (hexadecimal).

NumberLexeme $\triangleq \qquad Numeral^+$
$\qquad\qquad\qquad \mid (Numeral^* \& \{\text{"."}\} \& Numeral^+)$
$\qquad\qquad\qquad \mid \{\text{"\b"}, \text{"\B"}\} \& OneOf(\text{"01"})^+$
$\qquad\qquad\qquad \mid \{\text{"\o"}, \text{"\O"}\} \& OneOf(\text{"01234567"})^+$
$\qquad\qquad\qquad \mid \{\text{"\h"}, \text{"\H"}\} \& OneOf(\text{"0123456789abcdefABCDEF"})^+$

Number $\triangleq Tok(NumberLexeme)$

A *String* token represents a literal string. See Section 16.1.10 on page 307 to find out how special characters are typed in a string.

$String \;\triangleq\; Tok(\{\,``\,\texttt{"}\,"\} \;\&\; \text{STRING} \;\&\; \{``\,\texttt{"}\,"\})$

We next define the sets of tokens that represent prefix operators (like □), infix operators (like +), and postfix operators (like prime (′)). See Table 8 on page 273 to find out what symbols these ASCII strings represent.

$PrefixOp \;\triangleq\; Tok(\{\,$ "-", "~", "\lnot", "\neg", "[]", "<>", "DOMAIN",
$\qquad\qquad\qquad$ "ENABLED", "SUBSET", "UNCHANGED", "UNION"$\})$

$InfixOp \;\triangleq\;$
$\quad Tok(\{$

"!!",	"#",	"##",	"\$",	"\$\$",	"%",	"%%",
"&",	"&&",	"(+)",	"(-)",	"(.)",	"(/)",	"(\X)",
"*",	"**",	"+",	"++",	"-",	"-+->",	"--",
"-|",	"..",	"...",	"/",	"//",	"/=",	"/\\",
"::=",	":=",	":>",	"<",	"<:",	"<=>",	"=",
"=<",	"=>",	"=|",	">",	">=",	"?",	"??",
"@@",	"\\",	"\\/",	"^",	"^^",	"|",	"|-",
"|=",	"||",	"~>",	".",			

"\approx",	"\geq",	"\oslash",	"\sqsupseteq",
"\asymp",	"\gg",	"\otimes",	"\star",
"\bigcirc",	"\in",	"\prec",	"\subset",
"\bullet",	"\intersect",	"\preceq",	"\subseteq",
"\cap",	"\land",	"\propto",	"\succ",
"\cdot",	"\leq",	"\sim",	"\succeq",
"\circ",	"\ll",	"\simeq",	"\supset",
"\cong",	"\lor",	"\sqcap",	"\supseteq",
"\cup",	"\o",	"\sqcup",	"\union",
"\div",	"\odot",	"\sqsubset",	"\uplus",
"\doteq",	"\ominus",	"\sqsubseteq",	"\wr",
"\equiv",	"\oplus",	"\sqsupset"	$\})$

$PostfixOp \;\triangleq\; Tok(\{\,$ "^+", "^*", "^#", "'" $\})$

Formally, the grammar $TLAPlusGrammar$ of TLA⁺ is the smallest grammar satisfying the BNF productions below.

$TLAPlusGrammar \;\triangleq\;$
$\quad$ LET $P(G) \;\triangleq\;$

Here is the BNF grammar. Terms that begin with "G.", like $G.Module$, represent nonterminals. The terminals are sets of tokens, either defined above or described with the operators tok and Tok. The operators $AtLeast4$ and $CommaList$ are defined above.

$\land\; G.Module \;::=\;\quad AtLeast4(\text{"-"}) \;\&\; tok(\text{"MODULE"}) \;\&\; Name \;\&\; AtLeast4(\text{"-"})$
$\qquad\qquad\qquad\qquad \&\; (Nil \mid (tok(\text{"EXTENDS"}) \;\&\; CommaList(Name)))$
$\qquad\qquad\qquad\qquad \&\; (G.Unit)^*$
$\qquad\qquad\qquad\qquad \&\; AtLeast4(\text{"="})$

$\wedge$ *G.Unit* ::=   *G.VariableDeclaration*
            | *G.ConstantDeclaration*
            | (*Nil* | *tok*("LOCAL")) & *G.OperatorDefinition*
            | (*Nil* | *tok*("LOCAL")) & *G.FunctionDefinition*
            | (*Nil* | *tok*("LOCAL")) & *G.Instance*
            | (*Nil* | *tok*("LOCAL")) & *G.ModuleDefinition*
            | *G.Assumption*
            | *G.Theorem*
            | *G.Module*
            | *AtLeast4*("–")

$\wedge$ *G.VariableDeclaration* ::=
     Tok({"VARIABLE", "VARIABLES"}) & *CommaList*(*Identifier*)

$\wedge$ *G.ConstantDeclaration* ::=
     Tok({"CONSTANT", "CONSTANTS"}) & *CommaList*(*G.OpDecl*)

$\wedge$ *G.OpDecl* ::=   *Identifier*
            | *Identifier* & *tok*("(") & *CommaList*(*tok*("_")) & *tok*(")")
            | *PrefixOp* & *tok*("_")
            | *tok*("_") & *InfixOp* & *tok*("_")
            | *tok*("_") & *PostfixOp*

$\wedge$ *G.OperatorDefinition* ::=   (*G.NonFixLHS*
                              | *PrefixOp* & *Identifier*
                              | *Identifier* & *InfixOp* & *Identifier*
                              | *Identifier* & *PostfixOp*)
                   &   *tok*("==")
                   &   *G.Expression*

$\wedge$ *G.NonFixLHS* ::=
       Identifier
   &   (*Nil*
        | *tok*("(") & *CommaList*(*Identifier* | *G.OpDecl*) & *tok*(")"))

$\wedge$ *G.FunctionDefinition* ::=
       Identifier
   &   *tok*("[") & *CommaList*(*G.QuantifierBound*) & *tok*("]")
   &   *tok*("==")
   &   *G.Expression*

$\wedge$ *G.QuantifierBound* ::=   (*IdentifierOrTuple* | *CommaList*(*Identifier*))
                                  & *tok*("\in")
                                  & *G.Expression*

$\wedge$ *G.Instance* ::=   *tok*("INSTANCE")
                   & *Name*
                   & (*Nil* | *tok*("WITH") & *CommaList*(*G.Substitution*))

$\wedge$ *G.Substitution* ::=   (*Identifier* | *PrefixOp* | *InfixOp* | *PostfixOp*)
                       & *tok*("<-")
                       & *G.Argument*

$\wedge$ *G.Argument* ::=   *G.Expression*
                    | *G.GeneralPrefixOp*
                    | *G.GeneralInfixOp*
                    | *G.GeneralPostfixOp*

$\wedge$ *G.InstancePrefix* ::=
    (*Identifier*
     & (*Nil*
          | *tok*("(") & *CommaList*(*G.Expression*) & *tok*(")"))
     & *tok*("!")   )*

$\wedge$ *G.GeneralIdentifier* ::= *G.InstancePrefix* & *Identifier*
$\wedge$ *G.GeneralPrefixOp* ::= *G.InstancePrefix* & *PrefixOp*
$\wedge$ *G.GeneralInfixOp* ::= *G.InstancePrefix* & *InfixOp*
$\wedge$ *G.GeneralPostfixOp* ::= *G.InstancePrefix* & *PostfixOp*

$\wedge$ *G.ModuleDefinition* ::= *G.NonFixLHS* & *tok*("==") & *G.Instance*

$\wedge$ *G.Assumption* ::=
    Tok({"ASSUME", "ASSUMPTION", "AXIOM"}) & *G.Expression*

$\wedge$ *G.Theorem* ::= *tok*("THEOREM") & *G.Expression*

The comments give examples of each of the different types of expression.

$\wedge$ *G.Expression* ::=

     G.GeneralIdentifier $A(x+7)!B!Id$

    | *G.GeneralIdentifier* & *tok*("(") $A!Op(x+1, y)$
         & *CommaList*(*G.Argument*) & *tok*(")")

    | *G.GeneralPrefixOp* & *G.Expression*   SUBSET *S.foo*

| *G.Expression* & *G.GeneralInfixOp* & *G.Expression* $a + b$

| *G.Expression* & *G.GeneralPostfixOp* $x[1]'$

| *tok*("(") & *G.Expression* & *tok*(")") $(x + 1)$

| *Tok*({"\A", "\E"}) & *CommaList*(*G.QuantifierBound*) $\forall x \in S, \langle y, z \rangle \in T : F(x, y, z)$
        & *tok*(":") & *G.Expression*

| *Tok*({"\A", "\E", "\AA", "\EE"}) & *CommaList*(*Identifier*) $\exists x, y : x + y > 0$
        & *tok*(":") & *G.Expression*

| *tok*("CHOOSE")   CHOOSE $\langle x, y \rangle \in S : F(x, y)$
    & *IdentifierOrTuple*
    & (*Nil* | *tok*("\in") & *G.Expression*)
    & *tok*(":")
    & *G.Expression*

| *tok*("{") & (*Nil* | *CommaList*(*G.Expression*)) & *tok*("}") $\{1, 2, 2 + 2\}$

| *tok*("{") $\{x \in Nat : x > 0\}$
    & *IdentifierOrTuple* & *tok*("\in") & *G.Expression*
    & *tok*(":")
    & *G.Expression*
    & *tok*("}")

| *tok*("{") $\{F(x, y, z) : x, y \in S, z \in T\}$
    & *G.Expression*
    & *tok*(":")
    & *CommaList*(*G.QuantifierBound*)
    & *tok*("}")

| *G.Expression* & *tok*("[") & *CommaList*(*G.Expression*) & *tok*("]") $f[i + 1, j]$

| *tok*("[") $[i, j \in S, \langle p, q \rangle \in T \mapsto F(i, j, p, q)]$
    & *CommaList*(*G.QuantifierBound*)
    & *tok*("|->")
    & *G.Expression*
    & *tok*("]")

| *tok*("[") & *G.Expression* & *tok*("->") & *G.Expression* & *tok*("]") $[(S \cup T) \to U]$

| *tok*("[") & *CommaList*(*Name* & *tok*("|->") & *G.Expression*) $[a \mapsto x + 1, b \mapsto y]$
        & *tok*("]")

| $tok($ "[" $)$ & $CommaList($ $Name$ & $tok($ ":" $)$ & $G.Expression$ $)$ $[a : Nat,\ b : S \cup T]$
    & $tok($ "]" $)$

| $tok($ "[" $)$ $[f$ EXCEPT $![1, x].r = 4,\ ![\langle 2, y\rangle] = e]$
  & $G.Expression$
  & $tok($ "EXCEPT" $)$
  & $CommaList($ $tok($ "!" $)$
         & ($tok($ "." $)$ & $Name$
            | $tok($ "[" $)$ & $CommaList($G.Expression$)$ & $tok($ "]" $)$ $)^+$
       & $tok($ "=" $)$ & $G.Expression$ $)$
  & $tok($ "]" $)$

| $tok($ "<<" $)$ & $CommaList($G.Expression$)$ & $tok($ ">>" $)$ $\langle 1, 2, 1 + 2\rangle$

| $G.Expression$ & $(Tok($\{$ "\X", "\times" $\}$)$ & $G.Expression)^+$ $Nat \times (1 \mathrel{..} 3) \times Real$

| $tok($ "[" $)$ & $G.Expression$ & $tok($ "]_" $)$ & $G.Expression$ $[A \vee B]_{\langle x, y\rangle}$

| $tok($ "<<" $)$ & $G.Expression$ & $tok($ ">>_" $)$ & $G.Expression$ $\langle x' = y + 1\rangle_{(x*y)}$

| $Tok($\{$ "WF_", "SF_" $\}$)$ & $G.Expression$ $\mathrm{WF}_{vars}(Next)$
    & $tok($ "(" $)$ & $G.Expression$ & $tok($ ")" $)$

| $tok($ "IF" $)$ & $G.Expression$ & $tok($ "THEN" $)$ IF p THEN A ELSE B
    & $G.Expression$ & $tok($ "ELSE" $)$ & $G.Expression$

| $tok($ "CASE" $)$ CASE $p1$ $\rightarrow$ $e1$
& (LET $CaseArm$ $\triangleq$ $\Box\ p2$ $\rightarrow$ $e2$
       $G.Expression$ & $tok($ "->" $)$ & $G.Expression$ $\Box$ OTHER $\rightarrow$ $e3$
  IN $CaseArm$ & $(tok($ "[]" $)$ & $CaseArm)^*$)
& (Nil
    | $(tok($ "[]" $)$ & $tok($ "OTHER" $)$ & $tok($ "->" $)$ & $G.Expression)$))

| $tok($ "LET" $)$ LET x $\triangleq$ $y + 1$
& ($G.OperatorDefinition$ $f[t \in Nat]$ $\triangleq$ t^2
    | $G.FunctionDefinition$ IN $x + f[y]$
    | $G.ModuleDefinition$ $)^+$
& $tok($ "IN" $)$
& $G.Expression$

| $(tok($ "/\" $)$ & $G.Expression)^+$ $\wedge\ x = 1$
                                 $\wedge\ y = 2$

| $(tok($ "\/" $)$ & $G.Expression)^+$ $\vee\ x = 1$
                                 $\vee\ y = 2$

| *Number*   09001

| *String*   "foo"

| *tok*("@")   @ (Can be used only in an EXCEPT expression.)

IN *LeastGrammar*(*P*)

15.2 The Complete Grammar

We now complete our explanation of the syntax of TLA$^+$ by giving the details that are not described by the BNF grammar in the previous section. Section 15.2.1 gives the precedence rules, Section 15.2.2 gives the alignment rules for conjunction and disjunction lists, and Section 15.2.3 describes comments. Section 15.2.4 briefly discusses the syntax of temporal formulas. Finally, for completeness, Section 15.2.5 explains the handling of two anomalous cases that you're unlikely ever to encounter.

15.2.1 Precedence and Associativity

The expression $a + b * c$ is interpreted as $a + (b * c)$ rather than $(a + b) * c$. This convention is described by saying that the operator $*$ has *higher precedence* than the operator $+$. In general, operators with higher precedence are applied before operators of lower precedence. This applies to prefix operators (like SUBSET) and postfix operators (like $'$) as well as to infix operators like $+$ and $*$. Thus, $a + b'$ is interpreted as $a + (b')$, rather than as $(a + b)'$, because $'$ has higher precedence than $+$. Application order can also be determined by associativity. The expression $a - b - c$ is interpreted as $(a - b) - c$ because $-$ is a left-associative infix operator.

In TLA$^+$, the precedence of an operator is a range of numbers, like 9–13. The operator $\$$ has higher precedence than the operator $:>$ because the precedence of $\$$ is 9–13, and this entire range is greater than the precedence range of $:>$, which is 7–7. An expression is illegal (not syntactically well-formed) if the order of application of two operators is not determined because their precedence ranges overlap and they are not two instances of an associative infix operator. For example, the expression $a + b * c' \% d$ is illegal for the following reason. The precedence range of $'$ is higher than that of $*$, and the precedence range of $*$ is higher than that of both $+$ and $\%$, so this expression can be written as $a + (b * (c')) \% d$. However, the precedences of $+$ (10–10) and $\%$ (10–11) overlap, so we don't know if the expression is to be interpreted as $(a + (b * (c'))) \% d$ or $a + ((b * (c')) \% d)$, and it is therefore illegal.

TLA^{+} embodies the philosophy that it's better to require parentheses than to allow expressions that could easily be misinterpreted. Thus, $*$ and $/$ have overlapping precedence, making an expression like $a/b*c$ illegal. (This also makes $a*b/c$ illegal, even though $(a*b)/c$ and $a*(b/c)$ happen to be equal when $*$ and $/$ have their usual definitions.) Unconventional operators like $ have wide precedence ranges for safety. But, even when the precedence rules imply that parentheses aren't needed, it's often a good idea to use them anyway if you think there's any chance that a reader might not understand how an expression is parsed.

Table 6 on page 271 gives the precedence ranges of all operators and tells which infix operators are left associative. (No TLA^{+} operators are right associative.) Note that the symbols $\in$, $=$, and "." are used both as fixed parts of constructs and as infix operators. They are not infix operators in the following two expressions:

$$\{x \in S : p(x)\}\qquad [f \text{ EXCEPT } !.a = e]$$

so the precedence of the corresponding infix operators plays no role in parsing these expressions. Below are some additional precedence rules not covered by the operator precedence ranges.

Function Application

Function application is treated like an operator with precedence range 16–16, giving it higher precedence than any operator except period ("."), the record-field operator. Thus, $a + b.c[d]'$ is interpreted as $a + (((b.c)[d])')$.

Cartesian Products

In the Cartesian product construct, $\times$ (typed as \X or \times) acts somewhat like an associative infix operator with precedence range 10–13. Thus, $A \times B \subseteq C$ is interpreted as $(A \times B) \subseteq C$, rather than as $A \times (B \subseteq C)$. However, $\times$ is part of a special construct, not an infix operator. For example, the three sets $A \times B \times C$, $(A \times B) \times C$, and $A \times (B \times C)$ are all different:

$$
\begin{aligned}
A \times B \times C &= \{\langle a, b, c\rangle : a \in A,\ b \in B,\ c \in C\} \\
(A \times B) \times C &= \{\langle\langle a, b\rangle, c\rangle : a \in A,\ b \in B,\ c \in C\} \\
A \times (B \times C) &= \{\langle a, \langle b, c\rangle\rangle : a \in A,\ b \in B,\ c \in C\}
\end{aligned}
$$

The first is a set of triples; the last two are sets of pairs.

Undelimited Constructs

TLA^{+} has several expression-making constructs with no explicit right-hand terminator. They are: CHOOSE, IF/THEN/ELSE, CASE, LET/IN, and quantifier constructs. These constructs are treated as prefix operators with the lowest possible precedence, so an expression made with one of them extends as far as possible. More precisely, the expression is ended only by one of the following:

- The beginning of the next module unit. (Module units are produced by the *Unit* nonterminal in the BNF grammar of Section 11.1.4; they include definition and declaration statements.)

- A right delimiter whose matching left delimiter occurs before the beginning of the construct. Delimiter pairs are ( ), [ ], { }, and ⟨ ⟩.

- Any of the following lexemes, if they are not part of a subexpression: THEN, ELSE, IN, comma (,), colon (:), and →. For example, the subexpression $\forall x : P$ is ended by the THEN in the expression

    IF $\forall x : P$ THEN 0 ELSE 1

- The CASE separator □ (not the prefix temporal operator that is typed the same) ends all of these constructs except a CASE statement without an OTHER clause. That is, the □ acts as a delimiter except when it can be part of a CASE statement.

- Any symbol not to the right of the ∧ or ∨ prefixing a conjunction or disjunction list element containing the construct. (See Section 15.2.2 on the next page.)

Here is how some expressions are interpreted under this rule:

$$
\begin{array}{l}
\text{IF } x > 0 \text{ THEN } y + 1 \\
\qquad\qquad\text{ELSE } y - 1 \\
+\; 2
\end{array}
\quad\text{means}\quad
\begin{array}{l}
\text{IF } x > 0 \text{ THEN } y + 1 \\
\qquad\qquad\text{ELSE } (y - 1 + 2)
\end{array}
$$

$$
\begin{array}{l}
\quad\forall x \in S : P(x) \\
\lor\; Q
\end{array}
\quad\text{means}\quad
\forall x \in S : (P(x) \lor Q)
$$

As these examples show, indentation is ignored—except in conjunction and disjunction lists, discussed below. The absence of a terminating lexeme (an END) for an IF/THEN/ELSE or CASE construct usually makes an expression less cluttered, but sometimes it does require you to add parentheses.

Subscripts

TLA uses subscript notation in the following constructs: $[A]_e$, $\langle A \rangle_e$, $\mathrm{WF}_e(A)$, and $\mathrm{SF}_e(A)$. In TLA$^+$, these are written with a "_" character, as in `<<A>>_e`. This notation is, in principle, problematic. The expression `<<A>>_x /\ B`, which we expect to mean $(\langle A \rangle_x) \land B$, could conceivably be interpreted as $\langle A \rangle_{(x \land B)}$. The precise rule for parsing these constructs isn't important; you should put parentheses around the subscript except in the following two cases:

- The subscript is a *GeneralIdentifier* in the BNF grammar.

- The subscript is an expression enclosed by one of the following matching delimiter pairs: ( ), [ ], ⟨ ⟩, or { }—for example, ⟨x, y⟩ or ($x + y$).

Although `[A]_f[x]` is interpreted correctly as $[A]_{f[x]}$, it will be easier to read in the ASCII text (and will be formatted properly by TLATEX) if you write it as `[A]_(f[x])`.

15.2.2   Alignment

The most novel aspect of TLA$^+$ syntax is the aligned conjunction and disjunction lists. If you write such a list in a straightforward manner, then it will mean what you expect it to. However, you might wind up doing something weird through a typing error. So, it's a good idea to know what the exact syntax rules are for these lists. I give the rules here for conjunction lists; the rules for disjunction lists are analogous.

A conjunction list is an expression that begins with ∧, which is typed as `/\`. Let c be the column in which the `/` occurs. The conjunction list consists of a sequence of conjuncts, each beginning with a ∧. A conjunct is ended by any one of the following that occurs after the `/\`:

1. Another `/\` whose `/` character is in column c and is the first nonspace character on the line.

2. Any nonspace character in column c or a column to the left of column c.

3. A right delimiter whose matching left delimiter occurs before the beginning of the conjunction list. Delimiter pairs are ( ), [ ], { }, and ⟨ ⟩.

4. The beginning of the next module unit. (Module units are produced by the *Unit* nonterminal in the BNF grammar; they include definition and declaration statements.)

In case 1, the `/\` begins the next conjunct in the same conjunction list. In the other three cases, the end of the conjunct is the end of the entire conjunction list. In all cases, the character ending the conjunct does not belong to the conjunct. With these rules, indentation properly delimits expressions in a conjunction list—for example:

```
/\ IF e THEN P                  ∧ (IF e THEN P
        ELSE Q      means              ELSE  Q)
/\ R                            ∧ R
```

It's best to indent each conjunction completely to the right of its ∧ symbol. These examples illustrate precisely what happens if you don't:

```
/\ x'                 ∧ x' = y        /\ x'              (( ∧ x')
   = y      means                        = y    means      = y)
/\ y'=x               ∧ y' = x        /\ y'=x                ∧ (y' = x)
```

In the second example, $\wedge\, x'$ is interpreted as a conjunction list containing only one conjunct, and the second /\ is interpreted as an infix operator.

You can't use parentheses to circumvent the indentation rules. For example, this is illegal:

```
/\ (x'
   = y)
/\ y'=x
```

The rules imply that the first /\ begins a conjunction list that is ended before the =. That conjunction list is therefore $\wedge\, (x'$, which has an unmatched left parenthesis.

The conjunction/disjunction list notation is quite robust. Even if you mess up the alignment by typing one space too few or too many—something that's easy to do when the conjuncts are long—the formula is still likely to mean what you intended. Here's an example of what happens if you misalign a conjunct:

```
/\ A           ( ( ∧ A)   The bulleted list ∧ A of one conjunct; it equals A.

/\ B   means       ∧ B)   This ∧ is interpreted as an infix operator.

/\ C               ∧ C    This ∧ is interpreted as an infix operator.
```

While not interpreted the way you expected, this formula is equivalent to $A \wedge B \wedge C$, which is what you meant in the first place.

Most keyboards contain one key that is the source of a lot of trouble: the tab key (sometimes marked on the keyboard with a right arrow). On my computer screen, I can produce

```
A ==
        /\ x' = 1
        /\ y' = 2
```

by beginning the second line with eight space characters and the third with one tab character. In this case, it is unspecified whether or not the two / characters occur in the same column. Tab characters are an anachronism left over from the days of typewriters and of computers with memory capacity measured in kilobytes. I strongly advise you never to use them. But, if you insist on using them, here are the rules:

- A tab character is considered to be equivalent to one or more space characters, so it occupies one or more columns.

- Identical sequences of space and tab characters that occur at the beginning of a line occupy the same number of columns.

There are no other guarantees if you use tab characters.

15.2.3 Comments

Comments are described in Section 3.5 on page 32. A comment may appear between any two lexemes in a specification. There are two types of comments:

- A delimited comment is a string of the form "(*" $\circ s \circ$ "*)", where s is any string in which occurrences of "(*" and "*)" are properly matched. More precisely, a delimited comment is defined inductively to be a string of the form "(*" $\circ s_1 \circ \cdots \circ s_n \circ$ "*)", where each s_i is either (i) a string containing neither the substring "(*" nor the substring "*)", or (ii) a delimited comment. (In particular, "(**)" is a delimited comment.)

- An end-of-line comment is a string of the form "*" $\circ s \circ$ "⟨LF⟩", where s is any string not containing an end-of-line character ⟨LF⟩.

I like to write comments as shown here:

```
BufRcv == /\ InChan!Rcv              (*******************************)
          /\ q' = Append(q, in.val)  (* Receive message from channel *)
          /\ out                     (* 'in' and append to tail of q. *)
                                     (*******************************)
```

Grammatically, this piece of specification has four distinct comments, the first and last consisting of the same string (***···***). But a person reading it would regard them as a single comment, spread over four lines. This kind of commenting convention is not part of the TLA$^+$ language, but it is supported by the TLATEX typesetting program, as described in Section 13.4 on page 214.

15.2.4 Temporal Formulas

The BNF grammar treats □ and ◇ simply as prefix operators. However, as explained in Section 8.1 (page 88), the syntax of temporal formulas places restrictions on their use. For example, □$(x' = x+1)$ is not a legal formula. It's not hard to write a BNF grammar that specifies legal temporal formulas made from the temporal operators and ordinary Boolean operators like ¬ and ∧. However, such a BNF grammar won't tell you which of these two expressions is legal:

LET $F(P,Q) \triangleq P \wedge \Box Q$ LET $F(P,Q) \triangleq P \wedge \Box Q$
IN $\quad F(x = 1, \ x = y + 1)$ IN $\quad F(x = 1, \ x' = y + 1)$

The first is legal; the second isn't because it represents the illegal formula

$$(x = 1) \ \wedge \ \Box(x' = y + 1) \qquad \text{This formula is illegal.}$$

The precise rules for determining if a temporal formula is syntactically well-formed involve first replacing all defined operators by their definitions, using the procedure described in Section 17.4 below. I won't bother specifying these rules.

In practice, temporal operators are not used very much in TLA$^+$ specifications, and one rarely writes definitions of new ones such as

$$F(P, Q) \;\triangleq\; P \wedge \Box Q$$

The syntactic rules for expressions involving such operators are of academic interest only.

15.2.5 Two Anomalies

There are two sources of potential ambiguity in the grammar of TLA$^+$ that you are unlikely to encounter and that have *ad hoc* resolutions. The first of these arises from the use of $-$ as both an infix operator (as in $2 - 2$) and a prefix operator (as in $2 + -2$). This poses no problem when $-$ is used in an ordinary expression. However, there are two places in which an operator can appear by itself:

- As the argument of a higher-order operator, as in $HOp(+, -)$.

- In an INSTANCE substitution, such as

    INSTANCE M WITH $Plus \leftarrow +\,,\ Minus \leftarrow -$

In both these cases, the symbol - is interpreted as the infix operator. You must type -. to denote the prefix operator. You also have to type -. if you should ever want to define the prefix $-$ operator, as in

$$-.\, a \;\triangleq\; UMinus(a)$$

In ordinary expressions, you just type - as usual for both operators.

The second source of ambiguity in the TLA$^+$ syntax is an unlikely expression of the form $\{x \in S \;:\; y \in T\}$, which might be taken to mean either of the following:

LET $p(x) \;\triangleq\; y \in T$ IN $\{x \in S : p(x)\}$ This is a subset of S.

LET $p(y) \;\triangleq\; x \in S$ IN $\{p(y) : y \in T\}$ This is a subset of BOOLEAN (the set $\{\text{TRUE}, \text{FALSE}\}$).

It is interpreted as the first formula.

15.3   The Lexemes of TLA$^+$

So far, this chapter has described the sequences of lexemes that form syntactically correct TLA$^+$ modules. More precisely, because of the alignment rules, syntactic correctness depends not just on the sequence of lexemes, but also on the position of each lexeme—that is, on the row and columns in which the characters of the lexeme appear. To complete the definition of the syntax of TLA$^+$, this section explains how a sequence of characters is turned into a sequence of lexemes.

All characters that precede the beginning of the module are ignored. Ignoring a character does not change the row or column of any other character in the sequence. The module begins with a sequence of four or more dashes ("–" characters), followed by zero or more space characters, followed by the six-character string "MODULE". (This sequence of characters yields the first two lexemes of the module.) The remaining sequence of characters is then converted to a sequence of lexemes by iteratively applying the following rule until the module-ending == · · · == token is found:

> The next lexeme begins at the next text character that is not part of a comment, and consists of the largest sequence of consecutive characters that form a legal TLA$^+$ lexeme. (It is an error if no such lexeme exists.)

Space, tab, and the end-of-line character are not text characters. It is undefined whether characters such as form feed are considered text characters. (You should not use such characters outside comments.)

In the BNF grammar, a *Name* is a lexeme that can be used as the name of a record field. The semantics of TLA$^+$, in which $r.c$ is an abbreviation for $r["c"]$, would allow any string to be a *Name*. However, some restriction is needed—for example, allowing a string like "a+b" to be a *Name* would make it impossible in practice to decide if r.a+b meant $r["a+b"]$ or $r["a"] + b$. The one unusual restriction in the definition of *Name* on page 277 is the exclusion of strings beginning with (but not consisting entirely of) "WF_" and "SF_". With this restriction, such strings are not legal TLA$^+$ lexemes. Hence, the input WF_x(A) is broken into the five lexemes "WF_", "x", "(", "A", and ")", and it is interpreted as the expression WF$_x$(A).

Chapter 16

The Operators of TLA$^+$

This chapter describes the built-in operators of TLA$^+$. Most of these operators have been described in Part I. Here, you can find brief explanations of the operators, along with references to the longer descriptions in Part I. The explanations cover some subtle points not mentioned elsewhere. The chapter can serve as a reference manual for readers who have finished Part I or who are already familiar enough with the mathematical concepts that the brief explanations are all they need.

The chapter includes a formal semantics of the operators. The rigorous description of TLA$^+$ that a formal semantics provides is usually needed only by people building TLA$^+$ tools. If you're not building a tool and don't have a special fondness for formalism, you will probably want to skip all the subsections titled *Formal Semantics*. However, you may some day encounter an obscure question about the meaning of a TLA$^+$ operator that is answered only by the formal semantics.

This chapter also defines some of the "semantic" conditions on the syntax of TLA$^+$ that are omitted from the grammar of Chapter 15. For example, it tells you that $[a : Nat, a : \text{BOOLEAN}]$ is an illegal expression. Other semantic conditions on expressions arise from a combination of the definitions in this chapter and the conditions stated in Chapter 17. For example, this chapter defines $\exists x, x : p$ to equal $\exists x : (\exists x : p)$, and Chapter 17 tells you that the latter expression is illegal.

16.1 Constant Operators

We first define the constant operators of TLA$^+$. These are the operators of ordinary mathematics, having nothing to do with TLA or temporal logic. All the constant operators of TLA$^+$ are listed in Table 1 on page 268 and Table 2 on page 269.

An operator combines one or more expressions into a "larger" expression. For example, the set union operator $\cup$ combines two expressions e_1 and e_2 into the expression $e_1 \cup e_2$. Some operators don't have simple names like $\cup$. There's no name for the operator that combines the n expressions $e_1, \ldots, e_n$ to form the expression $\{e_1, \ldots, e_n\}$. We could name it $\{,\ldots,\}$ or $\{\_,\ldots,\_\}$, but that would be awkward. Instead of explicitly mentioning the operator, I'll refer to the *construct* $\{e_1, \ldots, e_n\}$. The distinction between an operator like $\cup$ and the nameless one used in the construct $\{e_1, \ldots, e_n\}$ is purely syntactic, with no mathematical significance. In Chapter 17, we'll abstract away from this syntactic difference and treat all operators uniformly. For now, we'll stay closer to the syntax.

Formal Semantics

A formal semantics for a language is a translation from that language into some form of mathematics. We assign a mathematical expression $[\![e]\!]$, called the *meaning* of e, to certain terms e in the language. Since we presumably understand the mathematics, we know what $[\![e]\!]$ means, and that tells us what e means.

Meaning is generally defined inductively. For example, the meaning $[\![e_1 \cup e_2]\!]$ of the expression $e_1 \cup e_2$ would be defined in terms of the meanings $[\![e_1]\!]$ and $[\![e_2]\!]$ of its subexpressions. This definition is said to define the semantics of the operator $\cup$.

Because much of TLA$^+$ is a language for expressing ordinary mathematics, much of its semantics is trivial. For example, the semantics of $\cup$ can be defined by

$$[\![e_1 \cup e_2]\!] \;\triangleq\; [\![e_1]\!] \cup [\![e_2]\!]$$

In this definition, the $\cup$ to the left of the $\triangleq$ is the TLA$^+$ symbol, while the one to the right is the set-union operator of ordinary mathematics. We could make the distinction between the two uses of the symbol $\cup$ more obvious by writing

$$[\![e_1 \texttt{ \textbackslash cup } e_2]\!] \;\triangleq\; [\![e_1]\!] \cup [\![e_2]\!]$$

But that wouldn't make the definition any less trivial.

Instead of trying to maintain a distinction between the TLA$^+$ operator $\cup$ and the operator of set theory that's written the same, we simply use TLA$^+$ as the language of mathematics in which to define the semantics of TLA$^+$. That is, we take as primitive certain TLA$^+$ operators that, like $\cup$, correspond to well-known mathematical operators. We describe the formal semantics of the constant operators of TLA$^+$ by defining them in terms of these primitive operators. We also describe the semantics of some of the primitive operators by stating the axioms that they satisfy.

16.1.1 Boolean Operators

The truth values of logic are written in TLA$^+$ as TRUE and FALSE. The built-in constant BOOLEAN is the set consisting of those two values:

$$\text{BOOLEAN} \ \triangleq \ \{\text{TRUE}, \text{FALSE}\}$$

TLA$^+$ provides the usual operators[1] of propositional logic:

$$\wedge \qquad \vee \qquad \neg \qquad \Rightarrow \text{(implication)} \qquad \equiv \qquad \text{TRUE} \qquad \text{FALSE}$$

They are explained in Section 1.1. Conjunctions and disjunctions can also be written as aligned lists:

$$\begin{array}{l} \wedge \ p_1 \\ \ \ \vdots \ \ \triangleq \ p_1 \wedge \ldots \wedge p_n \\ \wedge \ p_n \end{array} \qquad\qquad \begin{array}{l} \vee \ p_1 \\ \ \ \vdots \ \ \triangleq \ p_1 \vee \ldots \vee p_n \\ \vee \ p_n \end{array}$$

The standard quantified formulas of predicate logic are written in TLA$^+$ as

$$\forall\, x\, :\, p \qquad \exists\, x\, :\, p$$

I call these the *unbounded* quantifier constructions. The *bounded* versions are written as

$$\forall\, x \in S\, :\, p \qquad \exists\, x \in S\, :\, p$$

The meanings of these expressions are described in Section 1.3. TLA$^+$ allows some common abbreviations—for example:

$$\forall\, x, y\, :\, p \ \triangleq \ \forall\, x\, :\, (\forall\, y\, :\, p)$$
$$\exists\, x, y \in S,\, z \in T\, :\, p \ \triangleq \ \exists\, x \in S\, :\, (\exists\, y \in S\, :\, (\exists\, z \in T\, :\, p))$$

TLA$^+$ also allows bounded quantification over tuples, such as

$$\forall\, \langle x, y \rangle \in S\, :\, p$$

This formula is true iff, for any pair $\langle a, b \rangle$ in S, the formula obtained from p by substituting a for x and b for y is true.

Formal Semantics

Propositional and predicate logic, along with set theory, form the foundation of ordinary mathematics. In defining the semantics of TLA$^+$, we therefore take as primitives the operators of propositional logic and the simple unbounded quantifier constructs $\exists\, x : p$ and $\forall\, x : p$, where x is an identifier. Among the

---

[1] TRUE and FALSE are operators that take no arguments.

Boolean operators described above, this leaves only the general forms of the
quantifiers, given by the BNF grammar of Chapter 15, whose meanings must
be defined. This is done by defining those general forms in terms of the simple
forms.

The unbounded operators have the general forms

$$\forall\, x_1,\ldots,x_n : p \qquad \exists\, x_1,\ldots,x_n : p$$

where each x_i is an identifier. They are defined in terms of quantification over
a single variable by

$$\forall\, x_1,\ldots,x_n : p \;\triangleq\; \forall\, x_1 : (\forall\, x_2 : (\ldots \forall\, x_n : p)\ldots)$$

and similarly for $\exists$. The bounded operators have the general forms

$$\forall\, \mathbf{y}_1 \in S_1,\ldots,\mathbf{y}_n \in S_n : p \qquad \exists\, \mathbf{y}_1 \in S_1,\ldots,\mathbf{y}_n \in S_n : p$$

where each $\mathbf{y}_i$ has the form $x_1,\ldots,x_k$ or $\langle x_1,\ldots,x_k\rangle$, and each x_j is an identifier. The general forms of $\forall$ are defined inductively by

$$\forall\, x_1, \ldots, x_k \in S : p \quad\triangleq\quad \forall\, x_1, \ldots, x_k : \\ (x_1 \in S) \wedge \ldots \wedge (x_k \in S) \Rightarrow p$$

$$\forall\, \mathbf{y}_1 \in S_1, \ldots, \mathbf{y}_n \in S_n : p \;\triangleq\; \forall\, \mathbf{y}_1 \in S_1 : \ldots \forall\, \mathbf{y}_n \in S_n : p$$

$$\forall\, \langle x_1, \ldots, x_k\rangle \in S : p \quad\triangleq\quad \forall\, x_1, \ldots, x_k : (\langle x_1, \ldots, x_k\rangle \in S) \Rightarrow p$$

where the $\mathbf{y}_i$ are as before. In these expressions, S and the S_i lie outside the
scope of the quantifier's bound identifiers. The definitions for $\exists$ are similar. In
particular:

$$\exists\, \langle x_1, \ldots, x_k\rangle \in S : p \;\triangleq\; \exists\, x_1, \ldots, x_k : (\langle x_1, \ldots, x_k\rangle \in S) \wedge p$$

See Section 16.1.9 for further details about tuples.

16.1.2   The Choose Operator

A simple unbounded CHOOSE expression has the form

CHOOSE $x : p$

As explained in Section 6.6, the value of this expression is some arbitrary value
v such that p is true if v is substituted for x, if such a v exists. If no such v
exists, then the expression has a completely arbitrary value.

The bounded form of the CHOOSE expression is

CHOOSE $x \in S : p$

It is defined in terms of the unbounded form by

(16.1) CHOOSE $x \in S : p \stackrel{\Delta}{=}$ CHOOSE $x : (x \in S) \wedge p$

It is equal to some arbitrary value v in S such that p, with v substituted for x, is true—if such a v exists. If no such v exists, the CHOOSE expression has a completely arbitrary value.

A CHOOSE expression can also be used to choose a tuple. For example,

CHOOSE $\langle x, y \rangle \in S : p$

equals some pair $\langle v, w \rangle$ in S such that p, with v substituted for x and w substituted for y, is true—if such a pair exists. If no such pair exists, it has an arbitrary value, which need not be a pair.

The unbounded CHOOSE operator satisfies the following two rules:

(16.2) $(\exists x : P(x)) \equiv P(\text{CHOOSE } x : P(x))$

$(\forall x : P(x) = Q(x)) \Rightarrow ((\text{CHOOSE } x : P(x)) = (\text{CHOOSE } x : Q(x)))$

for any operators P and Q. We know nothing about the value chosen by CHOOSE except what we can deduce from these rules.

The second rule allows us to deduce the equality of certain CHOOSE expressions that we might expect to be different. In particular, for any operator P, if there exists no x satisfying $P(x)$, then CHOOSE $x : P(x)$ equals the unique value CHOOSE $x : \text{FALSE}$. For example, the *Reals* module defines division by

$a/b \stackrel{\Delta}{=}$ CHOOSE $c \in Real : a = b * c$

For any nonzero number a, there exists no number c such that $a = 0 * c$. Hence, $a/0$ equals CHOOSE $c : \text{FALSE}$, for any nonzero a. We can therefore deduce that $1/0$ equals $2/0$.

We would expect to be unable to deduce anything about the nonsensical expression $1/0$. It's a bit disquieting to prove that it equals $2/0$. If this upsets you, here's a way to define division that will make you happier. First define an operator *Choice* so that $Choice(v, P)$ equals CHOOSE $x : P(x)$ if there exists an x satisfying $P(x)$, and otherwise equals some arbitrary value that depends on v. There are many ways to define *Choice*; here's one:

$Choice(v, P(\_)) \stackrel{\Delta}{=}$ IF $\exists x : P(x)$ THEN CHOOSE $x : P(x)$
ELSE $(\text{CHOOSE } x : x.a = v).b$

You can then define division by

$a/b \stackrel{\Delta}{=}$ LET $P(c) \stackrel{\Delta}{=} (c \in Real) \wedge (a = b * c)$
IN $Choice(a, P)$

This definition makes it impossible to deduce any relation between $1/0$ and $2/0$. You can use *Choice* instead of CHOOSE whenever this kind of problem arises—if you consider $1/0$ equaling $2/0$ to be a problem. But there is seldom any practical reason for worrying about it.

Formal Semantics

We take the construct CHOOSE $x : p$, where x is an identifier, to be primitive. This form of the CHOOSE operator is known to mathematicians as *Hilbert's ε*. Its meaning is defined mathematically by the rules (16.2).[2]

An unbounded CHOOSE of a tuple is defined in terms of the simple unbounded CHOOSE construct by

$$\text{CHOOSE } \langle x_1, \ldots, x_n \rangle \; : \; p \;\; \triangleq$$
$$\text{CHOOSE } y \; : \; (\exists\, x_1, \ldots, x_n \; : \; (y = \langle x_1, \ldots, x_n \rangle)) \wedge p)$$

where y is an identifier that is different from the x_i and does not occur in p. The bounded CHOOSE construct is defined in terms of unbounded CHOOSE by (16.1), where x can be either an identifier or a tuple.

16.1.3   Interpretations of Boolean Operators

The meaning of a Boolean operator when applied to Boolean values is a standard part of traditional mathematics. Everyone agrees that TRUE $\wedge$ FALSE equals FALSE. However, because TLA$^+$ is untyped, an expression like $2 \wedge \langle 5 \rangle$ is legal. We must therefore decide what it means. There are three ways of doing this, which I call the *conservative*, *moderate*, and *liberal* interpretations.

In the conservative interpretation, the value of an expression like $2 \wedge \langle 5 \rangle$ is completely unspecified. It could equal $\sqrt{2}$. It need not equal $\langle 5 \rangle \wedge 2$. Hence, the ordinary laws of logic, such as the commutativity of $\wedge$, are valid only for Boolean values.

In the liberal interpretation, the value of $2 \wedge \langle 5 \rangle$ is specified to be a Boolean. It is not specified whether it equals TRUE or FALSE. However, all the ordinary laws of logic, such as the commutativity of $\wedge$, are valid. Hence, $2 \wedge \langle 5 \rangle$ equals $\langle 5 \rangle \wedge 2$. More precisely, any tautology of propositional or predicate logic, such as

$$(\forall\, x \; : \; p) \;\; \equiv \;\; \neg(\exists\, x \; : \; \neg p)$$

is valid, even if p is not necessarily a Boolean for all values of x.[3] It is easy to show that the liberal approach is sound.[4] For example, one way of defining operators that satisfy the liberal interpretation is to consider any non-Boolean value to be equivalent to FALSE.

The conservative and liberal interpretations are equivalent for most specifications, except for ones that use Boolean-valued functions. In practice, the

---

[2]Hilbert's ε is discussed at length in *Mathematical Logic and Hilbert's ε-Symbol* by A. C. Leisenring, published by Gordon and Breach, New York, 1969.

[3]Equality ($=$) is not an operator of propositional or predicate logic; this tautology need not be valid for non-Boolean values if $\equiv$ is replaced by $=$.

[4]A sound logic is one in which FALSE is not provable.

conservative interpretation doesn't permit you to use $f[x]$ as a Boolean expression even if f is defined to be a Boolean-valued function. For example, suppose we define the function *tnat* by

$$tnat \;\triangleq\; [n \in Nat \mapsto \text{TRUE}]$$

so $tnat[n]$ equals TRUE for all n in *Nat*. The formula

(16.3) $\forall\, n \in Nat \,:\, tnat[n]$

equals TRUE in the liberal interpretation, but not in the conservative interpretation. Formula (16.3) is equivalent to

$$\forall\, n \,:\, (n \in Nat) \Rightarrow tnat[n]$$

which asserts that $(n \in Nat) \Rightarrow tnat[n]$ is true for all n, including, for example, $n = 1/2$. For (16.3) to equal TRUE, the formula $(1/2 \in Nat) \Rightarrow tnat[1/2]$, which equals FALSE $\Rightarrow tnat[1/2]$, must equal TRUE. But the value of $tnat[1/2]$ is not specified; it might equal $\sqrt{2}$. The formula FALSE $\Rightarrow \sqrt{2}$ equals TRUE in the liberal interpretation; its value is unspecified in the conservative interpretation. Hence, the value of (16.3) is unspecified in the conservative interpretation. If we are using the conservative interpretation, instead of (16.3), we should write

$$\forall\, n \in Nat \,:\, (tnat[n] = \text{TRUE})$$

This formula equals TRUE in both interpretations.

The conservative interpretation is philosophically more satisfying, since it makes no assumptions about a silly expression like $2 \land \langle 5 \rangle$. However, as we have just seen, it would be nice if the not-so-silly formula FALSE $\Rightarrow \sqrt{2}$ equaled TRUE. We therefore introduce the *moderate interpretation*, which lies between the conservative and liberal interpretations. It assumes only that expressions involving FALSE and TRUE have their expected values—for example, FALSE $\Rightarrow \sqrt{2}$ equals TRUE, and FALSE $\land 2$ equals FALSE. In the moderate interpretation, (16.3) equals TRUE, but the value of $\langle 5 \rangle \land 2$ is still completely unspecified.

The laws of logic still do not hold unconditionally in the moderate interpretation. The formulas $p \land q$ and $q \land p$ are equivalent only if p and q are both Booleans, or if one of them equals FALSE. When using the moderate interpretation, we still have to check that all the relevant values are Booleans before applying any of the ordinary rules of logic in a proof. This can be burdensome in practice.

The semantics of TLA$^+$ asserts that the rules of the moderate interpretation are valid. The liberal interpretation is neither required nor forbidden. You should write specifications that make sense under the moderate interpretation. However, you (and the implementer of a tool) are free to use the liberal interpretation if you wish.

16.1.4   Conditional Constructs

TLA$^+$ provides two conditional constructs for forming expressions that are inspired by constructs from programming languages: IF/THEN/ELSE and CASE.

The IF/THEN/ELSE construct was introduced on page 16 of Section 2.2. Its general form is

$$\text{IF } p \text{ THEN } e_1 \text{ ELSE } e_2$$

It equals e_1 if p is true, and e_2 if p is false.

An expression can sometimes be simplified by using a CASE construct instead of nested IF/THEN/ELSE constructs. The CASE construct has two general forms:

(16.4) CASE $p_1 \rightarrow e_1 \;\square\; \ldots \;\square\; p_n \rightarrow e_n$

 CASE $p_1 \rightarrow e_1 \;\square\; \ldots \;\square\; p_n \rightarrow e_n \;\square\; \text{OTHER} \rightarrow e$

If some p_i is true, then the value of these expressions is some e_i such that p_i is true. For example, the expression

$$\text{CASE } n \geq 0 \rightarrow e_1 \;\square\; n \leq 0 \rightarrow e_2$$

equals e_1 if $n > 0$ is true, equals e_2 if $n < 0$ is true, and equals either e_1 or e_2 if $n = 0$ is true. In the latter case, the semantics of TLA$^+$ does not specify whether the expression equals e_1 or e_2. The CASE expressions (16.4) are generally used when the p_i are mutually disjoint, so at most one p_i can be true.

The two expressions (16.4) differ when p_i is false for all i. In that case, the value of the first is unspecified, while the value of the second is e, the OTHER expression. If you use a CASE expression without an OTHER clause, the value of the expression should matter only when $\exists\, i \in 1 \,..\, n : p_i$ is true.

Formal Semantics

The IF/THEN/ELSE and CASE constructs are defined as follows in terms of CHOOSE:

$$\text{IF } p \text{ THEN } e_1 \text{ ELSE } e_2 \;\triangleq$$
$$\quad \text{CHOOSE } v : (p \Rightarrow (v = e_1)) \wedge (\neg p \Rightarrow (v = e_2))$$

$$\text{CASE } p_1 \rightarrow e_1 \;\square\; \ldots \;\square\; p_n \rightarrow e_n \;\triangleq$$
$$\quad \text{CHOOSE } v : (p_1 \wedge (v = e_1)) \vee \ldots \vee (p_n \wedge (v = e_n))$$

$$\text{CASE } p_1 \rightarrow e_1 \;\square\; \ldots \;\square\; p_n \rightarrow e_n \;\square\; \text{OTHER} \rightarrow e \;\triangleq$$
$$\quad \text{CASE } p_1 \rightarrow e_1 \;\square\; \ldots \;\square\; p_n \rightarrow e_n \;\square\; \neg(p_1 \vee \ldots \vee p_n) \rightarrow e$$

16.1.5   The Let/In Construct

The LET/IN construct was introduced on page 60 of Section 5.6. The expression

LET $d \triangleq f$ IN e

equals e in the context of the definition $d \triangleq f$. For example,

LET $sq(i) \triangleq i * i$ IN $sq(1) + sq(2) + sq(3)$

equals $1 * 1 + 2 * 2 + 3 * 3$, which equals 14. The general form of the construct is

LET $\Delta_1 \ldots \Delta_n$ IN e

where each Δ_i has the syntactic form of any TLA$^+$ definition. Its value is e in the context of the definitions Δ_i. More precisely, it equals

LET Δ_1 IN (LET Δ_2 IN (... LET Δ_n IN e) ...)

Hence, the symbol defined in Δ_1 can be used in the definitions $\Delta_2, \ldots, \Delta_n$.

Formal Semantics

The formal semantics of the LET construct is defined below in Section 17.4 (page 325).

16.1.6   The Operators of Set Theory

TLA$^+$ provides the following operators on sets:

$\in$ $\notin$ $\cup$ $\cap$ $\subseteq$ $\setminus$ UNION   SUBSET

and the following set constructors:

$\{e_1, \ldots, e_n\}$ $\{x \in S : p\}$ $\{e : x \in S\}$

They are all described in Section 1.2 (page 11) and Section 6.1 (page 65). Equality is also an operator of set theory, since it formally means equality of sets. TLA$^+$ provides the usual operators $=$ and $\neq$.

The set construct $\{x \in S : p\}$ can also be used with x a tuple of identifiers. For example,

$\{\langle a, b \rangle \in Nat \times Nat : a > b\}$

is the set of all pairs of natural numbers whose first component is greater than its second—pairs such as $\langle 3, 1 \rangle$. In the set construct $\{e : x \in S\}$, the clause $x \in S$ can be generalized in exactly the same way as in a bounded quantifier such as $\forall x \in S : p$. For example,

$\{\langle a, b, c \rangle : a, b \in Nat, c \in Real\}$

is the set of all triples whose first two components are natural numbers and whose third component is a real number.

Formal Semantics

TLA$^+$ is based on Zermelo-Fränkel set theory, in which every value is a set. In set theory, $\in$ is taken as a primitive, undefined operator. We could define all the other operators of set theory in terms of $\in$, using predicate logic and the CHOOSE operator. For example, set union could be defined by

$$S \cup T \;\triangleq\; \text{CHOOSE } U \,:\, \forall x : (x \in U) \equiv (x \in S) \vee (x \in T)$$

(To reason about $\cup$, we would need axioms from which we can deduce the existence of the chosen set U.) Another approach we could take is to let certain of the operators be primitive and define the rest in terms of them. For example, $\cup$ can be defined in terms of UNION and the construct $\{e_1, \ldots, e_n\}$ by

$$S \cup T \;\triangleq\; \text{UNION } \{S, \, T\}$$

We won't try to distinguish a small set of primitive operators; instead, we treat $\cup$ and UNION as equally primitive. Operators that we take to be primitive are defined mathematically in terms of the rules that they satisfy. For example, $S \cup T$ is defined by

$$\forall x : (x \in (S \cup T)) \equiv (x \in S) \vee (x \in T)$$

However, there is no such defining rule for the primitive operator $\in$. We take only the simple forms of the constructs $\{x \in S : p\}$ and $\{e : x \in S\}$ as primitive, and we define the more general forms in terms of them.

$S = T \;\triangleq\; \forall x : (x \in S) \equiv (x \in T)$.

$e_1 \neq e_2 \;\triangleq\; \neg(e_1 = e_2)$.

$e \notin S \;\triangleq\; \neg(e \in S)$.

$S \cup T$ is defined by $\forall x : (x \in (S \cup T)) \equiv (x \in S) \vee (x \in T)$.

$S \cap T$ is defined by $\forall x : (x \in (S \cap T)) \equiv (x \in S) \wedge (x \in T)$.

$S \subseteq T \;\triangleq\; \forall x : (x \in S) \Rightarrow (x \in T)$.

$S \setminus T$ is defined by $\forall x : (x \in (S \setminus T)) \equiv (x \in S) \wedge (x \notin T)$.

SUBSET S is defined by $\forall T : (T \in \text{SUBSET } S) \equiv (T \subseteq S)$.

UNION S is defined by $\forall x : (x \in \text{UNION } S) \equiv (\exists T \in S : x \in T)$.

$\{e_1, \ldots, e_n\} \;\triangleq\; \{e_1\} \cup \ldots \cup \{e_n\}$,
       where $\{e\}$ is defined by
            $\forall x : (x \in \{e\}) \equiv (x = e)$
       For $n = 0$, this construct is the empty set $\{\}$, defined by
            $\forall x : x \notin \{\}$

$\{x \in S : p\}$

> where x is a bound identifier or a tuple of bound identifiers. The expression S is outside the scope of the bound identifier(s). For x an identifier, this is a primitive expression that is defined mathematically by

$$\forall y : (y \in \{x \in S : p\}) \equiv (y \in S) \wedge \widehat{p}$$

> where the identifier y does not occur in S or p, and $\widehat{p}$ is p with y substituted for x. For x a tuple, the expression is defined by

$$\{\langle x_1, \ldots, x_n \rangle \in S : p\} \triangleq$$
$$\{y \in S : (\exists x_1, \ldots, x_n : (y = \langle x_1, \ldots, x_n \rangle) \wedge p)\}$$

> where y is an identifier different from the x_i that does not occur in S or p. See Section 16.1.9 for further details about tuples.

$\{e : \mathbf{y}_1 \in S_1, \ldots, \mathbf{y}_n \in S_n\}$

> where each $\mathbf{y}_i$ has the form $x_1, \ldots, x_k$ or $\langle x_1, \ldots, x_k \rangle$, and each x_j is an identifier that is bound in the expression. The expressions S_i lie outside the scope of the bound identifiers. The simple form $\{e : x \in S\}$, for x an identifier, is taken to be primitive and is defined by

$$\forall y : (y \in \{e : x \in S\}) \equiv (\exists x \in S : e = y)$$

> The general form is defined inductively in terms of the simple form by

$$\{e : \mathbf{y}_1 \in S_1, \ldots, \mathbf{y}_n \in S_n\} \triangleq$$
$$\text{UNION} \{\{e : \mathbf{y}_1 \in S_1, \ldots, \mathbf{y}_{n-1} \in S_{n-1}\} : \mathbf{y}_n \in S_n\}$$

$$\{e : x_1, \ldots, x_n \in S\} \triangleq \{e : x_1 \in S, \ldots, x_n \in S\}$$

$$\{e : \langle x_1, \ldots, x_n \rangle \in S\} \triangleq$$
$$\{(\text{LET } z \triangleq \text{CHOOSE } \langle x_1, \ldots, x_n \rangle : y = \langle x_1, \ldots, x_n \rangle$$
$$x_1 \triangleq z[1]$$
$$\vdots$$
$$x_n \triangleq z[n] \quad \text{IN } e) : y \in S\}$$

> where the x_i are identifiers, and y and z are identifiers distinct from the x_i that do not occur in e or S. See Section 16.1.9 for further details about tuples.

16.1.7   Functions

Functions are described in Section 5.2 (page 48); the difference between functions and operators is discussed in Section 6.4 (page 69). In TLA$^+$, we write $f[v]$ for the value of the function f applied to v. A function f has a domain DOMAIN f, and the value of $f[v]$ is specified only if v is an element of DOMAIN f. We let $[S \to T]$ denote the set of all functions f such that DOMAIN $f = S$ and $f[v] \in T$, for all $v \in S$.

Functions can be described explicitly with the construct

(16.5) $[x \in S \mapsto e]$

This is the function f with domain S such that $f[v]$ equals the value obtained by substituting v for x in e, for any $v \in S$. For example,

$$[n \in \mathit{Nat} \mapsto 1/(n+1)]$$

is the function f with domain Nat such that $f[0] = 1$, $f[1] = 1/2$, $f[2] = 1/3$, etc. We can define an identifier fcn to equal the function (16.5) by writing

(16.6) $fcn[x \in S] \overset{\Delta}{=} e$

The identifier fcn can appear in the expression e, in which case this is a recursive function definition. Recursive function definitions were introduced in Section 5.5 (page 54) and discussed in Section 6.3 (page 67).

The EXCEPT construct describes a function that is "almost the same as" another function. For example,

(16.7) $[f \text{ EXCEPT } ![u] = a, ![v] = b]$

is the function $\widehat{f}$ that is the same as f, except that $\widehat{f}[u] = a$ and $\widehat{f}[v] = b$. More precisely, (16.7) equals

$$[x \in \text{DOMAIN } f \mapsto \text{IF } x = v \text{ THEN } b$$
$$\text{ELSE IF } x = u \text{ THEN } a \text{ ELSE } f[x]]$$

Hence, if neither u nor v is in the domain of f, then (16.7) equals f. If $u = v$, then (16.7) equals $[f \text{ EXCEPT } ![v] = b]$.

An exception clause can have the general form $![v_1] \cdots [v_n] = e$. For example,

(16.8) $[f \text{ EXCEPT } ![u][v] = a]$

is the function $\widetilde{f}$ that is the same as f, except that $\widetilde{f}[u][v]$ equals a. That is, $\widetilde{f}$ is the same as f, except that $\widetilde{f}[u]$ is the function that is the same as $f[u]$, except that $\widetilde{f}[u][v] = a$. The symbol @ occurring in an exception clause stands for the "original value". For example, an @ in the expression a of (16.8) denotes $f[u][v]$.

In TLA$^+$, a function of multiple arguments is one whose domain is a set of tuples; and $f[v_1, \ldots, v_n]$ is an abbreviation for $f[\langle v_1, \ldots, v_n \rangle]$. The $x \in S$ clause (16.5) and (16.6) can be generalized in the same way as in a bounded quantifier— for example, here are two different ways of writing the same function:

$$[m, n \in \mathit{Nat}, r \in \mathit{Real} \mapsto e] \qquad [\langle m, n, r \rangle \in \mathit{Nat} \times \mathit{Nat} \times \mathit{Real} \mapsto e]$$

This is a function whose domain is a set of triples. It is not the same as the function

$$[\langle m, n \rangle \in Nat \times Nat, \; r \in Real \mapsto e]$$

whose domain is the set $(Nat \times Nat) \times Real$ of pairs like $\langle \langle 1, 3 \rangle, 1/3 \rangle$, whose first element is a pair of natural numbers.

Formal Semantics

Mathematicians traditionally define a function to be a set of pairs. In TLA$^+$, pairs (and all tuples) are functions. We take as primitives the constructs

$$f[e] \qquad \text{DOMAIN} \; f \qquad [S \to T] \qquad [x \in S \mapsto e]$$

where x is an identifier. These constructs are defined mathematically by the rules they satisfy. The other constructs, and the general forms of the construct $[x \in S \mapsto e]$, are defined in terms of them. These definitions use the operator *IsAFcn*, which is defined as follows so that *IsAFcn(f)* is true iff f is a function:

$$IsAFcn(f) \;\triangleq\; f = [x \in \text{DOMAIN} \; f \mapsto f[x]]$$

The first rule, which is not naturally associated with any one construct, is that two functions are equal iff they have the same domain and assign the same value to each element in their domain:

$$
\begin{aligned}
\forall f, g \;:\; &IsAFcn(f) \wedge IsAFcn(g) \;\Rightarrow \\
&((f = g) \;\equiv\; \wedge \; \text{DOMAIN} \; f = \text{DOMAIN} \; g \\
&\qquad\qquad\qquad \wedge \; \forall x \in \text{DOMAIN} \; f \;:\; f[x] = g[x])
\end{aligned}
$$

The rest of the semantics of functions is given below. There is no separate defining rule for the DOMAIN operator.

$f[e_1, \ldots, e_n]$
> where the e_i are expressions. For $n = 1$, this is a primitive expression. For $n > 1$, it is defined by
>
> $$f[e_1, \ldots, e_n] \;=\; f[\langle e_1, \ldots, e_n \rangle]$$
>
> The tuple $\langle e_1, \ldots, e_n \rangle$ is defined in Section 16.1.9.

$[\mathbf{y}_1 \in S_1, \; \ldots, \mathbf{y}_n \in S_n \mapsto e]$
> where each $\mathbf{y}_i$ has the form $x_1, \; \ldots, x_k$ or $\langle x_1, \; \ldots, x_k \rangle$, and each x_j is an identifier that is bound in the expression. The expressions S_i lie outside the scope of the bound identifiers. The simple form $[x \in S \mapsto e]$, for x an identifier, is primitive and is defined by two rules:
>
> $$(\text{DOMAIN} \; [x \in S \mapsto e]) \;=\; S$$
>
> $$\forall y \in S \;:\; [x \in S \mapsto e][y] \;=\; \text{LET} \; x \triangleq y \; \text{IN} \; e$$

where y is an identifier different from x that does not occur in S or e. The general form of the construct is defined inductively in terms of the simple form by

$$[x_1 \in S_1, \ldots, x_n \in S_n \mapsto e] \quad \triangleq \quad [\langle x_1, \ldots, x_n \rangle \in S_1 \times \ldots \times S_n \mapsto e]$$

$$[\ldots, x_1, \ldots, x_k \in S_i, \ldots \mapsto e] \quad \triangleq \quad [\ldots, x_1 \in S_i, \ldots, x_k \in S_i, \ldots \mapsto e]$$

$$[\ldots, \langle x_1, \ldots, x_k \rangle \in S_i, \ldots \mapsto e] \triangleq$$

$$[\ldots, y \in S_i, \ldots \mapsto \text{LET } z \stackrel{\triangle}{=} \text{CHOOSE } \langle x_1, \ldots, x_k \rangle : y = \langle x_1, \ldots, x_k \rangle$$
$$x_1 \stackrel{\triangle}{=} z[1]$$
$$\vdots$$
$$x_k \stackrel{\triangle}{=} z[k] \quad \text{IN } e]$$

where y and z are identifiers that do not appear anywhere in the original expression. See Section 16.1.9 for details about tuples.

$[S \to T]$ is defined by

$$\forall f : f \in [S \to T] \equiv$$
$$IsAFcn(f) \wedge (S = \text{DOMAIN } f) \wedge (\forall x \in S : f[x] \in T)$$

where x and f do not occur in S or T, and $IsAFcn$ is defined above.

$[f \text{ EXCEPT } !\mathbf{a}_1 = e_1, \ldots, !\mathbf{a}_n = e_n]$
where each $\mathbf{a}_i$ has the form $[d_1] \ldots [d_k]$ and each d_j is an expression. For the simple case when $n = 1$ and $\mathbf{a}_1$ is $[d]$, this is defined by[5]

$$[f \text{ EXCEPT } ![d] = e] \quad \triangleq$$
$$[y \in \text{DOMAIN } f \mapsto \text{IF } y = d \text{ THEN LET } @ \stackrel{\triangle}{=} f[d] \text{ IN } e$$
$$\text{ELSE } f[y]]$$

where y does not occur in f, d, or e. The general form is defined inductively in terms of this simple case by

$$[f \text{ EXCEPT } !\mathbf{a}_1 = e_1, \ldots, !\mathbf{a}_n = e_n] \quad \triangleq$$
$$[[f \text{ EXCEPT } !\mathbf{a}_1 = e_1, \ldots, !\mathbf{a}_{n-1} = e_{n-1}] \text{ EXCEPT } !\mathbf{a}_n = e_n]$$

$$[f \text{ EXCEPT } ![d_1] \ldots [d_k] = e] \quad \triangleq$$
$$[f \text{ EXCEPT } ![d_1] = [@ \text{ EXCEPT } ![d_2] \ldots [d_k] = e]]$$

$f[\mathbf{y}_1 \in S_1, \ldots, \mathbf{y}_n \in S_n] \stackrel{\triangle}{=} e$ is defined to be an abbreviation for

$$f \stackrel{\triangle}{=} \text{CHOOSE } f : f = [\mathbf{y}_1 \in S_1, \ldots, \mathbf{y}_n \in S_n \mapsto e]$$

---

[5]Since @ is not actually an identifier, LET @ $\stackrel{\triangle}{=}$ ... isn't legal TLA⁺ syntax. However, its meaning should be clear.

16.1.8 Records

TLA$^+$ borrows from programming languages the concept of a record. Records were introduced in Section 3.2 (page 28) and further explained in Section 5.2 (page 48). As in programming languages, $r.h$ is the h field of record r. Records can be written explicitly as

$$[h_1 \mapsto e_1, \ldots, h_n \mapsto e_n]$$

which equals the record with n fields, whose h_i field equals e_i, for $i = 1, \ldots, n$. The expression

$$[h_1 : S_1, \ldots, h_n : S_n]$$

is the set of all such records with $e_i \in S_i$, for $i = 1, \ldots, n$. These expressions are legal only if the h_i are all different. For example, $[a : S, a : T]$ is illegal.

The EXCEPT construct, explained in Section 16.1.7 above, can be used for records as well as functions. For example,

$$[r \ \text{EXCEPT} \ !.a = e]$$

is the record $\widehat{r}$ that is the same as r, except that $\widehat{r}.a = e$. An exception clause can mix function application and record fields. For example,

$$[f \ \text{EXCEPT} \ ![v].a = e]$$

is the function $\widehat{f}$ that is the same as f, except that $\widehat{f}[v].a = e$.

In TLA$^+$, a record is a function whose domain is a finite set of strings, where $r.h$ means $r[\text{"}h\text{"}]$, for any expression r and record field h. Thus, the following two expressions describe the same record:

$$[fo \mapsto 7, \ ba \mapsto 8] \qquad [x \in \{ \text{"fo"}, \text{"ba"} \} \mapsto \text{IF} \ x = \text{"fo"} \ \text{THEN} \ 7 \ \text{ELSE} \ 8]$$

The name of a record field is syntactically an identifier. In the ASCII version of TLA$^+$, it is a string of letters, digits, and the underscore character (_) that contains at least one letter. Strings are described below in Section 16.1.10.

Formal Semantics

The record constructs are defined in terms of function constructs.

$$e.h \ \triangleq \ e[\text{"}h\text{"}]$$

$$[h_1 \mapsto e_1, \ldots, h_n \mapsto e_n] \ \triangleq$$
$$[y \in \{ \text{"}h_1\text{"}, \ldots, \text{"}h_n\text{"} \} \mapsto$$
$$\text{CASE} \ (y = \text{"}h_1\text{"}) \to e_1 \ \square \ \ldots \ \square \ (y = \text{"}h_n\text{"}) \to e_n]$$

where y does not occur in any of the expressions e_i. The h_i must all be distinct.

$$[h_1 : S_1, \ldots, h_n : S_n] \;\triangleq\; \{\, [h_1 \mapsto y_1, \ldots, h_n \mapsto y_n] \;:$$
$$y_1 \in S_1, \ldots, y_n \in S_n \}$$

where the y_i do not occur in any of the expressions S_j. The h_i must all be distinct.

$[r \text{ EXCEPT } !\mathbf{a}_1 = e_1, \ldots, !\mathbf{a}_n = e_n]$
where $\mathbf{a}_i$ has the form $b_1 \ldots b_k$ and each b_j is either (i) $[d]$, where d is an expression, or (ii) $.h$, where h is a record field. It is defined to equal the corresponding function EXCEPT construct in which each $.h$ is replaced by $[\text{``}h\text{''}]$.

16.1.9   Tuples

An n-tuple is written in TLA$^+$ as $\langle e_1, \ldots, e_n \rangle$. As explained in Section 5.4, an n-tuple is defined to be a function whose domain is the set $\{1, \ldots, n\}$, where $\langle e_1, \ldots, e_n \rangle[i] = e_i$, for $1 \le i \le n$. The Cartesian product $S_1 \times \cdots \times S_n$ is the set of all n-tuples $\langle e_1, \ldots, e_n \rangle$ such that $e_i \in S_i$, for $1 \le i \le n$.

In TLA$^+$, $\times$ is not an associative operator. For example,

$$\langle 1, 2, 3 \rangle \;\in\; Nat \times Nat \times Nat$$
$$\langle \langle 1, 2 \rangle, 3 \rangle \;\in\; (Nat \times Nat) \times Nat$$
$$\langle 1, \langle 2, 3 \rangle \rangle \;\in\; Nat \times (Nat \times Nat)$$

and the tuples $\langle 1, 2, 3 \rangle$, $\langle \langle 1, 2 \rangle, 3 \rangle$, and $\langle 1, \langle 2, 3 \rangle \rangle$ are not equal. More precisely, the triple $\langle 1, 2, 3 \rangle$ is unequal to either of the pairs $\langle \langle 1, 2 \rangle, 3 \rangle$ or $\langle 1, \langle 2, 3 \rangle \rangle$ because a triple and a pair have unequal domains. The semantics of TLA$^+$ does not specify if $\langle 1, 2 \rangle$ equals 1 or if 3 equals $\langle 2, 3 \rangle$, so we don't know whether or not $\langle \langle 1, 2 \rangle, 3 \rangle$ and $\langle 1, \langle 2, 3 \rangle \rangle$ are equal.

The 0-tuple $\langle \, \rangle$ is the unique function having an empty domain. The 1-tuple $\langle e \rangle$ is different from e. That is, the semantics does not specify whether or not they are equal. There is no special notation for writing a set of 1-tuples. The easiest way to denote the set of all 1-tuples $\langle e \rangle$ with $e \in S$ is $\{ \langle e \rangle \,:\, e \in S \}$.

In the standard *Sequences* module, described in Section 18.1 (page 339), an n-element sequence is represented as an n-tuple. The module defines several useful operators on sequences/tuples.

Formal Semantics

Tuples and Cartesian products are defined in terms of functions (defined in Section 16.1.7) and the set *Nat* of natural numbers (defined in Section 16.1.11).

$$\langle e_1, \ldots, e_n \rangle \;\triangleq\; [i \in \{ j \in Nat \,:\, (1 \le j) \land (j \le n) \} \mapsto e_i]$$
where i does not occur in any of the expressions e_j.

$$S_1 \times \cdots \times S_n \;\triangleq\; \{\, \langle y_1, \ldots, y_n \rangle \,:\, y_1 \in S_1, \ldots, y_n \in S_n \}$$

where the identifiers y_i do not occur in any of the expressions S_j.

16.1.10 Strings

TLA^+ defines a string to be a tuple of characters. (Tuples are defined in Section 16.1.9 above.) Thus, "abc" equals

$$\langle\, \text{"abc"}[1],\ \text{"abc"}[2],\ \text{"abc"}[3] \,\rangle$$

The semantics of TLA^+ does not specify what a character is. However, it does specify that different characters (those having different computer representations) are different. Thus "a"[1], "b"[1], and "A"[1] (the characters a, b, and A) are all different. The built-in operator STRING is defined to be the set of all strings.

Although TLA^+ doesn't specify what a character is, it's easy to define operators that assign values to characters. For example, here's the definition of an operator *Ascii* that assigns to every lower-case letter its ASCII representation.[6]

$$Ascii(char) \;\triangleq\; 96 + \text{CHOOSE } i \in 1 \mathinner{\ldotp\ldotp} 26 :$$
$$\text{"abcdefghijklmnopqrstuvwxyz"}[i] = char$$

This defines *Ascii*("a"[1]) to equal 97, the ASCII code for the letter a, and *Ascii*("z"[1]) to equal 122, the ASCII code for z. Section 11.1.4 on page 179 illustrates how a specification can make use of the fact that strings are tuples.

Exactly what characters may appear in a string is system-dependent. A Japanese version of TLA^+ might not allow the character a. The standard ASCII version contains the following characters:

$a\ b\ c\ d\ e\ f\ g\ h\ i\ j\ k\ l\ m\ n\ o\ p\ q\ r\ s\ t\ u\ v\ w\ x\ y\ z$
$A\ B\ C\ D\ E\ F\ G\ H\ I\ J\ K\ L\ M\ N\ O\ P\ Q\ R\ S\ T\ U\ V\ W\ X\ Y\ Z$
$0\ 1\ 2\ 3\ 4\ 5\ 6\ 7\ 8\ 9$
$\tilde{}\ @\ \#\ \$\ \%\ \hat{}\ \&\ *\ \_\ -\ +\ =\ (\ )\ \{\ \}\ [\ ]\ <\ >\ |\ /\ \backslash\ ,\ .\ ?\ :\ ;\ `\ '\ "$
$\langle \text{HT} \rangle$ (tab) $\langle \text{LF} \rangle$ (line feed) $\langle \text{FF} \rangle$ (form feed) $\langle \text{CR} \rangle$ (carriage return)

plus the space character. Since strings are delimited by a double-quote ("), some convention is needed for typing a string that contains a double-quote. Conventions are also needed to type characters like $\langle \text{LF} \rangle$ within a string. In the ASCII version of TLA^+, the following pairs of characters, beginning with a $\backslash$ character, are used to represent these special characters:

\"	"	\t	$\langle \text{HT} \rangle$	\f	$\langle \text{FF} \rangle$
\\	\	\n	$\langle \text{LF} \rangle$	\r	$\langle \text{CR} \rangle$

---

[6]This clever way of using CHOOSE to map from characters to numbers was pointed out to me by Georges Gonthier.

With this convention, `"a\\\"b\""` represents the string consisting of the following five characters: a \ " b ". In the ASCII version of TLA+, a \ character can appear in a string expression only as the first character of one of these six two-character sequences.

Formal Semantics

We assume a set *Char* of characters, which may depend on the version of TLA+. (The identifier *Char* is not a pre-defined symbol of TLA+.)

STRING $\triangleq$ *Seq(Char)*

>   where *Seq* is the operator defined in the *Sequences* module of Section 18.1
>   so that *Seq(S)* is the set of all finite sequences of elements of S.

"$c_1 \ldots c_n$" $\triangleq$ $\langle c_1, \ldots, c_n \rangle$

>   where each c_i is some representation of a character in *Char*.

16.1.11 Numbers

TLA+ defines a sequence of digits like 63 to be the usual natural number—that is, 63 equals $6 * 10 + 3$. TLA+ also allows the binary representation `\b111111`, the octal representation `\o77`, and the hexadecimal representation `\h3F` of that number. (Case is ignored in the prefixes and in the hexadecimal representation, so `\H3F` and `\h3f` are equivalent to `\h3F`.) Decimal numbers are also pre-defined in TLA+; for example, 3.14159 equals $314159/10^5$.

Numbers are pre-defined in TLA+, so 63 is defined even in a module that does not extend or instantiate one of the standard numbers modules. However, sets of numbers like *Nat* and arithmetic operators like + are not. You can write a module that defines + any way you want, in which case $40 + 23$ need not equal 63. Of course, $40 + 23$ does equal 63 for + defined by the standard numbers modules *Naturals*, *Integers*, and *Reals*, which are described in Section 18.4.

Formal Semantics

The set *Nat* of natural numbers, along with its zero element *Zero* and successor function *Succ*, is defined in module *Peano* on page 345. The meaning of a representation of a natural number is defined in the usual manner:

$$0 \triangleq Zero \qquad 1 \triangleq Succ[Zero] \qquad 2 \triangleq Succ[Succ[Zero]] \qquad \ldots$$

The *ProtoReals* module on pages 346–347 defines the set *Real* of real numbers to be a superset of the set *Nat*, and defines the usual arithmetic operators on real numbers. The meaning of a decimal number is defined in terms of these operators by

$$c_1 \cdots c_m . d_1 \cdots d_n \;\triangleq\; c_1 \cdots c_m \, d_1 \cdots d_n / 10^n$$

16.2 Nonconstant Operators

The nonconstant operators are what distinguish TLA$^+$ from ordinary mathematics. There are two classes of nonconstant operators: action operators, listed in Table 3 on page 269, and temporal operators, listed in Table 4 on page 269.

Section 16.1 above talks about the meanings of the built-in constant operators of TLA$^+$, without considering their arguments. We can do that for constant operators, since the meaning of $\subseteq$ in the expression $e_1 \subseteq e_2$ doesn't depend on whether or not the expressions e_1 and e_2 contain variables or primes. To understand the nonconstant operators, we need to consider their arguments. Thus, we can no longer talk about the meanings of the operators in isolation; we must describe the meanings of expressions built from those operators.

A *basic* expression is one that contains built-in TLA$^+$ operators, declared constants, and declared variables. We now describe the meaning of all basic TLA$^+$ expressions, including ones that contain nonconstant built-in operators. We start by considering basic constant expressions, ones containing only the constant operators we have already studied and declared constants.

16.2.1 Basic Constant Expressions

Section 16.1 above defines the meanings of the constant operators. This in turn defines the meaning of any expression built from these operators and declared constants. For example, if S and T are declared by

CONSTANTS $S,\ T(\_)$

then $\exists\, x : S \subseteq T(x)$ is a formula that equals TRUE if there is some value v such that every element of S is an element of $T(v)$, and otherwise equals FALSE. Whether $\exists\, x : S \subseteq T(x)$ equals TRUE or FALSE depends on what actual values we assign to S and to $T(v)$, for all v; so that's as far as we can go in assigning a meaning to the expression.

A formula is a Boolean-valued expression. There are some basic constant formulas that are true regardless of the values we assign to their declared constants—for example, the formula

$$(S \subseteq T) \equiv (S \cap T = S)$$

Such a formula is said to be *valid*.

Formal Semantics

Section 16.1 defines all the built-in constant operators in terms of a subset of them called the primitive operators. These definitions can be formulated as an inductive set of rules that define the meaning $[\![c]\!]$ of any basic constant expression c. For example, from the definition

$$e \notin S \;\triangleq\; \neg(e \in S)$$

we get the rule

$$[\![e \notin S]\!] \;=\; \neg([\![e]\!] \in [\![S]\!])$$

These rules define the meaning of a basic constant expression to be an expression containing only primitive constant operators and declared constants.

A basic constant expression e is a formula iff its meaning $[\![e]\!]$ is Boolean-valued, regardless of what values are substituted for the declared constants. As explained in Section 16.1.3, this will depend on whether we are using the liberal, moderate, or conservative interpretations of the Boolean operators.

If S and T are constants declared as above, then the meaning $[\![\exists x : S \subseteq T(x)]\!]$ of the expression $\exists x : S \subseteq T(x)$ is the expression itself. Logicians usually carry things further, assigning some meanings $[\![S]\!]$ and $[\![T]\!]$ to declared constants and defining $[\![\exists x : S \subseteq T(x)]\!]$ to equal $\exists x : [\![S]\!] \subseteq [\![T]\!](x)$. For simplicity, I have short-circuited that extra level of meaning.

We are taking as given the meaning of an expression containing only primitive constant operators and declared constants. In particular, we take as primitive the notion of validity for such expressions. Section 16.1 defines the meaning of any basic constant expression in terms of these expressions, so it defines what it means for a basic constant formula to be valid.

16.2.2   The Meaning of a State Function

A *state* is an assignment of values to variables. (In ZF set theory, on which the semantics of TLA$^+$ is based, *value* is just another term for *set*.) States were discussed in Sections 2.1 and 2.3.

A *state function* is an expression that is built from declared variables, declared constants, and constant operators. (State functions can also contain EN-ABLED expressions, which are described below.) State functions are discussed on page 25 in Section 3.1. A state function assigns a constant expression to every state. If state function e assigns to state s the constant expression v, then we say that v is the value of e in state s. For example, if x is a declared variable, T is a declared constant, and s is a state that assigns to x the value 42; then the value of $x \in T$ in state s is the constant expression $42 \in T$. A Boolean-valued state function is called a *state predicate*. A state predicate is *valid* iff it has the value TRUE in every state.

Formal Semantics

A state is an assignment of values to variables. Formally, a state s is a function whose domain is the set of all variable names, where $s[\text{"x"}]$ is the value that s assigns to variable x. We write $s[\![x]\!]$ instead of $s[\text{"x"}]$.

A *basic state function* is an expression that is built from declared variables, declared constants, constant operators, and ENABLED expressions, which are expressions of the form ENABLED e. An ENABLED-free basic state function is one with no ENABLED expressions. The meaning of a basic state function is a mapping from states to values. We let $s[\![e]\!]$ be the value that state function e assigns to a state s. Since a variable is a state function, we thus say both that state s assigns $s[\![x]\!]$ to variable x, and that the state function x assigns $s[\![x]\!]$ to state s.

Using the meanings assigned to the constant operators in Section 16.1 above, we inductively define $s[\![e]\!]$ for any ENABLED-free state function e to be an expression built from the primitive TLA$^+$ constant operators, declared constants, and the values assigned by s to each variable. For example, if x is a variable and S is a constant, then

$$s[\![x \notin S]\!] \;=\; \neg(s[\![x]\!] \in S)$$

It is easy to see that $s[\![c]\!]$ equals $[\![c]\!]$, for any constant expression c. (This expresses formally that a constant has the same value in all states.)

To define the meaning of all basic state function, not just ENABLED-free ones, we must define the meaning of an ENABLED expression. This is done below.

The formal semantics talks about state functions, not state predicates. Because TLA$^+$ is typeless, there is no formal distinction between a state predicate and a state function. By a state predicate, we mean a state function e such that $s[\![e]\!]$ is Boolean-valued for every reachable state s of some specification. See the discussion of actions on pages 313–314.

I described the meaning of a state function as a "mapping" on states. This mapping cannot be a function, because there is no set of all states. Since for any set S there is a state that assigns the value S to each variable, there are too many states to form a set. (See the discussion of Russell's paradox on page 66.) To be formal, we should define an operator M such that, if s is a state and e is a syntactically correct basic state function, then $M(s, e)$, which we write $s[\![e]\!]$, is the basic constant expression that is the meaning of e in state s.

Actually, this way of describing the semantics isn't right either. A state is a mapping from variables to values (sets), not to constant expressions. Since there are an uncountable number of sets and only a countable number of finite sequences of strings, there are values that can't be described by any expression. Suppose ξ is such a value, and let s be a state that assigns the value ξ to the variable x. Then $s[\![x = \{\}]\!]$ equals $\xi = \{\}$, which isn't a constant expression because ξ isn't an expression. So, to be *really* formal, we would have to define a semantic constant expression to be one made from primitive constant operators, declared constants, and arbitrary values. The meaning of a basic state function is a mapping from states to semantic constant expressions.

We won't bother with these details. Instead, we define a semi-formal semantics for basic expressions that is easier to understand. Mathematically sophisticated readers who understand the less formal exposition should be able to fill in the missing formal details.

16.2.3   Action Operators

A *transition function* is an expression built from state functions using the priming operator (′) and the other action operators of TLA$^+$ listed in Table 3 on page 269. A transition function assigns a value to every step, where a step is a pair of states. In a transition function, an unprimed occurrence of a variable x represents the value of x in the first (old) state, and a primed occurrence of x represents its value in the second (new) state. For example, if state s assigns the value 4 to x and state t assigns the value 5 to x, then the transition function $x' - x$ assigns to the step $s \to t$ the value $5 - 4$, which equals 1 (if $-$ has its usual definition).

An *action* is a Boolean-valued transition function, such as $x' > x$. We say that action A is true on step $s \to t$, or that $s \to t$ is an A step, iff A assigns the value TRUE to $s \to t$. An action is said to be *valid* iff it is true on any step.

The action operators of TLA$^+$ other than ′ have the following meanings, where A and B are actions and e is a state function:

$$[A]_e \ \triangleq \ A \lor (e' = e)$$

$$\langle A \rangle_e \ \triangleq \ A \land (e' \neq e)$$

ENABLED A is the state function that is true in state s iff there is some state t such that $s \to t$ is an A step.

UNCHANGED $e \ \triangleq \ e' = e$

$A \cdot B$ is the action that is true on step $s \to t$ iff there is a state u such that $s \to u$ is an A step and $u \to t$ is a B step.

Priming and the construct $[A]_v$ are introduced in Section 2.2 (page 15); the UNCHANGED operator is introduced on page 26 of Section 3.1; ENABLED is introduced on page 97 of Section 8.4; the construct $\langle A \rangle_v$ is defined on page 91 of Section 8.1; and the action-composition operator "·" is introduced in Section 7.3 (page 76).

Formal Semantics

A *basic transition function* is a basic expression that does not contain any temporal operators. The meaning of a basic transition function e is an assignment of a basic constant expression $\langle s, t \rangle [\![e]\!]$ to any pair of states $\langle s, t \rangle$. (We use here the more conventional notation $\langle s, t \rangle$ instead of $s \to t$.) A transition function is valid iff $\langle s, t \rangle [\![e]\!]$ is valid, for all states s and t.

If e is a basic state function, then we interpret e as a basic transition function by defining $\langle s, t \rangle [\![e]\!]$ to equal $s[\![e]\!]$. As indicated above, UNCHANGED and the constructs $[A]_e$ and $\langle A \rangle_e$ are defined in terms of priming. To define the meanings of the remaining action operators, we first define existential quantification over all states. Let $IsAState$ be an operator such that $IsAState(s)$ is true iff s is a state—that is, a function whose domain is the set of all variable names. (It's easy to define $IsAState$ using the operator $IsAFcn$, defined on page 303.) Existential quantification over all states is then defined by

$$\exists_{\text{state}} s : p \;\;\triangleq\;\; \exists s : IsAState(s) \wedge p$$

for any formula p. The meanings of all transition functions and all state functions (including ENABLED expressions) is then defined inductively by the definitions already given and the following definitions of the remaining action operators:

e' is the transition function defined by $\langle s, t \rangle [\![e']\!] \;\triangleq\; t[\![e]\!]$ for any state function e.

ENABLED A is the state function defined by

$$s[\![\text{ENABLED } A]\!] \;\;\triangleq\;\; \exists_{\text{state}} t : \langle s, t \rangle [\![A]\!]$$

for any transition function A.

$A \cdot B$ is the transition function defined by

$$\langle s, t \rangle [\![A \cdot B]\!] \;\;\triangleq\;\; \exists_{\text{state}} u : \langle s, u \rangle [\![A]\!] \wedge \langle u, t \rangle [\![B]\!]$$

for any transition functions A and B.

The formal semantics talks about transition functions, not actions. Since TLA$^+$ is typeless, there is no formal distinction between an action and an arbitrary transition function. We could define an action A to be a transition function such that $\langle s, t \rangle [\![A]\!]$ is a Boolean for all states s and t. However, what we usually mean by an action is a transition function A such that $\langle s, t \rangle [\![A]\!]$ is a Boolean whenever s and t are reachable states of some specification. For example, a specification with a variable b of type BOOLEAN might contain an action $b \wedge (y' = y)$. We can calculate the meaning of ENABLED $(b \wedge (y' = y))$ as follows:

> Types are explained on page 25.

$$
\begin{aligned}
&s[\![\text{ENABLED } (b \wedge (y' = y))]\!] \\
&\quad = \;\; \exists_{\text{state}} t : \langle s, t \rangle [\![b \wedge (y' = y)]\!] && \text{By definition of ENABLED.} \\
&\quad = \;\; \exists_{\text{state}} t : \langle s, t \rangle [\![b]\!] \wedge (\langle s, t \rangle [\![y']\!] = \langle s, t \rangle [\![y]\!]) && \text{By definition of } \wedge \text{ and } =. \\
&\quad = \;\; \exists_{\text{state}} t : s[\![b]\!] \wedge (t[\![y]\!] = s[\![y]\!]) && \text{By definition of } ', \text{ since } \langle s, t \rangle [\![e]\!] = s[\![e]\!], \\
&&& \text{for any state function } e.
\end{aligned}
$$

If $s[\![b]\!]$ is a Boolean, we can now continue the calculation as follows:

$$
\begin{aligned}
\exists_{\text{state}}\, t\, :\, & s[\![b]\!] \wedge (t[\![y]\!] = s[\![y]\!]) \\
= \;\; & s[\![b]\!] \wedge \exists_{\text{state}}\, t\, :\, (t[\![y]\!] = s[\![y]\!]) && \text{By predicate logic, since } t \text{ does not occur in } s[\![b]\!]. \\
= \;\; & s[\![b]\!] && \text{The existence of } t \text{ is obvious---for example, let it equal } s.
\end{aligned}
$$

Hence, $s[\![\text{ENABLED}\,(b \wedge (y' = y))]\!]$ equals $s[\![b]\!]$, if $s[\![b]\!]$ is a Boolean. However, if s is a state that assigns the value 2 to the variable b and the value -7 to the variable y, then

$$
s[\![\text{ENABLED}\,(b \wedge (y' = y))]\!] \;=\; \exists_{\text{state}}\, t\, :\, 2 \wedge (t[\![y]\!] = -7)
$$

The last expression may or may not equal 2. (See the discussion of the interpretation of the Boolean operators in Section 16.1.3 on page 296.) If the specification we are writing makes sense, it can depend on the meaning of ENABLED $(b \wedge (y' = y))$ only for states in which the value of b is a Boolean. We don't care about its value in a state that assigns to b the value 2, just as we don't care about the value of $3/x$ in a state that assigns the value "abc" to x. See the discussion of silly expressions in Section 6.2 (page 67).

16.2.4   Temporal Operators

As explained in Section 8.1, a temporal formula F is true or false for a behavior, where a behavior is a sequence of states. Syntactically, a temporal formula is defined inductively to be a state predicate or a formula having one of the forms shown in Table 4 on page 269, where e is a state function, A is an action, and F and G are temporal formulas. All the temporal operators in Table 4 are explained in Chapter 8—except for $\overset{+}{\leadsto}$, which is explained in Section 10.7 (page 156).

The formula $\Box F$ is true for a behavior σ iff the temporal formula F is true for σ and all suffixes of σ. To define the constructs $\Box[A]_e$ and $\Diamond\langle A\rangle_e$, we regard an action B to be a temporal formula that is true of a behavior σ iff the first two states of σ form a B step. Thus, $\Box[A]_e$ is true of σ iff every successive pair of states of σ is a $[A]_e$ step. All the other temporal operators of TLA$^+$, except $\exists$, $\forall$, and $\overset{+}{\leadsto}$, are defined as follows in terms of $\Box$:

$$
\begin{aligned}
\Diamond F & \triangleq \neg\Box\neg F \\
\text{WF}_e(A) & \triangleq \Box\Diamond\neg(\text{ENABLED}\,\langle A\rangle_e) \vee \Box\Diamond\langle A\rangle_e \\
\text{SF}_e(A) & \triangleq \Diamond\Box\neg(\text{ENABLED}\,\langle A\rangle_e) \vee \Box\Diamond\langle A\rangle_e \\
F \leadsto G & \triangleq \Box(F \Rightarrow \Diamond G)
\end{aligned}
$$

The temporal existential quantifier $\exists$ is a hiding operator, $\exists\, x : F$ meaning formula F with the variable x hidden. To define this more precisely, we first define $\natural\sigma$ to be the (possibly finite) sequence of states obtained by removing from σ all stuttering steps—that is, by removing any state that is the same as the previous one. We then define $\sigma \sim_x \tau$ to be true iff $\natural\sigma$ and $\natural\tau$ are the same except for the values that their states assign to the variable x. Thus, $\sigma \sim_x \tau$ is true iff σ can be obtained from τ (or vice-versa) by adding and/or removing stuttering steps and changing the values assigned to x by its states. Finally, $\exists\, x : F$ is defined to be true for a behavior σ iff F is true for some behavior τ such that $\sigma \sim_x \tau$.

The temporal universal quantifier $\forall$ is defined in terms of $\exists$ by

$$\forall x : F \;\triangleq\; \neg(\exists\, x : \neg F)$$

The formula $F \overset{+}{\Rightarrow} G$ asserts that G does not become false before F does. More precisely, we define a formula H to be true for a finite prefix ρ of a behavior σ iff H is true for some (infinite) behavior that extends ρ. (In particular, H is true of the empty prefix iff H satisfies some behavior.) Then $F \overset{+}{\Rightarrow} G$ is defined to be true for a behavior σ iff (i) $F \Rightarrow G$ is true for σ and (ii) for every finite prefix ρ of σ, if F is true for ρ then G is true for the prefix of σ that is one state longer than ρ.

Formal Semantics

Formally, a behavior is a function from the set *Nat* of natural numbers to states. (We think of a behavior σ as the sequence $\sigma[0], \sigma[1], \ldots$ of states.) The meaning of a temporal formula is a predicate on behaviors—that is, a mapping from behaviors to Booleans. We write $\sigma \models F$ for the value that the meaning of F assigns to the behavior σ. The temporal formula F is valid iff $\sigma \models F$ is true, for all behaviors σ.

Instead of writing σ_i as in Chapter 8, we use here the standard functional notation $\sigma[i]$.

Above, we have defined all the other temporal operators in terms of $\Box$, $\exists$, and $\overset{+}{\Rightarrow}$. Formally, since an action is not a temporal formula, the construct $\Box[A]_e$ is not an instance of the temporal operator $\Box$, so its meaning should be defined separately. The construct $\Diamond\langle A \rangle_e$, which is similarly not an instance of $\Diamond$, is then defined to equal $\neg\Box[\neg A]_e$.

To define the meaning of $\Box$, we first define σ^{+n} to be the behavior obtained by deleting the first n states of σ:

$$\sigma^{+n} \;\triangleq\; [\, i \in Nat \mapsto \sigma[i + n]\,]$$

We then define the meaning of $\Box$ as follows, for any temporal formula F, transition function A and state function e:

$$\sigma \models \Box F \;\triangleq\; \forall\, n \in Nat : \sigma^{+n} \models F$$
$$\sigma \models \Box[A]_e \;\triangleq\; \forall\, n \in Nat : \langle\sigma[n], \sigma[n+1]\rangle[\![A]_e]\!]$$

To formalize the definition of $\exists$ given above, we first define $\natural$ as follows, letting f be the function such that $\sigma[n] = \natural\sigma[f[n]]$, for all n:

$$\natural\sigma \;\triangleq\; \text{LET } f[n \in Nat] \;\triangleq\; \text{IF } n = 0 \text{ THEN } 0$$
$$\text{ELSE } \text{IF } \sigma[n] = \sigma[n-1]$$
$$\text{THEN } f[n-1]$$
$$\text{ELSE } f[n-1] + 1$$
$$S \;\triangleq\; \{f[n] : n \in Nat\}$$
$$\text{IN } \quad [n \in S \mapsto \sigma[\text{CHOOSE } i \in Nat : f[i] = n]]$$

Next, let $s_{x \leftarrow v}$ be the state that is the same as state s except that it assigns to the variable x the value v. We then define $\sim_x$ by

$$\sigma \sim_x \tau \;\triangleq\; \natural\sigma = [n \in \text{DOMAIN } \natural\tau \mapsto \tau_{x \leftarrow \natural\sigma[n][\![x]\!]}]$$

We next define existential quantification over behaviors. This is done much as we defined quantification over states on page 313 above; we first define *IsABehavior* so that *IsABehavior*(σ) is true iff σ is a behavior, and we then define

$$\exists_{\text{behavior}} \sigma : F \;\triangleq\; \exists \sigma : IsABehavior(\sigma) \land F$$

We can now define the meaning of $\exists$ by

$$\sigma \models \exists x : F \;\triangleq\; \exists_{\text{behavior}} \tau : (\sigma \sim_x \tau) \land (\tau \models F)$$

Finally, we define the meaning of $\xrightarrow{+}$ as follows:

$$\sigma \models F \xrightarrow{+} G \;\triangleq\;$$
$$\text{LET } PrefixSat(n, H) \;\triangleq\;$$
$$\exists_{\text{behavior}} \tau : \land \forall i \in 0 \,..\, (n-1) : \tau[i] = \sigma[i]$$
$$\land \tau \models H$$
$$\text{IN } \quad \land \sigma \models F \Rightarrow G$$
$$\land \forall n \in Nat : PrefixSat(n, F) \Rightarrow PrefixSat(n+1, G)$$

Chapter 17

The Meaning of a Module

Chapter 16 defines the meaning of the built-in TLA$^+$ operators. In doing so, it defines the meaning of a basic expression—that is, of an expression containing only built-in operators, declared constants, and declared variables. We now define the meaning of a module in terms of basic expressions. Since a TLA$^+$ specification consists of a collection of modules, this defines the semantics of TLA$^+$.

We also complete the definition of the syntax of TLA$^+$ by giving the remaining context-dependent syntactic conditions not described in Chapter 15. Here's a list of some illegal expressions that satisfy the grammar of Chapter 15, and where in this chapter you can find the conditions that make them illegal.

- $F(x)$, if F is defined by $F(x, y) \triangleq x + y$ (Section 17.1)

- $(x' + 1)'$ (Section 17.2)

- $x + 1$, if x is not defined or declared (Section 17.3)

- $F \triangleq 0$, if F is already defined (Section 17.5)

This chapter is meant to be read in its entirety. To try to make it as readable as possible, I have made the exposition somewhat informal. Wherever I could, I have used examples in place of formal definitions. The examples assume that you understand the approximate meanings of the TLA$^+$ constructs, as explained in Part I. I hope that mathematically sophisticated readers will see how to fill in the missing formalism.

17.1   Operators and Expressions

Because it uses conventional mathematical notation, TLA$^+$ has a rather rich syntax, with several different ways of expressing the same basic type of mathematical operation. For example, the following expressions are all formed by applying an operator to a single argument e:

$$Len(e) \qquad - e \qquad \{e\} \qquad e'$$

This section develops a uniform way of writing all these expressions, as well as more general kinds of expressions.

17.1.1   The Arity and Order of an Operator

An operator has an *arity* and an *order*. An operator's arity describes the number and order of its arguments. It's the arity of the *Len* operator that tells us that $Len(s)$ is a legal expression, while $Len(s,t)$ and $Len(+)$ are not. All the operators of TLA$^+$, whether built-in or defined, fall into three classes: 0th-, 1st-, and 2nd-order operators.[1] Here is how these classes, and their arities, are defined:

Len is defined in the *Sequences* module on page 341.

0. $E \triangleq x' + y$ defines E to be the 0th-order operator $x' + y$. A 0th-order operator takes no arguments, so it is an ordinary expression. We represent the arity of such an operator by the symbol $\_$ (underscore).

1. $F(x,y) \triangleq x \cup \{z,y\}$ defines F to be a 1st-order operator. For any expressions e_1 and e_2, it defines $F(e_1, e_2)$ to be an expression. We represent the arity of F by $\langle \_, \_ \rangle$.

   In general, a 1st-order operator takes expressions as arguments. Its arity is the tuple $\langle \_, \dots, \_ \rangle$, with one $\_$ for each argument.

2. $G(f(\_,\_), x, y) \triangleq f(x, \{x, y\})$ defines G to be a 2nd-order operator. The operator G takes three arguments: its first argument is a 1st-order operator that takes two arguments; its last two arguments are expressions. For any operator Op of arity $\langle \_, \_ \rangle$, and any expressions e_1 and e_2, this defines $G(Op, e_1, e_2)$ to be an expression. We say that G has arity $\langle \langle \_, \_ \rangle, \_, \_ \rangle$.

   In general, the arguments of a 2nd-order operator may be expressions or 1st-order operators. A 2nd-order operator has an arity of the form $\langle a_1, \dots, a_n \rangle$, where each a_i is either $\_$ or $\langle \_, \dots, \_ \rangle$. (We can consider a 1st-order operator to be a degenerate case of a 2nd-order operator.)

It would be easy enough to define 3rd- and higher-order operators. TLA$^+$ does not permit them because they are of little use and would make it harder to check level-correctness, which is discussed in Section 17.2 below.

---

[1]Even though it allows 2nd-order operators, TLA$^+$ is still what logicians call a first-order logic because it permits quantification only over 0th-order operators. A higher-order logic would allow us to write the formula $\exists x(\_) : exp$.

17.1.2 λ Expressions

When we define a 0th-order operator E by $E \triangleq exp$, we can write what the operator E equals—it equals the expression exp. We can explain the meaning of this definition by saying that it assigns the value exp to the symbol E. To explain the meaning of an arbitrary TLA$^+$ definition, we need to be able to write what a 1st- or 2nd-order operator equals—for example, the operator F defined by

$$F(x, y) \;\triangleq\; x \cup \{z, y\}$$

TLA$^+$ provides no way to write an expression that equals this operator F. (A TLA$^+$ expression can equal only a 0th-order operator.) We therefore generalize expressions to λ *expressions*, and we write the operator that F equals as the λ expression

$$\lambda \, x, y \,:\, x \cup \{z, y\}$$

The symbols x and y in this λ expression are called λ parameters. We use λ expressions only to explain the meaning of TLA$^+$ specifications; we can't write a λ expression in TLA$^+$.

We also allow 2nd-order λ expressions, where the operator G defined by

$$G(f(\_, \_), x, y) \;\triangleq\; f(y, \{x, z\})$$

is equal to the λ expression

(17.1) $\lambda \, f(\_, \_), x, y \,:\, f(y, \{x, z\})$

The general form of a λ expression is $\lambda \, p_1, \ldots, p_n : exp$, where exp is a λ expression, each parameter p_i is either an identifier id_i or has the form $id_i(\_, \ldots, \_)$, and the id_i are all distinct. We call id_i the *identifier* of the λ parameter p_i. We consider the $n = 0$ case, the λ expression $\lambda : exp$ with no parameters, to be the expression exp. This makes a λ expression a generalization of an ordinary expression.

A λ parameter identifier is a bound identifier, just like the identifier x in $\forall \, x : F$. As with any bound identifiers, renaming the λ parameter identifiers in a λ expression doesn't change the meaning of the expression. For example, (17.1) is equivalent to

$$\lambda \, abc(\_, \_), qq, m \,:\, abc(m, \{qq, z\})$$

For obscure historical reasons, this kind of renaming is called α *conversion*.

If Op is the λ expression $\lambda \, p_1, \ldots, p_n : exp$, then $Op(e_1, \ldots, e_n)$ equals the result of replacing the identifier of the λ parameter p_i in exp with e_i, for all i in $1 \, .. \, n$. For example,

$$(\lambda \, x, y \,:\, x \cup \{z, y\})\,(TT, w + z) \;=\; TT \cup \{z, (w + z)\}$$

This procedure for evaluating the application of a λ expression is called β *reduction*.

17.1.3   Simplifying Operator Application

To simplify the exposition, I assume that every operator application is written in the form $Op(e_1, \ldots, e_n)$. TLA$^+$ provides a number of different syntactic forms for operator application, so I have to explain how they are translated into this simple form. Here are all the different forms of operator application and their translations.

- Simple constructs with a fixed number of arguments, including infix operators like $+$, prefix operators like ENABLED , and constructs like WF, function application, and IF/THEN/ELSE. These operators and constructs pose no problem. We can write $+(a, b)$ instead of $a + b$, *IfThenElse*(p, e_1, e_2) instead of

    IF p THEN e_1 ELSE e_2

  and *Apply*(f, e) instead of $f[e]$. An expression like $a + b + c$ is an abbreviation for $(a + b) + c$, so it can be written $+(+(a, b), c)$.

- Simple constructs with a variable number of arguments—for example, $\{e_1, \ldots, e_n\}$ and $[h_1 \mapsto e_1, \ldots, h_n \mapsto e_n]$. We can consider each of these constructs to be repeated application of simpler operators with a fixed number of arguments. For example,

    $$\{e_1, \ldots, e_n\} = \{e_1\} \cup \ldots \cup \{e_n\}$$
    $$[h_1 \mapsto e_1, \ldots, h_n \mapsto e_n] = [h_1 \mapsto e_1] @@ \ldots @@ [h_n \mapsto e_n]$$

  where @@ is defined in the *TLC* module, on page 248. Of course, $\{e\}$ can be written *Singleton*(e) and $[h \mapsto e]$ can be written *Record*("h", e). Note that an arbitrary CASE expression can be written in terms of CASE expressions of the form

    CASE $p \rightarrow e \ \square \ q \rightarrow f$

  using the relation

    CASE $\dot{p}_1 \rightarrow e_1 \ \square \ \ldots \ \square \ p_n \rightarrow e_n =$
    CASE $p_1 \rightarrow e_1 \ \square \ (p_2 \vee \ldots \vee p_n) \rightarrow (\text{CASE } p_2 \rightarrow e_2 \ \square \ \ldots \ \square \ p_n \rightarrow e_n)$

- Constructs that introduce bound variables—for example,

    $\exists\, x \in S \ : \ x + z > y$

  We can rewrite this expression as

    ExistsIn$(S, \ \lambda\, x : x + z > y)$

  where *ExistsIn* is a 2nd-order operator of arity $\langle\, \_\,, \langle\, \_\, \rangle\, \rangle$. All the variants of the $\exists$ construct can be represented as expressions using either $\exists\, x \in S : e$ or $\exists\, x : e$. (Section 16.1.1 shows how these variants can be translated into expressions using only $\exists\, x : e$, but those translations don't maintain the scoping rules—for example, rewriting $\exists\,x \in S : e$ as $\exists\,x : (x \in S) \wedge e$ moves S inside the scope of the bound variable x.)

All other constructs that introduce bound variables, such as $\{x \in S : exp\}$, can similarly be expressed in the form $Op(e_1, \ldots, e_n)$ using λ expressions and 2nd-order operators Op. (Chapter 16 explains how to express constructs like $\{\langle x, y \rangle \in S : exp\}$, which have a tuple of bound identifiers, in terms of constructs with ordinary bound identifiers.)

- Operator applications such as $M(x)!\,Op(y, z)$ that arise from instantiation. We write this as $M!\,Op(x, y, z)$.

- LET expressions. The meaning of a LET expression is explained in Section 17.4 below. For now, we consider only LET-free λ expressions—ones that contain no LET expressions.

For uniformity, I will call an operator symbol an *identifier*, even if it is a symbol like $+$ that isn't an identifier according to the syntax of Chapter 15.

17.1.4   Expressions

We can now inductively define an expression to be either a 0th-order operator, or to have the form $Op(e_1, \ldots, e_n)$ where Op is an operator and each e_i is either an expression or a 1st-order operator. The expression must be *arity-correct*, meaning that Op must have arity $\langle a_1, \ldots, a_n \rangle$, where each a_i is the arity of e_i. In other words, e_i must be an expression if a_i equals $\_$; otherwise it must be a 1st-order operator with arity a_i. We require that Op not be a λ expression. (If it is, we can use β reduction to evaluate $Op(e_1, \ldots, e_n)$ and eliminate the λ expression Op.) Hence, a λ expression can appear in an expression only as an argument of a 2nd-order operator. This implies that only 1st-order λ expressions can appear in an expression.

We have eliminated all bound identifiers except the ones in λ expressions. We maintain the TLA$^+$ requirement that an identifier that already has a meaning cannot be used as a bound identifier. Thus, in any λ expression $\lambda\,p_1, \ldots, p_n : exp$, the identifiers of the parameters p_i cannot appear as parameter identifiers in any λ expression that occurs in exp.

Remember that λ expressions are used only to explain the semantics of TLA$^+$. They are not part of the language, and they can't be used in a TLA$^+$ specification.

17.2   Levels

TLA$^+$ has a class of syntactic restrictions that come from the underlying logic TLA and have no counterpart in ordinary mathematics. The simplest of these is that "double-priming" is prohibited. For example, $(x' + y)'$ is not syntactically well-formed, and is therefore meaningless, because the operator $'$ (priming) can be applied only to a state function, not to a transition function like $x' + y$. This class of restriction is expressed in terms of *levels*.

In TLA, an expression has one of four basic levels, which are numbered 0, 1, 2, and 3. These levels are described below, using examples that assume x, y, and c are declared by

> VARIABLES x, y CONSTANT c

and symbols like $+$ have their usual meanings.

0. A *constant*-level expression is a constant; it contains only constants and constant operators. Example: $c + 3$.

1. A *state*-level expression is a state function; it may contain constants, constant operators, and unprimed variables. Example: $x + 2 * c$.

2. A *transition*-level expression is a transition function; it may contain anything except temporal operators. Example: $x' + y > c$.

3. A *temporal*-level expression is a temporal formula; it may contain any TLA operator. Example: $\Box[x' > y + c]_{\langle x, y \rangle}$.

Chapter 16 assigns meanings to all basic expressions—ones containing only the built-in operators of TLA$^+$ and declared constants and variables. The meaning assigned to an expression depends as follows on its level.

0. The meaning of a constant-level basic expression is a constant-level basic expression containing only primitive operators.

1. The meaning of a state-level basic expression e is an assignment of a constant expression $s[\![e]\!]$ to any state s.

2. The meaning of a transition-level basic expression e is an assignment of a constant expression $\langle s, t \rangle [\![e]\!]$ to any transition $s \to t$.

3. The meaning of a temporal-level basic expression F is an assignment of a constant expression $\sigma \models F$ to any behavior σ.

An expression of any level can be considered to be an expression of a higher level, except that a transition-level expression is not a temporal-level expression.[2] For example, if x is a declared variable, then the state-level expression $x > 2$ is the

---

[2]More precisely, a transition-level expression that is not a state-level expression is not a temporal-level expression.

temporal-level formula such that $\sigma \models x$ is the value of $x > 2$ in the first state of σ, for any behavior σ.[3]

A set of simple rules inductively defines whether a basic expression is *level-correct* and, if so, what its level is. Here are some of the rules:

- A declared constant is a level-correct expression of level 0.

- A declared variable is a level-correct expression of level 1.

- If Op is declared to be a 1st-order constant operator, then the expression $Op(e_1, \ldots, e_n)$ is level-correct iff each e_i is level-correct, in which case its level is the maximum of the levels of the e_i.

- $e_1 \in e_2$ is level-correct iff e_1 and e_2 are, in which case its level is the maximum of the levels of e_1 and e_2.

- e' is level-correct and has level 2 iff e is level-correct and has level at most 1.[4]

- ENABLED e is level-correct and has level 1 iff e is level-correct and has level at most 2.

- $\exists\, x : e$ is level-correct and has level l iff e is level-correct and has level l, when x is considered to be a declared constant.

- $\exists\, x : e$ is level-correct and has level 3 iff e is level-correct and has any level other than 2, when x is considered to be a declared variable.

There are similar rules for the other TLA$^+$ operators.

A useful consequence of these rules is that level-correctness of a basic expression does not depend on the levels of the declared identifiers. In other words, an expression e is level-correct when c is declared to be a constant iff it is level-correct when c is declared to be a variable. Of course, the level of e may depend on the level of c.

We can abstract these rules by generalizing the concept of a level. So far, we have defined the level only of an expression. We can define the level of a 1st- or 2nd-order operator Op to be a rule for determining the level-correctness and level of an expression $Op(e_1, \ldots, e_n)$ as a function of the levels of the arguments e_i. The level of a 1st-order operator is a rule, so the level of a 2nd-order operator Op is a rule that depends in part on rules—namely, on the levels of the arguments that are operators. This makes a rigorous general definition of levels for 2nd-order operators rather complicated. Fortunately, there's a simpler, less general

---

[3]The expression $x + 2$ can be considered to be a temporal-level expression that, like the temporal-level expression $\Box(x + 2)$, is silly. (See the discussion of silliness in Section 6.2 on page 67.)

[4]If e is a constant expression, then e' equals e, so we could consider e' to have level 0. For simplicity, we consider e' to have level 2 even if e is a constant.

definition that handles all the operators of TLA$^+$. Even more fortunately, you don't have to know it, so I won't bother writing it down. All you need to know is that there exists a way of assigning a level to every built-in operator of TLA$^+$. The level-correctness and level of any basic expression is then determined by those levels and the levels of the declared identifiers that occur in the expression.

One important class of operator levels are the *constant* levels. Any expression built from constant-level operators and declared constants has constant level. The built-in constant operators of TLA$^+$, listed in Tables 1 and 2 (pages 268 and 269) all have constant level. Any operator defined solely in terms of constant-level operators and declared constants has constant level.

We now extend the definition of level-correctness from expressions to λ expressions. We define the λ expression $\lambda p_1, \ldots, p_n : exp$ to be level-correct iff exp is level-correct when the λ parameter identifiers are declared to be constants of the appropriate arity. For example, $\lambda p, q(\_) : exp$ is level-correct iff exp is level-correct with the additional declaration

    CONSTANTS $p, q(\_)$

This inductively defines level-correctness for λ expressions. The definition is reasonable because, as observed a few paragraphs ago, the level-correctness of exp doesn't depend on whether we assign level 0 or 1 to the λ parameters. One can also define the level of an arbitrary λ expression, but that would require the general definition of the level of an operator, which we want to avoid.

17.3   Contexts

Syntactic correctness of a basic expression depends on the arities of the declared identifiers. The expression $Foo = \{\}$ is syntactically correct if Foo is declared to be a variable, and hence of arity $\_$, but not if it's declared to be a (1st-order) constant of arity $\langle \_ \rangle$. The meaning of a basic expression also depends on the levels of the declared identifiers. We can't determine those arities and levels just by looking at the expression itself; they are implied by the context in which the expression appears. A nonbasic expression contains defined as well as declared operators. Its syntactic correctness and meaning depend on the definitions of those operators, which also depend on the context. This section defines a precise notion of a context.

For uniformity, built-in operators are treated the same as defined and declared operators. Just as the context might tell us that the identifier x is a declared variable, it tells us that $\in$ is declared to be a constant-level operator of arity $\langle \_, \_ \rangle$ and that $\notin$ is defined to equal $\lambda a, b : \neg(\in (a, b))$. We assume a standard context that specifies all the built-in operators of TLA$^+$.

To define contexts, let's first define declarations and definitions. A *declaration* assigns an arity and level to an operator name. A *definition* assigns a LET-free λ expression to an operator name. A *module definition* assigns the meaning of a module to a module name, where the meaning of a module is defined in Section 17.5 below.[5] A *context* consists of a set of declarations, definitions, and module definitions such that

C1. An operator name is declared or defined at most once by the context. (This means that it can't be both declared and defined.)

C2. No operator defined or declared by the context appears as the identifier of a λ parameter in any definition's expression.

C3. Every operator name that appears in a definition's expression is either a λ parameter's identifier or is declared (not defined) by the context.

C4. No module name is assigned meanings by two different module definitions.

Module and operator names are handled separately. The same string may be both a module name that is defined by a module definition and an operator name that is either declared or defined by an ordinary definition.

Here is an example of a context that declares the symbols $\cup$, a, b, and $\in$, defines the symbols c and *foo*, and defines the module *Naturals*:

$$(17.2) \ \{ \cup : \langle \_ , \_ \rangle , \quad a : \_ , \quad b : \_ , \quad \in : \langle \_ , \_ \rangle , \quad c \stackrel{\Delta}{=} \cup(a, b),$$
$$foo \stackrel{\Delta}{=} \lambda p, q(\_) : \ \in (p, \cup(q(b), a)), \quad Naturals \stackrel{m}{=} \dots \}$$

Not shown are the levels assigned to the operators $\cup$, a, b, and $\in$ and the meaning assigned to *Naturals*.

If $\mathcal{C}$ is a context, a $\mathcal{C}$-*basic λ expression* is defined to be a λ expression that contains only symbols declared in $\mathcal{C}$ (in addition to λ parameters). For example, $\lambda x : \ \in (x, \cup(a, b))$ is a $\mathcal{C}$-basic λ expression if $\mathcal{C}$ is the context (17.2). However, neither $\cap(a, b)$ nor $\lambda x : c(x, b)$ is a $\mathcal{C}$-basic λ expression because neither $\cap$ nor c is declared in $\mathcal{C}$. (The symbol c is defined, not declared, in $\mathcal{C}$.) A $\mathcal{C}$-basic λ expression is *syntactically correct* if it is arity- and level-correct with the arities and levels assigned by $\mathcal{C}$ to the expression's operators. Condition C3 states that if $Op \stackrel{\Delta}{=} exp$ is a definition in context $\mathcal{C}$, then exp is a $\mathcal{C}$-basic λ expression. We add to C3 the requirement that it be syntactically correct.

We also allow a context to contain a special definition of the form $Op \stackrel{\Delta}{=} ?$ that assigns to the name Op an "illegal" value ? that is not a λ expression. This definition indicates that, in the context, it is illegal to use the operator name Op.

17.4 The Meaning of a λ Expression

We now define the meaning $\mathcal{C}[\![e]\!]$ of a λ expression e in a context $\mathcal{C}$ to be a $\mathcal{C}$-basic λ expression. If e is an ordinary (nonbasic) expression, and $\mathcal{C}$ is the

---

[5]The meaning of a module is defined in terms of contexts, so these definitions appear to be circular. In fact, the definitions of context and of the meaning of a module together form a single inductive definition.

context that specifies the built-in TLA$^+$ operators and declares the constants and variables that occur in e, then this will define $\mathcal{C}[\![e]\!]$ to be a basic expression. Since Chapter 16 defines the meaning of basic expressions, this defines the meaning of an arbitrary expression. The expression e may contain LET constructs, so this defines the meaning of LET, the one operator whose meaning is not defined in Chapter 16.

Basically, $\mathcal{C}[\![e]\!]$ is obtained from e by replacing all defined operator names with their definitions and then applying β reduction whenever possible. Recall that β reduction replaces

$$(\lambda\ p_1,\ldots,p_n\ :\ exp)\,(e_1,\ldots,e_n)$$

with the expression obtained from exp by replacing the identifier of p_i with e_i, for each i. The definition of $\mathcal{C}[\![e]\!]$ does not depend on the levels assigned by the declarations of $\mathcal{C}$. So, we ignore levels in the definition. The inductive definition of $\mathcal{C}[\![e]\!]$ consists of the following rules:

- If e is an operator symbol, then $\mathcal{C}[\![e]\!]$ equals (i) e if e is declared in $\mathcal{C}$, or (ii) the λ expression of e's definition in $\mathcal{C}$ if e is defined in $\mathcal{C}$.

- If e is $Op(e_1,\ldots,e_n)$, where Op is declared in $\mathcal{C}$, then $\mathcal{C}[\![e]\!]$ equals the expression $Op(\mathcal{C}[\![e_1]\!],\ldots,\mathcal{C}[\![e_n]\!])$.

- If e is $Op(e_1,\ldots,e_n)$, where Op is defined in $\mathcal{C}$ to equal the λ expression d, then $\mathcal{C}[\![e]\!]$ equals the β reduction of $\overline{d}(\mathcal{C}[\![e_1]\!],\ldots,\mathcal{C}[\![e_n]\!])$, where $\overline{d}$ is obtained from d by α conversion (replacement of λ parameters) so that no λ parameter's identifier appears in both $\overline{d}$ and some $\mathcal{C}[\![e_i]\!]$.

- If e is $\lambda p_1,\ldots,p_n : exp$, then $\mathcal{C}[\![e]\!]$ equals $\lambda p_1,\ldots,p_n : \mathcal{D}[\![exp]\!]$, where $\mathcal{D}$ is the context obtained by adding to $\mathcal{C}$ the declarations that, for each i in $1 \mathrel{..} n$, assign to the ith λ parameter's identifier the arity determined by p_i.

- If e is where d is a λ expression and exp an expression, then $\mathcal{C}[\![e]\!]$ equals $\mathcal{D}[\![exp]\!]$, where $\mathcal{D}$ is the context obtained by adding to $\mathcal{C}$ the definition that assigns $\mathcal{C}[\![d]\!]$ to Op.

- If e is

    LET $Op(p_1,\ldots,p_n) \triangleq$ INSTANCE ...   IN exp

  then $\mathcal{C}[\![e]\!]$ equals $\mathcal{D}[\![exp]\!]$, where $\mathcal{D}$ is the new current context obtained by "evaluating" the statement

    $Op(p_1,\ldots,p_n) \triangleq$ INSTANCE ...

  in the current context $\mathcal{C}$, as described in Section 17.5.5 below.

The last two conditions define the meaning of any LET construct, because

- The operator definition $Op(p_1, \ldots, p_n) \triangleq d$ in a LET means

$$Op \triangleq \lambda p_1, \ldots, p_n : d$$

- A function definition $Op[x \in S] \triangleq d$ in a LET means

$$Op \triangleq \text{CHOOSE } Op : Op = [x \in S \mapsto d]$$

- The expression LET $Op_1 \triangleq d_1 \ldots Op_n \triangleq d_n$ IN exp is defined to equal

$$\text{LET } Op_1 \triangleq d_1 \text{ IN } (\text{LET } \ldots \text{ IN } (\text{LET } Op_n \triangleq d_n \text{ IN } exp) \ldots)$$

The λ expression e is defined to be legal (syntactically well-formed) in the context $\mathcal{C}$ iff these rules define $\mathcal{C}[\![e]\!]$ to be a legal $\mathcal{C}$-basic expression.

17.5 The Meaning of a Module

The meaning of a module depends on a context. For an external module, which is not a submodule of another module, the context consists of declarations and definitions of all the built-in operators of TLA$^+$, together with definitions of some other modules. Section 17.7 below discusses where the definitions of those other modules come from.

The meaning of a module in a context $\mathcal{C}$ consists of six sets:

Dcl   A set of declarations. They come from CONSTANT and VARIABLE declarations and declarations in extended modules (modules appearing in an EXTENDS statement).

GDef A set of global definitions. They come from ordinary (non-LOCAL) definitions and global definitions in extended and instantiated modules.

LDef A set of local definitions. They come from LOCAL definitions and LOCAL instantiations of modules. (Local definitions are not obtained by other modules that extend or instantiate the module.)

MDef A set of module definitions. They come from submodules of the module and of extended modules.

Ass   A set of assumptions. They come from ASSUME statements and from extended modules.

Thm   A set of theorems. They come from THEOREM statements, from theorems in extended modules, and from the assumptions and theorems of instantiated modules, as explained in Section 17.5.5 below.

The λ expressions of definitions in *GDef* and *LDef*, as well as the expressions in *Ass* and *Thm*, are $(\mathcal{C} \cup Dcl)$-basic λ expressions. In other words, the only operator symbols they contain (other than λ parameter identifiers) are ones declared in $\mathcal{C}$ or in *Dcl*.

The meaning of a module in a context $\mathcal{C}$ is defined by an algorithm for computing these six sets. The algorithm processes each statement in the module in turn, from beginning to end. The meaning of the module is the value of those sets when the end of the module is reached.

Initially, all six sets are empty. The rules for handling each possible type of statement are given below. In these rules, the *current context* $\mathcal{CC}$ is defined to be the union of $\mathcal{C}$, *Dcl*, *GDef*, *LDef*, and *MDef*.

When the algorithm adds elements to the context $\mathcal{CC}$, it uses α conversion to ensure that no defined or declared operator name appears as a λ parameter's identifier in any λ expression in $\mathcal{CC}$. For example, if the definition $foo \triangleq \lambda x : x + 1$ is in *LDef*, then adding a declaration of x to *Dcl* requires α conversion of this definition to rename the λ parameter identifier x. This α conversion is not explicitly mentioned.

17.5.1   Extends

An EXTENDS statement has the form

   EXTENDS $M_1, \ldots, M_n$

where each M_i is a module name. This statement must be the first one in the module. The statement sets the values of *Dcl*, *GDef*, *MDef*, *Ass*, and *Thm* equal to the union of the corresponding values for the module meanings assigned by $\mathcal{C}$ to the module names M_i.

This statement is legal iff the module names M_i are all defined in $\mathcal{C}$, and the resulting current context $\mathcal{CC}$ does not assign more than one meaning to any symbol. More precisely, if the same symbol is defined or declared by two or more of the M_i, then those duplicate definitions or declarations must all have been obtained through a (possibly empty) chains of EXTENDS statements from the same definition or declaration. For example, suppose M_1 extends the *Naturals* module, and M_2 extends M_1. Then the three modules *Naturals*, M_1, and M_2 all define the operator $+$. The statement

   EXTENDS *Naturals*, M_1, M_2

can still be legal, because each of the three definitions is obtained by a chain of EXTENDS statements (of length 0, 1, and 2, respectively) from the definition of $+$ in the *Naturals* module.

When decomposing a large specification into modules, we often want a module M to extend modules $M_1, \ldots, M_n$, where the M_i have declared constants and/or variables in common. In this case, we put the common declarations in a module P that is extended by all the M_i.

17.5.2 Declarations

A declaration statement has one of the forms

$$\text{CONSTANT } c_1, \ldots, c_n \qquad \text{VARIABLE } v_1, \ldots, v_n$$

where each v_i is an identifier and each c_i is either an identifier or has the form $Op(\_, \ldots, \_)$, for some identifier Op. This statement adds to the set Dcl the obvious declarations. It is legal iff none of the declared identifiers is defined or declared in $\mathcal{CC}$.

17.5.3 Operator Definitions

A global operator definition[6] has one of the two forms

$$Op \triangleq exp \qquad Op(p_1, \ldots, p_n) \triangleq exp$$

where Op is an identifier, exp is an expression, and each p_i is either an identifier or has the form $P(\_, \ldots, \_)$, where P is an identifier. We consider the first form an instance of the second with $n = 0$.

This statement is legal iff Op is not declared or defined in $\mathcal{CC}$ and the λ expression $\lambda p_1, \ldots, p_n : exp$ is legal in context $\mathcal{CC}$. In particular, no λ parameter in this λ expression can be defined or declared in $\mathcal{CC}$. The statement adds to $GDef$ the definition that assigns to Op the λ expression $\mathcal{CC}[\![\lambda p_1, \ldots, p_n : exp]\!]$.

A local operator definition has one of the two forms

$$\text{LOCAL } Op \triangleq exp \qquad \text{LOCAL } Op(p_1, \ldots, p_n) \triangleq exp$$

It is the same as a global definition, except that it adds the definition to $LDef$ instead of $GDef$.

17.5.4 Function Definitions

A global function definition has the form

$$Op[fcnargs] \triangleq exp$$

---

[6]An operator definition statement should not be confused with a definition clause in a LET expression. The meaning of a LET expression is described in Section 17.4.

where *fcnargs* is a comma-separated list of elements, each having the form $Id_1, \ldots, Id_n \in S$ or $\langle Id_1, \ldots, Id_n \rangle \in S$. It is equivalent to the global operator definition

$$Op \;\triangleq\; \textsc{choose}\; Op \,:\, Op = [fcnargs \mapsto exp]$$

A local function definition, which has the form

$$\textsc{local}\; Op[fcnargs] \;\triangleq\; exp$$

is equivalent to the analogous local operator definition.

17.5.5   Instantiation

We consider first a global instantiation of the form

(17.3) $\;I(p_1, \ldots, p_m) \;\triangleq\; \textsc{instance}\; N \;\textsc{with}\; q_1 \leftarrow e_1, \ldots, q_n \leftarrow e_n$

For this to be legal, N must be a module name defined in $\mathcal{CC}$. Let *NDcl*, *NDef*, *NAss*, and *NThm* be the sets *Dcl*, *GDef*, *Ass*, and *Thm* in the meaning assigned to N by $\mathcal{CC}$. The q_i must be distinct identifiers declared by *NDcl*. We add a WITH clause of the form $Op \leftarrow Op$ for any identifier Op that is declared in *NDcl* but is not one of the q_i, so the q_i constitute all the identifiers declared in *NDcl*.

Neither I nor any of the identifiers of the definition parameters p_i may be defined or declared in $\mathcal{CC}$. Let $\mathcal{D}$ be the context obtained by adding to $\mathcal{CC}$ the obvious constant-level declaration for each p_i. Then e_i must be syntactically well-formed in the context $\mathcal{D}$, and $\mathcal{D}[\![e_i]\!]$ must have the same arity as q_i, for each $i \in 1 \mathinner{\ldotp\ldotp} n$.

The instantiation must also satisfy the following level-correctness condition. Define module N to be a *constant* module iff every declaration in *NDcl* has constant level, and every operator appearing in every definition in *NDef* has constant level. If N is *not* a constant module, then for each i in $1 \mathinner{\ldotp\ldotp} n$:

- If q_i is declared in *NDcl* to be a constant operator, then $\mathcal{D}[\![e_i]\!]$ has constant level.

- If q_i is declared in *NDcl* to be a variable (a 0th-order operator of level 1), then $\mathcal{D}[\![e_i]\!]$ has level 0 or 1.

The reason for this condition is explained in Section 17.8 below.

For each definition $Op \triangleq \lambda r_1, \ldots, r_p : e$ in *NDef*, the definition

(17.4) $\;I!Op \triangleq \lambda\, p_1, \ldots, p_m, r_1, \ldots, r_p : \overline{e}$

is added to *GDef*, where $\overline{e}$ is the expression obtained from e by substituting e_i for q_i, for all $i \in 1 \mathinner{\ldotp\ldotp} n$. Before doing this substitution, α conversion must be applied to ensure that $\mathcal{CC}$ is a correct context after the definition of $I\,!\,Op$ is added to $GDef$. The precise definition of $\overline{e}$ is a bit subtle; it is given in Section 17.8 below. We require that the λ expression in (17.4) be level-correct. (If N is a nonconstant module, then level-correctness of this λ expression is implied by the level condition on parameter instantiation described in the preceding paragraph.) Legality of the definition of Op in module N and of the WITH substitutions implies that the λ expression is arity-correct in the current context. Remember that $I\,!\,Op(c_1,\ldots,c_m,d_1,\ldots,d_n)$ is actually written in TLA$^+$ as $I(c_1,\ldots,c_m)\,!\,Op(d_1,\ldots,d_n)$.

Also added to $GDef$ is the special definition $I \triangleq ?$. This prevents I from later being defined or declared as an operator name.

If $NAss$ equals the set $\{A_1,\ldots,A_k\}$ of assumptions, then for each theorem T in $NThm$, we add to Thm the theorem

$$\overline{A_1} \wedge \ldots \wedge \overline{A_k} \;\Rightarrow\; \overline{T}$$

(As above, $\overline{T}$ and the $\overline{A_j}$ are obtained from T and the A_j by substituting e_i for q_i, for each i in $1 \mathinner{\ldotp\ldotp} k$.)

A global INSTANCE statement can also have the two forms

$$I \;\triangleq\; \text{INSTANCE } N \text{ WITH } q_1 \leftarrow e_1,\, \ldots,\, q_n \leftarrow e_n$$
$$\text{INSTANCE } N \text{ WITH } q_1 \leftarrow e_1,\, \ldots,\, q_n \leftarrow e_n$$

The first is just the $m = 0$ case of (17.3); the second is similar to the first, except the definitions added to $GDef$ do not have $I\,!$ prepended to the operator names. The second form also has the legality condition that none of the defined symbols in N may be defined or declared in the current context, except in the following case. An operator definition may be included multiple times through chains of INSTANCE and EXTENDS statements if it is defined in a module[7] having no declarations. For example, suppose the current context contains a definition of $+$ obtained through extending the *Naturals* module. Then an INSTANCE N statement is legal even though N also extends *Naturals* and therefore defines $+$. Because the *Naturals* module declares no parameters, instantiation cannot change the definition of $+$.

In all forms of the INSTANCE statement, omitting the WITH clause is equivalent to the case $n = 0$ of these statements. (Remember that all the declared identifiers of module N are either explicitly or implicitly instantiated.)

A local INSTANCE statement consists of the keyword LOCAL followed by an INSTANCE statement of the form described above. It is handled in a similar fashion to a global INSTANCE statement, except that all definitions are added to $LDef$ instead of $GDef$.

---

[7]An operator J!Op is defined in the module that contains the $J \triangleq$ INSTANCE ... statement.

17.5.6   Theorems and Assumptions

A theorem has one of the forms

$$\text{THEOREM } exp \qquad \text{THEOREM } Op \stackrel{\Delta}{=} exp$$

where *exp* is an expression, which must be legal in the current context CC. The first form adds the theorem $CC[\![exp]\!]$ to the set *Thm*. The second form is equivalent to the two statements

$$Op \stackrel{\Delta}{=} exp$$
$$\text{THEOREM } Op$$

An assumption has one of the forms

$$\text{ASSUME } exp \qquad \text{ASSUME } Op \stackrel{\Delta}{=} exp$$

The expression *exp* must have constant level. An assumption is similar to a theorem except that $CC[\![exp]\!]$ is added to the set *Ass*.

17.5.7   Submodules

A module can contain a submodule, which is a complete module that begins with

$$\boxed{\qquad\qquad\qquad \text{MODULE } N \qquad\qquad\qquad}$$

for some module name N, and ends with

$$\boxed{\qquad\qquad\qquad\qquad\qquad\qquad\qquad\qquad\qquad}$$

This is legal iff the module name N is not defined in CC and the module is legal in the context CC. In this case, the module definition that assigns to N the meaning of the submodule in context CC is added to *MDef*.

A submodule can be used in an INSTANCE statement that appears either later in the current module or in a module that extends the current module. Submodules of a module M are *not* added to the set *MDef* of a module that instantiates M.

17.6   Correctness of a Module

Section 17.5 above defines the meaning of a module to consist of the six sets *Dcl*, *GDef*, *LDef*, *MDef*, *Ass*, and *Thm*. Mathematically, we can view the meaning of a module to be the assertion that all the theorems in *Thm* are consequences of the assumptions in *Ass*. More precisely, let A be the conjunction of all the assumptions in *Ass*. The module asserts that, for every theorem T in *Thm*, the formula $A \Rightarrow T$ is valid.[8]

An assumption or theorem of the module is a $(\mathcal{C} \cup Dcl)$-basic expression. For an outermost module (not a submodule), $\mathcal{C}$ declares only the built-in operators of TLA$^+$, and *Dcl* declares the declared constants and variables of the module. Therefore, each formula $A \Rightarrow T$ asserted by the module is a basic expression. We say that the module is *semantically correct* if each of these expressions $A \Rightarrow T$ is a valid formula in the context *Dcl*. Chapter 16 defines what it means for a basic expression to be a valid formula.

By defining the meaning of a theorem, we have defined the meaning of a TLA$^+$ specification. Any mathematically meaningful question we can ask about a specification can be framed as the question of whether a certain formula is a valid theorem.

17.7 Finding Modules

For a module M to have a meaning in a context $\mathcal{C}$, every module N extended or instantiated by M must have its meaning defined in $\mathcal{C}$—unless N is a submodule of M or of a module extended by M. In principle, module M is interpreted in a context containing declarations and definitions of the built-in TLA$^+$ operator names and module definitions of all modules needed to interpret M. In practice, a tool (or a person) begins interpreting M in a context $\mathcal{C}_0$ initially containing only declarations and definitions of the built-in TLA$^+$ operator names. When the tool encounters an EXTENDS or INSTANCE statement that mentions a module named N not defined in the current context $\mathcal{CC}$ of M, the tool finds the module named N, interprets it in the context $\mathcal{C}_0$, and then adds the module definition for N to $\mathcal{C}_0$ and to $\mathcal{CC}$.

The definition of the TLA$^+$ language does not specify how a tool finds a module named N. A tool will most likely look for the module in a file named N`.tla`.

The meaning of a module depends on the meanings of the modules that it extends or instantiates. The meaning of each of those modules in turn may depend on the meanings of other modules, and so on. Thus, the meaning of a module depends on the meanings of some set of modules. A module M is syntactically incorrect if this set of modules includes M itself.

---

[8]In a temporal logic like TLA, the formula $F \Rightarrow G$ is not in general equivalent to the assertion that G is a consequence of assumption F. However, the two are equivalent if F is a constant formula, and TLA$^+$ allows only constant assumptions.

17.8   The Semantics of Instantiation

Section 17.5.5 above defines the meaning of an INSTANCE statement in terms of substitution. I now define precisely how that substitution is performed and explain the level-correctness rule for instantiating nonconstant modules.

Suppose that module M contains the statement

$$I \; \triangleq \; \text{INSTANCE } N \text{ WITH } q_1 \leftarrow e_1, \ldots, q_n \leftarrow e_n$$

where the q_i are all the declared identifiers of module N, and that N contains the definition

$$F \triangleq e$$

where no λ parameter identifier in e is defined or declared in the current context of M. The INSTANCE statement then adds to the current context of M the definition

(17.5) $I!F \triangleq \overline{e}$

where $\overline{e}$ is obtained from e by substituting e_i for q_i, for all i in $1 .. n$.

A fundamental principle of mathematics is that substitution preserves validity; substituting in a valid formula yields a valid formula. So, we want to define $\overline{e}$ so that, if F is a valid formula in N, then $I!F$ is a valid formula in M.

A simple example shows that the level rule for instantiating nonconstant modules is necessary to preserve the validity of F. Suppose F is defined to equal $\Box[c' = c]_c$, where c is declared in N to be a constant. Then F is a temporal formula asserting that no step changes c. It is valid because a constant has the same value in every state of a behavior. If we allowed an instantiation that substitutes a variable x for the constant c, then $I!F$ would be the formula $\Box[x' = x]_x$. This is not a valid formula because it is false for any behavior in which the value of x changes. Since x is a variable, such a behavior obviously exists. Preserving validity requires that we not allow substitution of a nonconstant for a declared constant when instantiating a nonconstant module. (Since $\Box$ and $'$ are nonconstant operators, this definition of F can appear only in a nonconstant module.)

In ordinary mathematics, there is one tricky problem in making substitution preserve validity. Consider the formula

(17.6) $(n \in Nat) \Rightarrow (\exists\, m \in Nat : m \geq n)$

This formula is valid because it is true for any value of n. Now, suppose we substitute $m + 1$ for n. A naive substitution that simply replaces n by $m + 1$ would yield the formula

(17.7) $(m + 1 \in Nat) \Rightarrow (\exists\, m \in Nat : m \geq m + 1)$

Since the formula $\exists\, m \in Nat : m \geq m + 1$ is equivalent to FALSE, (17.7) is obviously not valid. Mathematicians call this problem *variable capture*; m is "captured" by the quantifier $\exists\, m$. Mathematicians avoid it by the rule that, when substituting for an identifier in a formula, one does not substitute for bound occurrences of the identifier. This rule requires that m be removed from (17.6) by α conversion before $m + 1$ is substituted for n.

Section 17.5.5 defines the meaning of the INSTANCE statement in a way that avoids variable capture. Indeed, formula (17.7) is illegal in TLA$^+$ because the subexpression $m + 1 \in Nat$ is allowed only in a context in which m is defined or declared, in which case m cannot be used as a bound identifier, so the subexpression $\exists\, m \ldots$ is illegal. The α conversion necessary to produce a syntactically well-formed expression makes this kind of variable capture impossible.

The problem of variable capture occurs in a more subtle form in certain nonconstant operators of TLA$^+$, where it is not prevented by the syntactic rules. Most notable of these operators is ENABLED. Suppose x and y are declared variables of module N, and F is defined by

$$F \;\triangleq\; \text{ENABLED}\ (x' = 0 \,\wedge\, y' = 1)$$

Then F is equivalent to TRUE, so it is valid in module N. (For any state s, there exists a state t in which $x = 0$ and $y = 1$.) Now suppose z is a declared variable of module M, and let the instantiation be

$$I \;\triangleq\; \text{INSTANCE}\ N\ \text{WITH}\ x \leftarrow z,\ y \leftarrow z$$

With naive substitution, $I!F$ would equal

$$\text{ENABLED}\ (z' = 0 \,\wedge\, z' = 1)$$

which is equivalent to FALSE. (For any state s, there is no state t in which $z = 0$ and $z = 1$ are both true.) Hence, $I!F$ would not be a theorem, so instantiation would not preserve validity.

Naive substitution in a formula of the form ENABLED A does not preserve validity because the primed variables in A are really bound identifiers. The formula ENABLED A asserts that *there exist* values of the primed variables such that A is true. Substituting z' for x' and y' in the ENABLED formula is really substitution for a bound identifier. It isn't ruled out by the syntactic rules of TLA$^+$ because the quantification is implicit.

To preserve validity, we must define $\overline{e}$ in (17.5) so it avoids capture of identifiers implicitly bound in ENABLED expressions. Before performing the substitution, we first replace the primed occurrences of variables in ENABLED expressions with new variable symbols. That is, for each subexpression of e of the form ENABLED A and each declared variable q of module N, we replace every primed occurrence of q in A with a new symbol, which we write $\$q$, that does not appear in A. This new symbol is considered to be bound by the ENABLED operator. For example, the module

```
┌──────────────────────────── MODULE N ────────────────────────────┐
│ VARIABLE u                                                        │
│                                                                   │
│ G(v, A)  ≜  ENABLED (A ∨ ({u, v}' = {u, v}))                      │
│ H  ≜  (u' = u) ∧ G(u, u' ≠ u)                                     │
└───────────────────────────────────────────────────────────────────┘
```

has as its global definitions the set

$$\{ G \;\triangleq\; \lambda v, A : \text{ENABLED}\,(A \vee (\{u, v\}' = \{u, v\})),$$
$$\quad H \;\triangleq\; (u' = u) \wedge \text{ENABLED}\,((u' \neq u) \vee (\{u, u\}' = \{u, u\})) \}$$

The statement

$$I \;\triangleq\; \text{INSTANCE } N \text{ WITH } u \leftarrow x$$

adds the following definitions to the current module:

$$I!G \;\triangleq\; \lambda v, A : \text{ENABLED}\,(A \vee (\{\$u, v\}' = \{x, v\}))$$
$$I!H \;\triangleq\; (x' = x) \wedge \text{ENABLED}\,((\$u' \neq x) \vee (\{\$u, \$u\}' = \{x, x\}))$$

Observe that $I!H$ does not equal $(x' = x) \wedge I!G(x, x' \neq x)$, even though H equals $(u' = u) \wedge G(u, u' \neq u)$ in module N and the instantiation substitutes x for u.

As another example, consider the module

```
┌──────────────────────────── MODULE N ────────────────────────────┐
│ VARIABLES u, v                                                    │
│                                                                   │
│ A   ≜  (u' = u) ∧ (v' ≠ v)                                        │
│ B(d) ≜  ENABLED d                                                 │
│ C   ≜  B(A)                                                       │
└───────────────────────────────────────────────────────────────────┘
```

The instantiation

$$I \;\triangleq\; \text{INSTANCE } N \text{ WITH } u \leftarrow x, \; v \leftarrow x$$

adds the following definitions to the current module:

$$I!A \;\triangleq\; (x' = x) \wedge (x' \neq x)$$
$$I!B \;\triangleq\; \lambda d : \text{ENABLED } d$$
$$I!C \;\triangleq\; \text{ENABLED}\,((\$u' = x) \wedge (\$v' \neq x))$$

Observe that $I!C$ is not equivalent to $I!B(I!A)$. In fact, $I!C \equiv \text{TRUE}$ and $I!B(I!A) \equiv \text{FALSE}$.

We say that instantiation *distributes* over an operator Op if

$$\overline{Op(e_1, \ldots, e_n)} \;=\; Op(\overline{e_1}, \ldots, \overline{e_n})$$

for any expressions e_i, where the overlining operator $(^-)$ denotes some arbitrary instantiation. Instantiation distributes over all constant operators—for example, $+$, $\subseteq$, and $\exists$.[9] Instantiation also distributes over most of the nonconstant operators of TLA$^+$, like priming $(')$ and $\Box$.

If an operator Op implicitly binds some identifiers in its arguments, then instantiation would not preserve validity if it distributed over Op. Our rules for instantiating in an ENABLED expression imply that instantiation does not distribute over ENABLED. It also does not distribute over any operator defined in terms of ENABLED—in particular, the built-in operators WF and SF.

There are two other TLA$^+$ operators that implicitly bind identifiers: the action composition operator "$\cdot$", defined in Section 16.2.3, and the temporal operator $\xrightarrow{+}$, introduced in Section 10.7. The rule for instantiating an expression $A \cdot B$ is similar to that for ENABLED A—namely, bound occurrences of variables are replaced by a new symbol. In the expression $A \cdot B$, primed occurrences of variables in A and unprimed occurrences in B are bound. We handle a formula of the form $F \xrightarrow{+} G$ by replacing it with an equivalent formula in which the quantification is made explicit.[10] Most readers won't care, but here's how that equivalent formula is constructed. Let $\mathbf{x}$ be the tuple $\langle x_1, \ldots, x_n \rangle$ of all declared variables; let $b, \widehat{x_1}, \ldots, \widehat{x_n}$ be symbols distinct from the x_i and from any bound identifiers in F or G; and let $\widehat{e}$ be the expression obtained from an expression e by substituting the variables $\widehat{x_i}$ for the corresponding variables x_i. Then $F \xrightarrow{+} G$ is equivalent to

$$(17.8) \quad \forall\, b : (\ \land (b \in \text{BOOLEAN}) \land \Box[b' = \text{FALSE}]_b$$
$$\land\ \exists\, \widehat{x_1}, \ldots, \widehat{x_n} : \widehat{F} \land \Box(b \Rightarrow (\mathbf{x} = \widehat{\mathbf{x}}))\ )$$
$$\Rightarrow \exists\, \widehat{x_1}, \ldots, \widehat{x_n} : \widehat{G} \land (\mathbf{x} = \widehat{\mathbf{x}}) \land \Box[b \Rightarrow (\mathbf{x}' = \widehat{\mathbf{x}}')]_{\langle b, \mathbf{x}, \widehat{\mathbf{x}} \rangle}$$

Here's a complete statement of the rules for computing $\overline{e}$, for an arbitrary expression e.

1. Remove all $\xrightarrow{+}$ operators by replacing each subformula of the form $F \xrightarrow{+} G$ with the equivalent formula (17.8).

2. Recursively perform the following replacements, starting from the innermost subexpressions of e, for each declared variable x of N:

   - For each subexpression of the form ENABLED A, replace each primed occurrence of x in A by a new symbol $\$x$ that is different from any identifier and from any other symbol that occurs in A.

---

[9]Recall the explanation on pages 320–321 of how we consider $\exists$ to be a second-order operator. Instantiation distributes over $\exists$ because TLA$^+$ does not permit variable capture when substituting in λ expressions.

[10]Replacing ENABLED and "$\cdot$" expressions by equivalent formulas with explicit quantifiers before substituting would result in some surprising instantiations. For example, if N contains the definition $E(A) \triangleq$ ENABLED A, then $I \triangleq$ INSTANCE N would effectively obtain the definition $I!E(A) \triangleq A$.

- For each subexpression of the form $B \cdot C$, replace each primed occurrence of x in B and each unprimed occurrence of x in C by a new symbol $\$x$ that is different from any identifier and from any other symbol that occurs in B or C.

For example, applying these rules to the inner ENABLED expression and to the "$\cdot$" expression converts

$$\text{ENABLED}\,((\text{ENABLED}\,(x' = x))' \wedge ((y' = x) \cdot (x' = y)))$$

to

$$\text{ENABLED}\,((\text{ENABLED}\,(\$x' = x))' \wedge ((\$y' = x) \cdot (x' = \$y)))$$

and applying them again to the outer ENABLED expression yields

$$\text{ENABLED}\,((\text{ENABLED}\,(\$x' = \$xx))' \wedge ((\$y' = x) \cdot (\$xx' = \$y)))$$

where $\$xx$ is some new symbol different from x, $\$x$, and $\$y$.

3. Replace each occurrence of q_i with e_i, for all i in $1 \mathbin{..} n$.

Chapter 18

The Standard Modules

Several standard modules are provided for use in TLA$^+$ specifications. Some of the definitions they contain are subtle—for example, the definitions of the set of real numbers and its operators. Others, such as the definition of $1 \,..\, n$, are obvious. There are two reasons to use standard modules. First, specifications are easier to read when they use basic operators that we're already familiar with. Second, tools can have built-in knowledge of standard operators. For example, the TLC model checker (Chapter 14) has efficient implementations of some standard modules; and a theorem-prover might implement special decision procedures for some standard operators. The standard modules of TLA$^+$ are described here, except for the *RealTime* module, which appears in Chapter 9.

18.1   Module *Sequences*

The *Sequences* module was introduced in Section 4.1 on page 35. Most of the operators it defines have already been explained. The exceptions are

$SubSeq(s, m, n)$ The subsequence $\langle s[m], s[m+1], \ldots, s[n] \rangle$ consisting of the mth through n^{th} elements of s. It is undefined if $m < 1$ or $n > Len(s)$, except that it equals the empty sequence if $m > n$.

$SelectSeq(s, Test)$ The subsequence of s consisting of the elements $s[i]$ such that $Test(s[i])$ equals TRUE. For example,

$$PosSubSeq(s) \;\triangleq\; \text{LET } IsPos(n) \;\triangleq\; n > 0$$
$$\text{IN} \quad SelectSeq(s, IsPos)$$

defines $PosSubSeq(\langle 0, 3, -2, 5 \rangle)$ to equal $\langle 3, 5 \rangle$.

The *Sequences* module uses operators on natural numbers, so we might expect it to extend the *Naturals* module. However, this would mean that any module that extends *Sequences* would then also extend *Naturals*. Just in case someone wants to use sequences without extending the *Naturals* module, the *Sequences* module contains the statement

LOCAL INSTANCE *Naturals*

This statement introduces the definitions from the *Naturals* module, just as an ordinary INSTANCE statement would, but it does not export those definitions to another module that extends or instantiates the *Sequences* module. The LOCAL modifier can also precede an ordinary definition; it has the effect of making that definition usable within the current module, but not in a module that extends or instantiates it. (The LOCAL modifier cannot be used with parameter declarations.)

Everything else that appears in the *Sequences* module should be familiar. The module is in Figure 18.1 on the next page.

18.2   Module *FiniteSets*

As described in Section 6.1 on page 66, the *FiniteSets* module defines the two operators *IsFiniteSet* and *Cardinality*. The definition of *Cardinality* is discussed on page 70. The module itself is in Figure 18.2 on the next page.

18.3   Module *Bags*

A *bag*, also called a multiset, is a set that can contain multiple copies of the same element. A bag can have infinitely many elements, but only finitely many copies of any single element. Bags are sometimes useful for representing data structures. For example, the state of a network in which messages can be delivered in any order could be represented as a bag of messages in transit. Multiple copies of an element in the bag represent multiple copies of the same message in transit.

The *Bags* module defines a bag to be a function whose range is a subset of the positive integers. An element e belongs to bag B iff e is in the domain of B, in which case bag B contains $B[e]$ copies of e. The module defines the following operators. In our customary style, we leave unspecified the value obtained by applying an operator on bags to something other than a bag.

IsABag(B)   True iff B is a bag.

BagToSet(B)   The set of elements of which bag B contains at least one copy.

────────────────────── MODULE *Sequences* ──────────────────────

Defines operators on finite sequences, where a sequence of length n is represented as a function whose domain is the set $1 .. n$ (the set $\{1, 2, \ldots, n\}$). This is also how TLA$^+$ defines an n-tuple, so tuples are sequences.

LOCAL INSTANCE *Naturals*   Imports the definitions from *Naturals*, but doesn't export them.

$Seq(S) \triangleq \text{UNION } \{[1 .. n \rightarrow S] : n \in Nat\}$ The set of all finite sequences of elements in S.

$Len(s) \triangleq \text{CHOOSE } n \in Nat : \text{DOMAIN } s = 1 .. n$ The length of sequence s.

$s \circ t \triangleq$ The sequence obtained by concatenating sequences s and t.

$\quad [i \in 1 .. (Len(s) + Len(t)) \mapsto \text{IF } i \leq Len(s) \text{ THEN } s[i]$
$\qquad\qquad\qquad\qquad\qquad\qquad\qquad \text{ELSE } t[i - Len(s)]]$

$Append(s, e) \triangleq s \circ \langle e \rangle$ The sequence obtained by appending element e to the end of sequence s.

$Head(s) \triangleq s[1]$ The usual head (first)
$Tail(s) \triangleq [i \in 1 .. (Len(s) - 1) \mapsto s[i + 1]]$ and tail (rest) operators.

$SubSeq(s, m, n) \triangleq [i \in 1 .. (1 + n - m) \mapsto s[i + m - 1]]$ The sequence $\langle s[m], s[m + 1], \ldots, s[n] \rangle$.

$SelectSeq(s, Test(\_)) \triangleq$ The subsequence of s consisting of all elements $s[i]$ such that $Test(s[i])$ is true.

$\quad \text{LET } F[i \in 0 .. Len(s)] \triangleq$ $F[i]$ equals $SelectSeq(SubSeq(s, 1, i), Test)$.
$\qquad\qquad \text{IF } i = 0 \text{ THEN } \langle \rangle$
$\qquad\qquad\qquad\quad \text{ELSE IF } Test(s[i]) \text{ THEN } Append(F[i - 1], s[i])$
$\qquad\qquad\qquad\qquad\qquad\qquad \text{ELSE } F[i - 1]$
$\quad \text{IN } F[Len(s)]$

Figure 18.1: The standard *Sequences* module.

────────────────────── MODULE *FiniteSets* ──────────────────────

LOCAL INSTANCE *Naturals*   Imports the definitions from *Naturals* and *Sequences*, but doesn't
LOCAL INSTANCE *Sequences*   export them.

$IsFiniteSet(S) \triangleq$ A set is finite iff there is a finite sequence containing all its elements.

$\quad \exists\, seq \in Seq(S) : \forall s \in S : \exists n \in 1 .. Len(seq) : seq[n] = s$

$Cardinality(S) \triangleq$ Cardinality is defined only for finite sets.

$\quad \text{LET } CS[T \in \text{SUBSET } S] \triangleq \text{IF } T = \{\} \text{ THEN } 0$
$\qquad\qquad\qquad\qquad\qquad\qquad \text{ELSE } 1 + CS[T \setminus \{\text{CHOOSE } x : x \in T\}]$
$\quad \text{IN } CS[S]$

Figure 18.2: The standard *FiniteSets* module.

$SetToBag(S)$	The bag that contains one copy of every element in the set S.
$BagIn(e, B)$	True iff bag B contains at least one copy of e. $BagIn$ is the $\in$ operator for bags.
$EmptyBag$	The bag containing no elements.
$CopiesIn(e, B)$	The number of copies of e in bag B; it is equal to 0 iff $BagIn(e, B)$ is false.
$B1 \oplus B2$	The union of bags $B1$ and $B2$. The operator $\oplus$ satisfies $$CopiesIn(e, B1 \oplus B2) =$$ $$CopiesIn(e, B1) + CopiesIn(e, B2)$$ for any e and any bags $B1$ and $B2$.
$B1 \ominus B2$	The bag $B1$ with the elements of $B2$ removed—that is, with one copy of an element removed from $B1$ for each copy of the same element in $B2$. If $B2$ has at least as many copies of e as $B1$, then $B1 \ominus B2$ has no copies of e.
$BagUnion(S)$	The bag union of all elements of the set S of bags. For example, $BagUnion(\{B1, B2, B3\})$ equals $B1 \oplus B2 \oplus B3$. $BagUnion$ is the analog of UNION for bags.
$B1 \sqsubseteq B2$	True iff, for all e, bag $B2$ has at least as many copies of e as bag $B1$ does. Thus, $\sqsubseteq$ is the analog for bags of $\subseteq$.
$SubBag(B)$	The set of all subbags of bag B. $SubBag$ is the analog of SUBSET for bags.
$BagOfAll(F, B)$	The bag analog of the construct $\{F(x) : x \in B\}$. It is the bag that contains, for each element e of bag B, one copy of $F(e)$ for every copy of e in B. This defines a bag iff, for any value v, the set of e in B such that $F(e) = v$ is finite.
$BagCardinality(B)$	If B is a finite bag (one such that $BagToSet(B)$ is a finite set), then this is its cardinality—the total number of copies of elements in B. Its value is unspecified if B is not a finite bag.

The module appears in Figure 18.3 on the next page. Note the local definition of *Sum*, which makes *Sum* defined within the *Bags* module but not in any module that extends or instantiates it.

$\overline{}$ MODULE $Bags$ $\overline{}$

LOCAL INSTANCE $Naturals$ Import definitions from $Naturals$, but don't export them.

$IsABag(B) \;\triangleq\; B \in [\text{DOMAIN } B \to \{n \in Nat \,:\, n > 0\}]$ True iff B is a bag.

$BagToSet(B) \;\triangleq\; \text{DOMAIN } B$ The set of elements at least one copy of which is in B.

$SetToBag(S) \;\triangleq\; [e \in S \mapsto 1]$ The bag that contains one copy of every element of the set S.

$BagIn(e, B) \;\triangleq\; e \in BagToSet(B)$ The $\in$ operator for bags.

$EmptyBag \;\triangleq\; SetToBag(\{\})$

$CopiesIn(e, B) \;\triangleq\; \text{IF } BagIn(e, B) \text{ THEN } B[e] \text{ ELSE } 0$ The number of copies of e in B.

$B1 \oplus B2 \;\triangleq\;$ The union of bags $B1$ and $B2$.
     $[e \in (\text{DOMAIN } B1) \cup (\text{DOMAIN } B2) \mapsto CopiesIn(e, B1) + CopiesIn(e, B2)]$

$B1 \ominus B2 \;\triangleq\;$ The bag $B1$ with the elements of $B2$ removed.
     LET $B \;\triangleq\; [e \in \text{DOMAIN } B1 \mapsto CopiesIn(e, B1) - CopiesIn(e, B2)]$
     IN $[e \in \{d \in \text{DOMAIN } B \,:\, B[d] > 0\} \mapsto B[e]]$

LOCAL $Sum(f) \;\triangleq\;$ The sum of $f[x]$ for all x in DOMAIN f.
     LET $DSum[S \in \text{SUBSET DOMAIN } f] \;\triangleq\;$ LET $elt \;\triangleq\;$ CHOOSE $e \in S \,:\, \text{TRUE}$
                          IN   IF $S = \{\}$ THEN 0
                                      ELSE $f[elt] + DSum[S \setminus \{elt\}]$
    IN $DSum[\text{DOMAIN } f]$

$BagUnion(S) \;\triangleq\;$ The bag union of all elements of the set S of bags.
     $[e \in \text{UNION } \{BagToSet(B) \,:\, B \in S\} \mapsto Sum([B \in S \mapsto CopiesIn(e, B)])]$

$B1 \sqsubseteq B2 \;\triangleq\; \wedge (\text{DOMAIN } B1) \subseteq (\text{DOMAIN } B2)$ The subset operator for bags.
            $\wedge \forall e \in \text{DOMAIN } B1 \,:\, B1[e] \leq B2[e]$

$SubBag(B) \;\triangleq\;$ The set of all subbags of bag B.
     LET $AllBagsOfSubset \;\triangleq\;$ The set of bags SB such that $BagToSet(SB) \subseteq BagToSet(B)$.
            UNION $\{[SB \to \{n \in Nat \,:\, n > 0\}] \,:\, SB \in \text{SUBSET } BagToSet(B)\}$
    IN $\{SB \in AllBagsOfSubset \,:\, \forall e \in \text{DOMAIN } SB \,:\, SB[e] \leq B[e]\}$

$BagOfAll(F(\_), B) \;\triangleq\;$ The bag analog of the set $\{F(x) : x \in B\}$ for a set B.
     $[e \in \{F(d) \,:\, d \in BagToSet(B)\} \mapsto$
        $Sum([d \in BagToSet(B) \mapsto \text{IF } F(d) = e \text{ THEN } B[d] \text{ ELSE } 0])]$

$BagCardinality(B) \;\triangleq\; Sum(B)$ The total number of copies of elements in bag B.

Figure 18.3: The standard $Bags$ module.

18.4   The Numbers Modules

The usual sets of numbers and operators on them are defined in the three modules *Naturals*, *Integers*, and *Reals*. These modules are tricky because their definitions must be consistent. A module M might extend both the *Naturals* module and another module that extends the *Reals* module. The module M thereby obtains two definitions of an operator such as $+$, one from *Naturals* and one from *Reals*. These two definitions of $+$ must be the same. To make them the same, we have them both come from the definition of $+$ in a module *ProtoReals*, which is locally instantiated by both *Naturals* and *Reals*.

The *Naturals* module defines the following operators:

$+$	$*$	$<$	$\leq$	Nat	$\div$ integer division
$-$ binary minus	$\hat{}$ exponentiation	$>$	$\geq$	$..$	$\%$ modulus

Except for $\div$, these operators are all either standard or explained in Chapter 2. Integer division ($\div$) and modulus ($\%$) are defined so that the following two conditions hold, for any integer a and positive integer b:

$$a \% b \in 0 .. (b-1) \qquad a = b * (a \div b) + (a \% b)$$

The *Integers* module extends the *Naturals* module and also defines the set *Int* of integers and unary minus ($-$). The *Reals* module extends *Integers* and introduces the set *Real* of real numbers and ordinary division ($/$). In mathematics, (unlike programming languages), integers are real numbers. Hence, *Nat* is a subset of *Int*, which is a subset of *Real*.

The *Reals* module also defines the special value *Infinity*. *Infinity*, which represents a mathematical ∞, satisfies the following two properties:

$$\forall\, r \in Real \;:\; -Infinity < r < Infinity \qquad -(-Infinity) = Infinity$$

The precise details of the number modules are of no practical importance. When writing specifications, you can just assume that the operators they define have their usual meanings. If you want to prove something about a specification, you can reason about numbers however you want. Tools like model checkers and theorem provers that care about these operators will have their own ways of handling them. The modules are given here mainly for completeness. They can also serve as models if you want to define other basic mathematical structures. However, such definitions are rarely necessary for writing specifications.

The set *Nat* of natural numbers, with its zero element and successor function, is defined in the *Peano* module, which appears in Figure 18.4 on the next page. It simply defines the naturals to be a set satisfying Peano's axioms. This definition is separated into its own module for the following reason. As explained in Section 16.1.9 (page 306) and Section 16.1.10 (page 307), the meanings of tuples and strings are defined in terms of the natural numbers. The *Peano* module,

Peano's axioms are discussed in many books on the foundations of mathematics.

─────────────────── MODULE *Peano* ───────────────────

This module defines *Nat* to be an arbitrary set satisfying Peano's axioms with zero element *Zero* and successor function *Succ*. It does not use strings or tuples, which in TLA$^+$ are defined in terms of natural numbers.

$PeanoAxioms(N, Z, Sc) \triangleq$ Asserts that N satisfies Peano's axioms with zero element Z and
  $\land\ Z \in N$ successor function *Sc*.
  $\land\ Sc \in [N \to N]$
  $\land\ \forall\, n \in N : (\exists\, m \in N : n = Sc[m]) \equiv (n \neq Z)$
  $\land\ \forall\, S \in \text{SUBSET } N : (Z \in S) \land (\forall\, n \in S : Sc[n] \in S) \Rightarrow (S = N)$

ASSUME $\exists\, N, Z, Sc : PeanoAxioms(N, Z, Sc)$ Asserts the existence of a set satisfying Peano's axioms.

$Succ \triangleq \text{CHOOSE } Sc : \exists\, N, Z : PeanoAxioms(N, Z, Sc)$
$Nat \triangleq \text{DOMAIN } Succ$
$Zero \triangleq \text{CHOOSE } Z : PeanoAxioms(Nat, Z, Succ)$

Figure 18.4: The *Peano* module.

which defines the natural numbers, does not use tuples or strings. Hence, there is no circularity.

As explained in Section 16.1.11 on page 308, numbers like 42 are defined in TLA$^+$ so that 0 equals *Zero* and 1 equals *Succ[Zero]*, where *Zero* and *Succ* are defined in the *Peano* module. We could therefore replace *Zero* by 0 and *Succ[Zero]* by 1 in the *ProtoReals* module. But doing so would obscure how the definition of the reals depends on the definition of the natural numbers in the *Peano* module.

Most of the definitions in modules *Naturals*, *Integers*, and *Reals* come from module *ProtoReals* in Figure 18.5 on the following two pages. The definition of the real numbers in module *ProtoReals* uses the well-known mathematical result that the reals are uniquely defined, up to isomorphism, as an ordered field in which every subset bounded from above has a least upper bound. The details will be of interest only to mathematically sophisticated readers who are curious about the formalization of ordinary mathematics. I hope that those readers will be as impressed as I am by how easy this formalization is—once you understand the mathematics.

Given the *ProtoReals* module, the rest is simple. The *Naturals*, *Integers*, and *Reals* modules appear in Figures 18.6–18.8 on page 348. Perhaps the most striking thing about them is the ugliness of an operator like *R!+*, which is the version of + obtained by instantiating *ProtoReals* under the name *R*. It demonstrates that you should not define infix operators in a module that may be used with a named instantiation.

────────────────────────────── MODULE *ProtoReals* ──────────────────────────────

This module provides the basic definitions for the *Naturals*, *Integers*, and *Reals* module. It does this by defining the real numbers to be a complete ordered field containing the naturals.

EXTENDS *Peano*

$IsModelOfReals(R, Plus, Times, Leq) \triangleq$

> Asserts that R satisfies the properties of the reals with $a + b = Plus[a, b]$, $a * b = Times[a, b]$, and $(a \leq b) = (\langle a, b \rangle \in Leq)$. (We will have to quantify over the arguments, so they must be values, not operators.)

LET $IsAbelianGroup(G, Id, \_ + \_) \triangleq$ Asserts that G is an Abelian group with identity Id and

> $\wedge Id \in G$ group operation +.
> $\wedge \forall a, b \in G : a + b \in G$
> $\wedge \forall a \in G : Id + a = a$
> $\wedge \forall a, b, c \in G : (a + b) + c = a + (b + c)$
> $\wedge \forall a \in G : \exists minusa \in G : a + minusa = Id$
> $\wedge \forall a, b \in G : a + b = b + a$
>
> $a + b \triangleq Plus[a, b]$
> $a * b \triangleq Times[a, b]$
> $a \leq b \triangleq \langle a, b \rangle \in Leq$

IN $\wedge Nat \subseteq R$ The first two conjuncts assert that *Nat*
> $\wedge \forall n \in Nat : Succ[n] = n + Succ[Zero]$ is embedded in R.
> $\wedge IsAbelianGroup(R, Zero, +)$ The next three conjuncts assert that R
> $\wedge IsAbelianGroup(R \setminus \{Zero\}, Succ[Zero], *)$ is a field.
> $\wedge \forall a, b, c \in R : a * (b + c) = (a * b) + (a * c)$
> $\wedge \forall a, b \in R : \wedge (a \leq b) \vee (b \leq a)$ The next two conjuncts assert that R is
> $\wedge (a \leq b) \wedge (b \leq a) \equiv (a = b)$ an ordered field.
> $\wedge \forall a, b, c \in R : \wedge (a \leq b) \wedge (b \leq c) \Rightarrow (a \leq c)$
> $\wedge (a \leq b) \Rightarrow \wedge (a + c) \leq (b + c)$
> $\wedge (Zero \leq c) \Rightarrow (a * c) \leq (b * c)$
>
> $\wedge \forall S \in$ SUBSET $R :$ The last conjunct asserts that every
>   LET $SBound(a) \triangleq \forall s \in S : s \leq a$ subset S of R bounded from above has
>   IN $(\exists a \in R : SBound(a)) \Rightarrow$ a least upper bound *sup*.
> $(\exists sup \in R : \wedge SBound(sup)$
> $\wedge \forall a \in R : SBound(a) \Rightarrow (sup \leq a))$

THEOREM $\exists R, Plus, Times, Leq : IsModelOfReals(R, Plus, Times, Leq)$

$RM \triangleq$ CHOOSE $RM : IsModelOfReals(RM.R, RM.Plus, RM.Times, RM.Leq)$

$Real \triangleq RM.R$

Figure 18.5a: The *ProtoReals* module (beginning).

We define *Infinity*, $\leq$, and $-$ so $-Infinity \leq r \leq Infinity$, for any $r \in Real$, and $-(-Infinity) = Infinity$.

$Infinity \triangleq$ CHOOSE $x : x \notin Real$

$MinusInfinity \triangleq$ CHOOSE $x : x \notin Real \cup \{Infinity\}$

> *Infinity* and *MinusInfinity* (which will equal $-Infinity$) are chosen to be arbitrary values not in *Real*.

$a + b \triangleq RM.Plus[a, b]$

$a * b \triangleq RM.Times[a, b]$

$a \leq b \triangleq$ CASE $(a \in Real) \wedge (b \in Real) \qquad\qquad\quad \rightarrow \quad \langle a, b \rangle \in RM.Leq$
$\qquad\qquad\quad \square\ (a = Infinity) \wedge (b \in Real \cup \{MinusInfinity\}) \quad \rightarrow \quad$ FALSE
$\qquad\qquad\quad \square\ (a \in Real \cup \{MinusInfinity\}) \wedge (b = Infinity) \quad \rightarrow \quad$ TRUE
$\qquad\qquad\quad \square\ a = b \qquad\qquad\qquad\qquad\qquad\qquad\qquad\quad \rightarrow \quad$ TRUE

$a - b \triangleq$ CASE $(a \in Real) \wedge (b \in Real) \qquad\quad \rightarrow \quad$ CHOOSE $c \in Real : c + b = a$
$\qquad\qquad\quad \square\ (a \in Real) \wedge (b = Infinity) \qquad \rightarrow \quad MinusInfinity$
$\qquad\qquad\quad \square\ (a \in Real) \wedge (b = MinusInfinity) \ \rightarrow \quad Infinity$

$a / b \triangleq$ CHOOSE $c \in Real : a = b * c$

$Int \triangleq Nat \cup \{Zero - n : n \in Nat\}$

We define a^b (exponentiation) for $a > 0$, or $b > 0$, or $a \neq 0$ and $b \in Int$, by the four axioms

$\quad a^1 = a \qquad a^{m+n} = a^m * a^n$ if $a \neq 0$ and $m, n \in Int \qquad 0^b = 0$ if $b > 0 \qquad a^{b*c} = (a^b)^c$ if $a > 0$

plus the continuity condition that $0 < a$ and $0 < b \leq c$ imply $a^b \leq a^c$.

$a^b \triangleq$ LET $RPos \triangleq \{r \in Real \setminus \{Zero\} : Zero \leq r\}$
$\qquad\qquad\quad exp \triangleq$ CHOOSE $f \in [(RPos \times Real) \cup (Real \times RPos)$
$\qquad\qquad\qquad\qquad\qquad\qquad\qquad \cup ((Real \setminus \{Zero\}) \times Int) \rightarrow Real] :$
$\qquad\qquad\qquad\qquad \wedge\ \forall\, r \in Real : \wedge f[r, Succ[Zero]] = r$
$\qquad\qquad\qquad\qquad\qquad\qquad\qquad\quad \wedge \forall\, m, n \in Int : (r \neq Zero) \Rightarrow$
$\qquad\qquad\qquad\qquad\qquad\qquad\qquad\qquad\qquad\qquad (f[r, m + n] = f[r, m] * f[r, n])$
$\qquad\qquad\qquad\qquad \wedge\ \forall\, r \in RPos : \wedge f[Zero, r] = Zero$
$\qquad\qquad\qquad\qquad\qquad\qquad\qquad\quad \wedge \forall\, s, t \in Real : f[r, s * t] = f[f[r, s], t]$
$\qquad\qquad\qquad\qquad\qquad\qquad\qquad\quad \wedge \forall\, s, t \in RPos : (s \leq t) \Rightarrow (f[r, s] \leq f[r, t])$
$\qquad\quad$ IN $\quad exp[a, b]$

Figure 18.5b: The *ProtoReals* module (end).

───────────────── MODULE *Naturals* ─────────────────

LOCAL $R \triangleq$ INSTANCE *ProtoReals*

$Nat \triangleq R!Nat$

$a + b \triangleq a \; R!+ \; b$ $R!+$ is the operator + defined in module *ProtoReals*.

$a - b \triangleq a \; R!- \; b$

$a * b \triangleq a \; R!* \; b$

$a^b \triangleq a \; R!\char`\^ \; b$     $a^b$ is written in ASCII as `a^b`.

$a \leq b \triangleq a \; R!\leq \; b$

$a \geq b \triangleq b \leq a$

$a < b \triangleq (a \leq b) \wedge (a \neq b)$

$a > b \triangleq b < a$

$a \mathrel{..} b \triangleq \{i \in R!Int : (a \leq i) \wedge (i \leq b)\}$

$a \div b \triangleq$ CHOOSE $n \in R!Int : \exists r \in 0 \mathrel{..} (b-1) : a = b * n + r$ We define $\div$ and % so that

$a \mathrel{\%} b \triangleq a - b * (a \div b)$ $a = b * (a \div b) + (a \mathrel{\%} b)$
for all integers a and b with $b > 0$.

Figure 18.6: The standard *Naturals* module.

───────────────── MODULE *Integers* ─────────────────

EXTENDS *Naturals*   The *Naturals* module already defines operators like + to work on all real numbers.

LOCAL $R \triangleq$ INSTANCE *ProtoReals*

$Int \triangleq R!Int$

$-.\, a \triangleq 0 - a$ Unary − is written −. when being defined or used as an operator argument.

Figure 18.7: The standard *Integers* module.

───────────────── MODULE *Reals* ─────────────────

EXTENDS *Integers*   The *Integers* module already defines operators like + to work on all real numbers.

LOCAL $R \triangleq$ INSTANCE *ProtoReals*

$Real \triangleq R!Real$

$a/b \triangleq a \; R!/ \; b$ $R!/$ is the operator / defined in module *ProtoReals*.

$Infinity \triangleq R!Infinity$

Figure 18.8: The standard *Reals* module.

Index